Lecture Notes in Artificial Intelligence 7788

Subseries of Lecture Notes in Computer Science

LNAI Series Editors

Randy Goebel
   University of Alberta, Edmonton, Can(
Yuzuru Tanaka
   Hokkaido University, Sapporo, Japan
Wolfgang Wahlster
   DFKI and Saarland University, Saarbrücken, Germany

T0216464

LNAI Founding Series Editor

Joerg Siekmann
   DFKI and Saarland University, Saarbrücken, Germany

Maria Paola Bonacina   Mark E. Stickel (Eds.)

Automated Reasoning and Mathematics

Essays in Memory of William W. McCune

 Springer

Series Editors

Randy Goebel, University of Alberta, Edmonton, Canada
Jörg Siekmann, University of Saarland, Saarbrücken, Germany
Wolfgang Wahlster, DFKI and University of Saarland, Saarbrücken, Germany

Volume Editors

Maria Paola Bonacina
Università degli Studi di Verona, Dipartimento di Informatica
Strada Le Grazie 15, 37134 Verona, Italy
E-mail: mariapaola.bonacina@univr.it

Mark E. Stickel
SRI International, Artificial Intelligence Center
333 Ravenswood Avenue, Menlo Park, CA 94025, USA
E-mail: stickel@ai.sri.com

The picture of William W. McCune on page V was taken from the website
http://www.cs.unm.edu/ mccune/photo.html.

The cover illustration is licensed under the Creative Commons Attribution-Share
Alike 3.0 United States license. Attribution: http://www.ForestWander.com

ISSN 0302-9743   e-ISSN 1611-3349
ISBN 978-3-642-36674-1   e-ISBN 978-3-642-36675-8
DOI 10.1007/978-3-642-36675-8
Springer Heidelberg Dordrecht London New York

Library of Congress Control Number: 2013932090

CR Subject Classification (1998): I.2.3, F.4.1-2, F.3.1

LNCS Sublibrary: SL 7 – Artificial Intelligence

Typesetting: Camera-ready by author, data conversion by Scientific Publishing Services, Chennai, India

Printed on acid-free paper

Springer is part of Springer Science+Business Media (www.springer.com)

William W. McCune (1953–2011)

Preface

In Memory of William W. McCune (1953–2011)

This volume is a tribute to the memory of William (Bill) McCune, whose sudden death on May 4, 2011, left the field of automated reasoning deprived of one of the founders of practical theorem proving and model building. While he was an accomplished computer scientist all around, Bill McCune was especially a fantastic system builder and software engineer. He developed a series of systems of astounding power, robustness, useability, and portability, including the theorem provers *Otter*, *EQP*, and *Prover9*, the parallel prover *ROO*, the proof checker *Ivy*, the prototype SAT-solver *ANL-DP*, and the model builders *Mace2* and *Mace4*.

Bill McCune's Scientific and Professional Contributions

Originary of New Hampshire, Bill did his undergraduate studies in mathematics at the University of Vermont, and completed his education with an MS and a PhD in computer science from Northwestern University, with adviser Larry Henschen. After the PhD, he started his career as Assistant Computer Scientist at the Mathematics and Computer Science (MCS) Division of the Argonne National Laboratory (ANL), at Argonne, near Chicago. He was soon promoted to Computer Scientist, and later became Senior Computer Scientist. Bill spent most of his professional life in the red-brick building hosting the MCS Division on the vast and quiet ANL campus. He stayed there until the unexpected demise of the automated reasoning program in 2006, when he joined the Department of Computer Science of the University of New Mexico as Research Professor.

ANL is a research laboratory mostly funded by the federal government of the United States through its Department of Energy. A primary mission of its MCS Division is to advance research in computing and mathematics to enable the solving of the mathematical and computational challenges posed by research in physics and other natural sciences. Perhaps surprisingly, and due in part to the lack of specialization in the early days of computer science, Argonne, was, and still remains from a historic point of view, the cradle of automated reasoning. J. Alan Robinson worked on *resolution* and *unification* during summer jobs at Argonne during 1962–1964, writing the milestone article on "A Machine-Oriented Logic" in the summer of 1963. Approximately in the same years, Larry Wos started at ANL a research program in automated reasoning, where *paramodulation*, or resolution with equality built-in, *demodulation*, or the *ad hoc* use of equations for rewriting, and the *set of support strategy* were invented in the years 1965–1969. Work on automated reasoning at Argonne continued during the 1970s and early 1980s. When Bill McCune started at ANL in 1984, he joined Larry Wos, Ewing L. (Rusty) Lusk, and Ross Overbeek. The Argonne theorem provers in that period were ITP, LMA/ITP, and AURA, which already featured an early version of the *given-clause algorithm* later popularized by Otter.

A positive consequence of Argonne's great inventions was the persuasion that time was ripe to focus on implementation and system building. A less positive one was the notion that research in theory was almost over. In reality, in the mid-1980s the quest for building equality into resolution was not over. The *Wos–Robinson conjecture*, namely, whether paramodulation is complete without paramodulating into variables and adding instances of reflexivity – the *functionally reflexive axioms* – was not settled. Also, it was not known how to control the termination of demodulation, that was called k-demodulation, because it depended on an *ad hoc* bound k on the number of allowed rewriting steps.

In the same years when Larry Wos and G. Robinson invented paramodulation, Don Knuth and his student Peter B. Bendix devised a *completion procedure*, featuring *superposition*, or paramodulation between the left sides of two rewrite rules, and *simplification* of a rewrite rule by another. A key idea was that equations were oriented into rewrite rules by a *well-founded ordering* on terms. In 1981, Gérard Huet proved that Knuth–Bendix completion is correct: if it does not fail by generating an equation that cannot be oriented, it generates in the limit a *confluent rewriting system*, and it semi-decides the validity of equational theorems, as suggested independently also by Dallas Lankford. At the IEEE Symposium on Logic in Computer Science (LICS) of 1986, Leo Bachmair, Nachum Dershowitz, and Jieh Hsiang reobtained these results in a more general framework based on *well-founded proof orderings*. At the 8th International Conference on Automated Deduction (CADE-8), held at Oxford in July 1986, Jieh Hsiang and Michäel Rusinowitch showed that *ordered resolution* and *ordered paramodulation*, restricted by a well-founded ordering, are refutationally complete *without functionally reflexive axioms*. During 1987–1989, Jieh Hsiang and Michäel Rusinowitch, on one hand, and Leo Bachmair, Nachum Dershowitz, and David A. Plaisted, on the other, came up independently with *unfailing*, or *ordered*, completion, which works for equations with no need of orienting them once and for all into rewrite rules. Throughout the 1980s, Nachum Dershowitz, David A. Plaisted, and others worked systematically on well-founded orderings, their properties, and termination of rewriting for theorem proving.

Bill McCune's greatness at that time was that he deeply understood the rewriting and completion research developed elsewhere, and united it with the best results of the Argonne tradition in a new theorem prover named *Otter*. Otter stands for *organized techniques for theorem-proving and effective research*, and it is also the name of a rare semiaquatic mammal, that inhabits rivers and unites a playful, shy nature with the determination of a skilled hunter. The release of Otter at CADE-9 in 1988 was a turning point in the history of automated reasoning. Never before had the computer science community seen a theorem prover of such awesome power. Otter proved theorems in full first-order logic with equality with amazing speed, relative to the technology of the day. It was almost surely sound,[1] and endowed with both complete and incomplete

---

[1] Though Bill used to joke that he would not jump off a bridge if a soundness bug were exposed in Otter or any other of his systems.

strategies, with the latter often most useful in practice. Otter quickly became the touchstone for an entire field.

Already in the early 1990s, Otter matured into a robust, reliable, and portable prover, with *options* and *parameters* that the experimenter could set to define the adopted strategy. Over time, Bill added features such as the *pick-given ratio* to mix clause evaluation functions, *hints* and a *hot list* to guide the search, and an *auto(matic) mode* enabling the prover to choose by itself the strategy based on the input's syntax. Several of these enhancements came from Bill's readiness to learn from experiments, including those carried out by others, and to integrate users' suggestions or requests with his own apparently infallible sense of what was practical to implement. Notwithstanding its wealth of features, Otter was remarkably easy to use, and therefore a significant user community grew world-wide, including scientists not working in theorem proving, and especially mathematicians and logicians. Indeed, Larry Wos and Bill McCune shared a keen interest in applying theorem proving to mathematics, especially algebra, geometry, and logic, also building on historic ties that the MCS Division of ANL had with the Department of Mathematics of the University of Chicago.

However, perhaps Otter's greatest impact was due to Bill's generous and far-looking decision to make its source code publicly available. It is impossible to describe completely a reasoning program in research papers. There is always some amount of knowledge, often a surprising amount, that is written only in the code, and therefore remains hidden, if the code is not public or is too hard to read. Bill's code was admirably readable and well organized. Other researchers, including those whose systems eventually overtook Otter in speed or in variety of inference rules, also learnt from Bill's code data structures, algorithms, and indexing schemes, which are fundamental for implementing theorem provers.

Although Bill developed Otter for several years, he had a clear sense that it may not be wise to try to put too many features in one system. For instance, he refused to implement in Otter *reasoning modulo associativity and commutativity* (AC). Rather, he built another theorem prover, called *EQP*, for *equational prover*, that had *AC-matching* and *AC-unification*, but worked only for equational theories. Bill always considered EQP as a prototype to be used by himself and a few others, rather than a system for all like Otter. EQP was written with a specific goal in mind: proving the *Robbins conjecture*, an open problem in mathematics whose existence Larry Wos had discovered in his continuous quest for challenges for theorem provers.

The Robbins conjecture dated back to 1933, when a mathematician, named E. V. Huntington, demonstrated by hand that a certain equation, later called *Huntington axiom*, was sufficient, together with associativity and commutativity of addition, to axiomatize Boolean algebra. In the same year, another mathematician, Herbert Robbins, conjectured that the same was true of another equation, later called *Robbins axiom*. A proof was not found, and algebras presented by Robbins axiom, together with associativity and commutativity of addition, became known as Robbins algebras. In 1990-1992, S. Winker proposed two equations, later called *first Winker condition* and *second Winker condition*, and proved by hand that if each of them is added to the Robbins axiom, the Huntington axiom follows. This led to decomposing the problem into proving that Robbins axiom implies the second Winker condition, the second Winker condition implies the first, and the first implies Huntington axiom. While the first step was relatively easy, the other two remained beyond the possibilities of theorem provers.

In 1996, EQP proved them both, thereby solving a problem that had challenged mathematicians for over 60 years. The first successful run for the hardest lemma, the one showing that the second Winker condition implies the first, in February 1996, took an impressive 12.5 days of computation on a 486DX2/66. The experiment was repeated on an RS/6000 taking 7.2 days. In the same year, Bill also succeeded in having EQP prove that Robbins axiom implies Huntington axiom in one run, in order to have a single mechanical proof. A human version of the proof was extracted from the mechanical one for readability and persuasion. The field reacted with awe, and this momentous achievement brought unprecedented visibility to automated deduction, artificial intelligence, and the whole of computer science. Bill McCune and EQP made it onto the pages of the *New York Times*. One of the very early dreams of artificial intelligence, namely, machines capable of proving mathematical theorems that human experts could not prove, was no longer only a dream: it was a reality. The Robbins algebra proof is an ideal example of a successful blending of multiple strands of research in automated reasoning: the extensive experimentation characteristic of Argonne, new theory about equality reasoning, and associative-commutative unification.

While obtaining these outstanding results in theorem proving, Bill was also among the first to understand the importance of the dual problem of *model building*, or *theorem disproving*. A precipitating event was the solution of open problems of *quasigroup existence* in 1992. The problems were posed by mathematician Frank Bennett, and solved initially by Masayuki Fujita, with ICOT's Model Generation Theorem Prover (MGTP), and by John Slaney, with his model builder FINDER. The opportunity and excitement of solving open problems stimulated a lot of activity by researchers, including the second editor of this volume, Hantao Zhang, and Bill McCune, on new, advanced implementations of the Davis–Putnam–Loveland–Logemann (DPLL) procedure, known since 1960-1962, to decide propositional satisfiability (SAT), and find models. A Japanese–American Workshop on Automated Theorem Proving that focused on finite domain theorem proving was held at Duke University in March 1994. Frank Bennett, Masayuki Fujita, Bill McCune, Hantao Zhang, and the second editor of this volume, were among the attendees. A few months later, in June 1994, Bill McCune announced his DPLL-based SAT-solver ANL-DP, which was already being applied to solving quasigroup existence problems.

Bill was too involved with first-order reasoning to delve into SAT, and ANL-DP remained a prototype. Since DPLL works by trying to build a model, and reports unsatisfiability when it has found that none exists, ANL-DP was preparatory work for Bill's next great system, the SAT-based finite model finder *Mace*, whose most successful version was *Mace2*. The mathematical community that

Bill supported benefited enormously from his providing a model builder as well as a theorem prover. As we shall see, several chapters of this volume report results that depend on both. For all these achievements, in 2000 Bill McCune received *the Herbrand Award*, the highest honor in automated reasoning and one of the highest in computer science.

Bill McCune was not the kind who would rest on his laurels. Although he maintained Otter through August 2003, which is the date of that glorious theorem prover's last manual, Bill knew that Otter had become too old to continue developing it. Thus, he embarked in designing a brand new theorem prover for first-order logic with equality, called *Prover9*, and with the ever-optimistic, forward-looking subtitle *"the future of theorem proving."* The years 2005–2010 were devoted to Prover9 and *Mace4*, a new Mace, no longer SAT-based, but using a more general constraint solving approach. Prover9 and Mace4 inherited all the great qualities of their predecessors Otter and Mace2, as witnessed by the fact that they are still very much in use today.

Not only was Bill McCune a marvelous system builder, he also wanted to make it easy for others to build reasoning programs. In addition to making his own code available, already in 1992–1993, he had the idea of building a software library to assemble theorem provers. The version dating from those years was named OPS, for *Otter parts store*. Later it evolved into a new library, called LADR, or *Library for Automated Deduction Research*. Bill was probably also the first to think of a web-based interface to let anyone play with an automated reasoner: *Son of BirdBrain* gave access to Otter and Mace2, and *Son of BirdBrain II* gave access to Prover9 and Mace4.

Since he worked for most of his career in a research laboratory, Bill did not have PhD students. However, he mentored through cooperation several junior colleagues. Also, he served the scientific community as the first Secretary of the Association for Automated Reasoning (AAR) in 1986–1993, as co-organizer of CADE-9 at Argonne in 1988, as Program Chair of CADE-14 in Townsville, Australia, in 1997, and as Trustee of CADE Inc. in 1997–2000. Bill McCune was remembered with heart-felt speeches by several colleagues at CADE-23 in Wrocław, Poland, in August 2011, and at the 8th International Workshop on First-Order Theorem Proving (FTP) in Bern, Switzerland, in July 2011.

Some Recollections of Bill McCune by Maria Paola Bonacina

It is likely that my first interaction with Bill was by e-mail, when he was AAR Secretary, I was a graduate student, and I wanted to become a member of the AAR. As part of my PhD work at SUNY Stony Brook, I implemented a distributed version of Otter, called *Aquarius*, later presented at the Third International Symposium on Design and Implementation of Symbolic Computation Systems (DISCO), in Gmunden, Austria, in September 1993. The availability of Bill's code helped the implementation part of my thesis enormously. It meant that I could focus on implementing the methodology for distributed deduction that I was proposing in the thesis, called *clause diffusion*, reusing Bill's code for everything else. It was my first opportunity to appreciate the clarity and robustness of his code. Also, Bill's turn-around time by e-mail was fantastic.

My adviser, Jieh Hsiang, suggested inviting Bill to my PhD defense committee. Bill kindly accepted, and travelled from Argonne to Stony Brook for the defense in December 1992. Later Larry Wos told me that Bill went back to Argonne so excited about distributed deduction, that he implemented right away a prototype with a master–slave architecture for equational reasoning modulo AC. Indeed, that was when Bill was starting to expand the OPS with code that would lead later to the development of EQP. After my defense, Bill invited me to visit Argonne in January–February 1993, before starting a postdoc at INRIA Lorraine in Nancy. Thus, I had a wonderful opportunity to work side by side with Bill McCune, and discuss research with Larry Wos, Rusty Lusk, and Ross Overbeek. At that time, Rusty and Ross were interested in parallel programming, and, together with other colleagues at Argonne, had developed *the p4 Parallel Programming System*. It was a perfect match, and Bill and I worked together on implementing a new clause-diffusion prover, using p4 and the OPS code for equational reasoning modulo AC. Upon Bill's suggestion, the prover was called *Peers*, because in clause diffusion there was no master–slave concept, and all deductive processes cooperated as peers. The Peers prover was presented at CADE-12 in Nancy in June 1994.

When I started as Assistant Professor in the Department of Computer Science of the University of Iowa in the fall of 1993, I was asked to serve as Colloquium Chair, and encouraged to invite leading figures in my field, including mentors. Because Jieh Hsiang had moved from Stony Brook to the National Taiwan University, my first invitation was for Bill McCune. During my years at the University of Iowa, our scientific paths diverged somewhat. Bill delved into the implementation of EQP and the final attack to the Robbins problem. I pursued other topics, including a new version of clause diffusion, called *modified clause diffusion*. However, I continued to benefit from Bill's code: my next distributed theorem prover, *Peers-mcd*, implemented modified clause diffusion using MPI (message passing interface), and EQP as sequential base. In July 1997, Peers-mcd was presented at CADE-14, and at the second International Symposium on Parallel Symbolic Computation (PASCO), in Maui, Hawaii, where the experimental report included instances of super-linear speed-up, made possible by distributed search, in the proofs for the Robbins problem.

I have fond memories of working and discussing with Bill, especially during my visits at Argonne in January–February 1993 and June 1998. In addition to being such a great computer scientist, Bill McCune was a very fine gentleman: he had an admirable self-control, used very few words, and was very kind. I never saw him upset or losing his temper about anything. It seemed impossible that such a thing could ever happen. In January–February 1993, Bill had just gotten his beloved dog, and was very concerned that the puppy would cry if left alone all day. Thus, he would quit the office early and continue work at home. His usage of time was extremely effective. He used to say, "There is a whole lot we can do in 15 minutes!" — and it was true. Bill was very open towards people. When I arrived at Argonne in January 1993 he asked me how long I had been in the States, and when I answered, "Three years and half," he said, "You're one of us!" When I visited again in June 1998, and we found that ANL had tightened security checks on visitors without a US passport, Bill commented that it was nonsense. He could not see how the trustworthiness of people had anything to do with their passport. During my visits at Argonne we used to have lunch at the cafeteria on campus. A couple of times we took extended walks to go see a historic linear accelerator of the Physics Division of ANL.

I never knew how Bill really felt about the termination of the automated reasoning program at ANL. I did not dare to inquire too much. In earlier conversations, Bill had told me that he did not necessarily see himself spending the rest of his career at Argonne. He said at some point he might have moved into teaching, preferably at some university or college in New England. When his position at ANL was terminated, for some time it looked like he would stay in Illinois as a sort of freelance researcher. Then, he joined the Department of Computer Science of the University of New Mexico. I invited him to visit my new Department in Verona, but at that time he was not too keen on long flights. We mentioned me visiting his new department in Albuquerque, but regrettably it did not happen.

I would like to dedicate the work I personally put in the editing of this volume to the memory of my mother, Annita Fusari, who unexpectedly passed away on May 21, 2011, a couple of weeks after Bill McCune.

Some Recollections of Bill McCune by Mark E. Stickel

Although I was not one of Bill McCune's collaborators or long-term visitors, our joint attendance at workshops and conferences was always a welcome opportunity for discussion.

A memorable encounter was at the Workshop on Empirically Successful First Order Reasoning (ESFOR) at the International Joint Conference on Automated Reasoning (IJCAR) 2004 in Cork, Ireland. Bill and I enthused about Stephan Schulz's *feature vector indexing* (described and refined in his chapter in this volume) over beers the evening after his presentation. Bill was quick to test the idea, reported six months later to Stephan and me that he had tried it, and praised it highly. Indexing is crucial to the performance of theorem provers and, as a top system builder, Bill paid close attention to it. Back in 1989 he was also quick to implement my then new *path indexing* method, alongside with *discrimination tree indexing*, in Otter. Through comprehensive testing, he created test sets of terms that were long used by other researchers to evaluate indexing methods.

Also in 1989, Argonne asserted the near indispensability of subsumption in automated reasoning in the article "Subsumption, a Sometimes Undervalued Procedure" by Ross Overbeek, Larry Wos, and Ewing Lusk, that appeared in 1991 in the volume edited by Jean-Louis Lassez and Gordon Plotkin in honor of Alan Robinson. Approaches like my *Prolog technology theorem prover* (PTTP), which is based on logic programming, generally lack subsumption for controlling redundancy. As another instance of his openness to appreciating different theorem-proving paradigms, Bill once told me that he *liked* PTTP, a great encouragement and then a cherished memory.

PTTP can also be criticized for lacking another feature Argonne found to be nearly indispensable: making inferences from clauses in order of ascending weight in the given-clause algorithm, instead of using first-in-first-out order (i.e., level saturation) or depth-first search from the goal as in PTTP. There is considerable justification for these criticisms; however, PTTP and other systems that lack subsumption and weight-ordered search occasionally surprised the field by finding proofs contrary to expectation. This revealed that an overreliance on weight-ordered search may be a weakness. I do not know if PTTP influenced his thinking, but Bill also saw the weakness and a solution. His pick-given ratio allows the system to choose clauses for inference either by weight or first-in-first-out ordering in alternation according to a user supplied ratio. This can be used to avoid the problem of large clauses, especially those derived early but necessary for a proof, being postponed too long.

Bill and I were both system builders who learned from each other's systems. I often consult Otter or Prover9 code to see how Bill did things, and Bill looked at my implementation of the DPLL procedure when developing ANL-DP and my implementation of AC-unification when developing EQP. We both strived to build a library of code that others could use for building systems. I would have liked to use his LADR, but my preference for programming in Lisp is too strong. We shared the attitude inherited from the Argonne tradition that new problems or application areas often require considerable user input. Fully automated proof is not always feasible and opportunities for user control of the proving process should be provided in abundance. This is illustrated, for example, in the vital role of Robert (Bob) Veroff's hint mechanism in chapters in this volume. As a system builder myself, I wish to emphasize that Bill was peerless in making his systems valuable to a large community of users, especially mathematicians, by excellent design, implementation, documentation, outreach, and support.

Outline of This Volume

We had the idea of this volume back in May 2011, when the field of automated reasoning was still under shock at the news of Bill McCune's sudden passing. We were encouraged by Larry Wos, Deepak Kapur, and Bob Veroff: we thank them for their support. A first call for papers appeared in September 2011 and was repeated a few months later: we thank Carsten Schürmann for helping with the publicity. We received 15 submissions and accepted 13 of them. Each paper had at least two and up to four reviewers, who wrote accurate and detailed reviews: we thank them all for their precious cooperation. All accepted papers were thoroughly revised, and in some cases extended, by their authors.

The volume is organized as follows. The first article is a recollection of Bill McCune by Larry Wos, his main colleague and friend at ANL. Larry describes how Bill approached proving theorems in mathematics, especially with Otter, and how Bill and Larry cooperated in the search for shorter or otherwise more elegant proofs. We thank Gail Pieper for helping Larry with the editing. Then there are four articles on core topics where Bill gave fundamental contributions: *strategies, indexing, superposition-based theorem proving,* and *model building.*

Leonardo De Moura and Grant Olney Passmore discuss the strategy challenge in SMT-solving, where SMT stands for satisfiability modulo theories. This article debunks the notion that SMT-solvers are push-button tools, whereas first-order theorem provers depend heavily on heuristics, and user-defined parameters and options. Both kinds of system need heuristic search, feature search strategies, and may involve user-defined settings, based on the problem's difficulty. Indeed, SMT-solvers *are* theorem provers and model builders. The importance of strategies is such that the authors propose enabling users to exert strategic control, so that they can program, in a sense, their strategies. Thus, this work also advances Bill's vision of enabling others to build reasoners.

Stephan Schulz presents a simple and efficient indexing technique for *clause susbumption*, called *feature vector indexing*. This article focuses on *clause indexing*, when most of the literature emphasizes *term indexing* for unification and matching. As Bill McCune was among the initiators of the research in indexing, this article continues a central topic in his research program. Thomas Hillenbrand and Christoph Weidenbach contribute an article on *superposition-based decision procedures* for theories of *bounded domains*, including, for example, the theory of *bitvectors*. This article connects two fundamental themes in Bill McCune's research: superposition-based deduction and reasoning about finite domains.

The latter topic leads to the article on *finite model generation* by Jian Zhang and Hantao Zhang. After touching briefly on SAT-based model generators, such as Bill McCune's Mace2 and Hantao Zhang's ModGen, the authors analyze model builders based on constraint solving, comparing Mace4 with their SEM. The article presents the paradigm of *backtracking search* for constraint solving applied to model finding, and then treats in greater detail two main issues of this paradigm: *heuristics to choose assignments* and inference rules for *constraint propagation*. Apparently, there has not been much exchange between this kind of work and SMT-solving: this volume might help establish connections.

The core of the volume is devoted to the *application of automated reasoning to mathematics*, which Bill pursued throughout his career. Ranganathan Padmanabhan was Bill's companion of investigations in the realms of ternary Boolean algebras, cubic curves, lattices, ortholattices, and more. Together they wrote a book on *Automated Deduction in Equational Logic and Cubic Curves* that appeared in the LNAI series of Springer as volume 1249. Ranganathan Padmanabhan contributed two articles in *geometry*. The one with Bob Veroff reports on proving new theorems about *cubic curves* with Prover9, and discusses the pros and cons of building theory properties in the inference system versus using generic inference rules, an ever-returning issue in automated deduction. The one with Eric Ens describes using Prover9 to prove theorems connecting *projective planes* and *abelian groups*.

This leads us from geometry to *algebra*: Michael Kinyon, Petr Vojtěchovský, and Bob Veroff contributed an article on applying Prover9 to reason about algebraic structures, including *loops* and *quasigroups*, by using proof hints and proof sketches. The next two articles take us from algebra to *logic*: Rob Arthan and Paulo Oliva investigate *continuous logic*, finding counter-examples with Mace4 and proofs with Prover9; Branden Fitelson studies an axiomatic approach to proving a theorem in *sentential logic* called *Gibbard's collapse theorem*. Prover9 is used to prove it, while Mace4 is used to show that the axioms in the proposed axiomatization are pair-wise independent, by exhibiting counter-models to the conjecture that one depends on the others.

The third part of the volume collects articles on applications of automated reasoning that Bill's work contributed to make possible: *program verification, data mining*, and *computer formalized mathematics*. Deepak Kapur, Zhihai Zhang, Hengjun Zhao, Matthias Horbach, Qi Lu, and Thanvu Nguyen investigate deductive methods for *invariant generation* by quantifier elimination. Zachary Ernst and Seth Kurtenbach explore how to apply data mining and statistical learning techniques, which paid off in fields such as *computational biology* or *computational linguistics*, to theorem proving, which also deals with dazzling amounts of data. The volume is closed by an article on a grand project that Bill McCune cared for, as witnessed by his engagement with the *QED manifesto: computer formalization of mathematics*. Josef Urban and Jiří Vyskočil survey recent work on interfacing the Mizar library and proof assistant for formalized mathematics with automated theorem provers and model builders.

As computer scientists designing algorithms, we are trained to refrain from brute-force solutions, and seek to instill in our programs as much intelligence as possible. However, as computer scientists we are fascinated with computers, and with the unavoidably brute-force character, in a way, of mechanical solutions: there is simultaneously intelligence and brute force in a machine playing chess *à la* Deep Blue, answering questions *à la* Watson, and proving theorems *à la* Otter. The balance between the two is a constantly renewed challenge of artificial intelligence. Thus, we would like to close this preface with a quote from the article by Zachary Ernst and Seth Kurtenbach: "We do not expect a mathematician to work from scratch re-proving everything each time, so why would we want that from a theorem prover?" Indeed, when it comes to machines, humans sometimes set the threshold unreasonably high, perhaps because there is still a certain reluctance to renounce the assumption of a human monopoly on intelligence, and admit that there is intelligence in machines, and there is intelligence in non-human animals. Bill McCune did engineer wonderful intelligent machines, and the best way to honor his legacy is to continue pursuing this research.

January 2013

Maria Paola Bonacina
Mark E. Stickel
Editors

Referees

Table of Contents

The Legacy of a Great Researcher

Larry Wos*

Argonne National Laboratory

Abstract. This article presents examples of the monumental contributions of Bill McCune to mathematics, logic, computer science, and especially automated reasoning. The examples are presented in the form of short stories and recollections of the author during his long association with Bill. In addition to Bill's accomplishments as a researcher, the author provides personal memories giving glimpses into Bill's complex personality and his generosity as a collaborator.

1 Perspective and Genesis

Perhaps you have wondered what would result if you had the opportunity to spend thousands of hours with a great mind. For more than two decades, I had that opportunity as I shared research time with my esteemed colleague William (Bill) McCune. We shared many ideas, conjectures, and, yes, guesses. Each of us had two main goals. The first goal was to formulate enhancements for an automated reasoning program, enhancements that would substantially add to its power. The second goal was to employ the program in a way that would contribute to various areas of mathematics and logic.

From the viewpoint of making contributions to mathematics and logic, Bill and I had a marvelous automated assistant; indeed, in 1988, if memory serves, he designed and implemented the automated reasoning program called OTTER. (We did have access to a program designed at Argonne before OTTER was produced.) In but four months, even though Bill was also involved in research of different aspects, he wrote more than 30,000 lines of code, producing a program that, from then until now, has exhibited the smallest number of bugs. Yes, his effort was and is monumental; indeed, when you obtain a conclusion, a set of conclusions, a proof, you can assume with almost total certainty that all is in order. Also important is the robustness of OTTER, permitting you to have it search for desired objects for weeks, if needed, without stopping.

In this article, I shall tell a number of short stories that provide ample evidence of Bill's inventive mind, his accurate insights, and his impeccable professionalism. His successes in the context of enhancements have played a key role in much of what has occurred in the past ten years. As I shall highlight here, Bill answered (with one of his programs) open questions in areas that include group theory, lattice theory, Boolean algebra, combinatory logic, and—so impressive—Robbins algebra; for various open questions, see Chapter 7 of [14].

---

* This work was supported by the U.S. Department of Energy, under Contract DE-AC02-06CH11357.

M.P. Bonacina and M.E. Stickel (Eds.): McCune Festschrift, LNAI 7788, pp. 1–14, 2013.

As many of you may know, Bill provided at Argonne National Laboratory a means for researchers to easily copy OTTER. And, possibly because that program was correctly viewed as extremely powerful, many, many copies were taken. As but one bit of evidence of his professionalism, he enabled me to place (on disc) a copy of OTTER, with a manual, in the back of one of my books, thus materially adding to the usefulness of my books. Even his users manual is well written [5]. An examination of history would reveal that OTTER provided the basis for a large number of programs that followed its birth.

This narrative will almost certainly not follow the chronology of history. Nor will it echo my view of the significance of Bill's achievements. Instead, the order of the topics will reflect, though probably somewhat hidden, some form of intensity of my recollections, recollections coupled with memories of excitement, curiosity, and, yes, surprising results in many cases. The field of automated reasoning is deeply indebted to Bill for all that he provided.

2   Combinatory Logic

Our foray into combinatory logic began with our colleague Ross Overbeek, who had read a charming book on combinatory logic by Raymond Smullyan, *To Mock a Mockingbird* [12]. Ross began the study of two combinators, B and W, that respectively satisfy the following, where expressions by convention are assumed to be left associated unless otherwise indicated.

```
Bxyz = x(yz)
Wxy = xyy
```

The object of Ross's study was to determine whether the fragment with basis consisting just of B and W satisfied the strong fixed point property, in the context of the following definition: A combinator $\mathbf{F}$ is a fixed point combinator if and only if $\mathbf{F}x = x(\mathbf{F}x)$ for all x. (As I recall, Smullyan called such combinators "sages".) The $\mathbf{F}$ in this case must be expressed solely in terms of B and W. To his disappointment during his study, Ross found that the open question concerning B and W had been answered by R. Statman, who found (in February 1986) a fixed point combinator for B and W expressed in eight letters.

```
B(WW)(BW(BBB))   (Statman's fixed point combinator)
```

Upon learning of Statman's result, Ross asked me to find a way for an automated reasoning program to find a fixed point combinator $\mathbf{F}$ expressed solely in terms of B and W. In other words, you could view his request as amounting to answering, with a program of ours and no knowledge about whether the answer was yes or no, the following open question: Does the fragment based on B and W alone satisfy the strong fixed point property?

I began thinking of how paramodulation could be used. (Paramodulation, as many of you probably know, is an inference rule that generalizes equality substitution; its use builds in a powerful treatment of equality.) Shortly after I initiated my effort, Bill entered my office. Upon learning of my activity, he asked if he could join me, a request I gladly agreed to. What Bill did was have the reasoning program ITP make two searches, one forward and one backward, believing that insufficient memory was present to attack the question directly. (ITP, designed and implemented by Ross and Ewing Lusk, and contributed to heavily by Bill, preceded the birth of OTTER [3].)

Bill accumulated two sets of deductions. Then, on a weekend, he assigned the program the task of comparing pairs of equations, one from the forward search and one from the backward search, to see whether unit conflict could be found. In particular, the forward search yielded (positive) equalities, and the backward negative equalities.

On the following Monday, to the amazement of both Bill and me, five pairs were discovered by ITP. That result showed that, rather than one fixed point combinator for the B-and-W fragment, there were five. When Bill and I wrote to Smullyan, he was indeed surprised and impressed at our discovery—mostly Bill's, to be precise [11].

For the curious reader, the discovery of the five fixed point combinators for the B-and-W fragment eventually led to the discovery of an infinite set of fixed point combinators for this fragment. Not too long after that discovery, an infinite class of infinite sets of such fixed point combinators was proved (by me) to exist. Yes, these discoveries can be traced directly to Bill's use of ITP and that profitable weekend. And this episode led to a second, one that illustrates Bill's insight.

Indeed, some time later, Bill called me at home and made the following observation: "Did you know that, if you took the five fixed point combinators, each of length eight, and demodulated each with B, you get the same expression?" The magical expression is the following.

`(W(B(Bx)))` cubed

Eventually, this expression would be called a kernel for the B-and-W fragment, after the *kernel strategy* was formulated [13]. I would never have thought of this strategy were it not for Bill's insightful observation. That strategy proved to be most powerful when seeking fixed point combinators.

Sometime during the study with Bill—I cannot pinpoint exactly when that was—I proved that the B-and-L fragment was too weak to satisfy the strong fixed point property. I am fairly certain that result also answered a question that had been open. Bill, not too much later, generalized what I had done and showed that various other fragments also failed to satisfy the fixed point property. That bit of research provides yet more evidence of Bill's research strength, nicely illustrating his ability to think as a mathematician.

3   Boolean Algebra

During Bill's position as a staff member at Argonne, I believe 1983 to 2006, I was certainly not the only researcher who benefited from collaboration with him. For example, with Ranganathan Padmanabhan he answered a number of open questions, many of which were published in a marvelous monograph (in 1996) by the two of them [8]. One of the questions that was answered with OTTER concerns a theorem called by Bill DUAL-BA-3. The following equations (which rely on the notation used in that monograph) capture that theorem, where $x@$ denotes the complement of x and where the two inequalities arise from, respectively, negating the theorem to be proved and negating its dual.

```
x = x.
x * (y + z) = (x * y) + (x * z).
x + (y * z) = (x + y) * (x + z).
x + x@ = 1.
x * x@ = 0.
(x * y@) + ((x * x) + (y@ * x)) = x.
(x * x@) + ((x * z) + (x@ * z)) = z.
(x * y@) + ((x * y) + (y@ * y)) = x.
(x + y@) * ((x + x) * (y@ + x)) = x.
(x + x@) * ((x + z) * (x@ + z)) = z.
(x + y@) * ((x + y) * (y@ + y)) = x.
(A * B) + B != B | $ANS(A2).
(A + B) * B != B | $ANS(A4).
```

Bill called me at home, having in mind my interest in improving on given proofs. First, he told me that he had a proof of the theorem; then he informed me that the proof consisted of 816 equations, the longest proof I had ever heard of produced by OTTER. The proof relied on a Knuth-Bendix approach and, therefore, featured the use of demodulation. After I expressed amazement, he asked whether I could "elegantize" the proof.

I in turn asked whether there was one or more equations in his proof that he wished to avoid. That was not his goal. Instead, being well aware of my interest in proof shortening, he asked me to produce a shorter proof, far shorter; I agreed to try. But before I hung up, he interrupted to refine his request, saying that he wished me to find a proof of length 100 or less—which I felt was either a joke or essentially absurd. Bill, however, knew I enjoyed challenges, especially with such a nice number as the goal. (He told me that he intended to place the proof in a monograph, I believe written with Padmanabhan, but wished to avoid the cited long proof, which took more than nineteen pages to display.)

Indeed, over the next couple of weeks—inspired by Bill, no doubt—with Bill's goal in mind, I developed methodology, much of which I still use, for proof shortening.

I had a fine start because of one of Bill's enhancements to OTTER, namely, *ancestor subsumption*. Ancestor subsumption is a procedure that compares derivation lengths to the same conclusion and prefers the strictly shorter. Yes, Bill's professionalism is exhibited: he designed and implemented this procedure solely because of my interest in finding "short" proofs.

Of course, you have anticipated what is now to be said. With that superb procedure and the methodology I was able to formulate, Bill did get his 100-step proof. (Bill was gratified in that the proof required only a bit more than three pages to display.) Quite a while later, I found a 94-step proof, which, as was true of the 100-step proof, relied on demodulation. Therefore, you could properly object to measuring its length when demodulation steps are not counted. So, I note that I did find a proof of length 147 in which demodulation is not used. (If you enjoy a challenge, from what I can recall, I know of no proof of length strictly less than 147 that relies on forward reasoning, with paramodulation, avoids demodulation, and proves DUAL-BA-3.) For those who enjoy history, along the way OTTER produced a proof of level 107. Bill's request was made, I believe, in May 2002; the 100-step proof was found perhaps in June; in November of that year, OTTER produced a 94-step proof.

At this point, some of you may be wondering why I have not yet cited one of Bill's greatest successes. I shall shortly. For now, I turn to another study Bill conducted in Boolean algebra. Specifically, he had formulated a technique for generating thousands of candidates, when seeking a single axiom, with the set filtered to avoid those that could not possibly succeed. Although I am not certain, I believe he employed this technique in his search in 2002 for single axioms for Boolean algebra in terms of **not** and **or**. Whether such is the case or not, Bill found the following ten (given in the notation he used), each of length 22.

```
~ (~ (x + y) + ~ z) + ~ (~ (~ u + u) + (~ z + x)) = z.
% 13345 685 sec
~ (~ (~ (x + y) + z) + ~ (x + ~ (~ z + ~ (z + u)))) = z.
% 20615    6 sec
~ (~ (x + y) + ~ z) + ~ (x + ~ (z + ~ (~ z + u))) = z.
% 20629   19 sec
~ (~ (x + y) + ~ (~ (x + z) + ~ (~ y + ~ (y + u)))) = y
% 20775   18 sec
~ (x + ~ y) + ~ (~ (x + z) + ~ (y + ~ (~ y + u))) = y.
% 20787   80 sec
~ (~ (x + y) + ~ (~ (~ y + ~ (z + y)) + ~ (x + u))) = y.
% 24070   28 sec
~ (x + ~ y) + ~ (~ (y + ~ (z + ~ y)) + ~ (x + u)) = y
% 24086   44 sec
~ (~ (x + y) + ~ (~ (~ y + ~ (z + y)) + ~ (u + x))) = y.
% 24412   40 sec
~ (x + ~ y) + ~ (~ (y + ~ (z + ~ y)) + ~ (u + x)) = y.
% 24429   36 sec
~ (~ (~ (x + y) + z) + ~ (~ (~ z + ~ (u + z)) + y)) = z.
% 24970   47 sec
```

Open question: For a far greater challenge that, if met, would merit a publication, one might study the open question concerning the possible existence of a single axiom for Boolean algebra in terms of disjunction and negation that has length strictly less than 22, the length of Bill's single axioms.

And now for the long-awaited highlight of Bill's study of Boolean algebra; even the New York Times was impressed, enough to write an article on the success under the byline of Gina Kolata [2]. Bill designed another automated reasoning program he called EQP, a program with built-in commutative/associative unification. Perhaps one reason he did so, perhaps the main one, was his intention of answering the decades-old Robbins algebra problem. A Robbins algebra is axiomatized with the following three axioms, where (in the notation Bill used in this study) + denotes union and the function n denotes complement.

```
x + y = y + x.                    %  commutativity
(x + y) + z = x + (y + z).        %  associativity
n(n(n(x) + y) + n(x + y)) = y.    %  Robbins
```

Whether a Robbins algebra is a Boolean algebra was unknown for decades. S. Winker, I believe in the late 1970s, brought the problem to the Argonne researchers in automated reasoning. Not too long afterwards, the problem was attacked by various people throughout the world without success. Winker did supply a number of conditions that, if satisfied along with the three given axioms, sufficed to enable a deduction of the properties of a Boolean algebra. Stated a bit differently, if one of Winker's conditions was adjoined to the three given axioms, then you could prove that the resulting algebra is a Boolean algebra.

Bill's approach, with EQP, was to try to prove, from the three Robbins algebra axioms, one of Winker's conditions. Now, as far as I know, all the people attempting to answer the open question focusing on Robbins algebra believed that the key axiom (known as the Robbins axiom) was a misprint. Nevertheless, the question proved fascinating.

In almost eight days of computing, EQP deduced one of Winker's conditions, and, therefore, the question was no longer open: Indeed, every Robbins algebra is a Boolean algebra. Bill found the proof in late 1996, I believe [7]. Adding piquancy to the story was what occurred after Bill's monumental success. Specifically, Bill called Robbins to inform him of the result, commenting that the third axiom, as we thought we knew, had been published with an error. Robbins replied that, no, it was not an error, that he had conjectured that the three axioms—commutativity of union, associativity of union, and the third axiom (which focuses on a complicated expression involving complement and union)—might axiomatize Boolean algebra. Robbins was elated at Bill's information—how nice for the originator, then 81 years old, to learn of this result.

By way of what might be termed a post mortem to the story, Bill followed the tradition of the solid mathematician. Indeed, to find the earlier-cited ten single axioms for Boolean algebra in terms of **or** and **not**, he needed a target (as is typical) to show that each totally axiomatized the algebra. He did not choose to attempt to deduce the usual set of axioms for Boolean algebra. Instead, building on his success with Robbins algebra, he chose as target to deduce the Robbins basis, the set of three equations (given earlier) that characterize a Boolean algebra. A most charming action to take.

4   Logic Calculi and More Enhancements

Various areas of logic are often referred to as calculi. Bill and I spent some time with equivalential calculus. A possible obstacle that Bill noted was that, early in a run, depending on the hypothesis, deduced formulas could be complex (long in symbol count). If you permit your program to retain many complex formulas, the program can drown in deduced conclusions. On the other hand, if you place a tight bound on all deduced and retained information, then some key item might not be retained. In particular, if you are using a program to study, say, XHK or XHN, captured respectively by the following two clauses, and if you assign a large value to retained items, you will find early in the output rather lengthy deductions. (Predicates such as P can be thought of as meaning "is provable".)

```
P(e(x,e(e(y,z),e(e(x,z),y)))).  % XHK
P(e(x,e(e(y,z),e(e(z,x),y)))).  % XHN
```

With OTTER, if you were to assign the value, say, 35 to max_weight, you would find that too many conclusions were being retained. But if instead you assigned the value, say, 15, then you might be prevented from reaching your goal, perhaps of deducing some known single axiom for this area of logic. Yes, XHK and XHN each are single axioms.

Bill formulated and then encoded a feature that permits the program to retain, early in a run, very complex conclusions, but, shortly afterwards, discard such new conclusions. The following two commands show how it works.

```
assign(change_limit_after,100).
assign(new_max_weight,10).
```

The first of the two commands has OTTER, after 100 clauses have been chosen to initiate application of inference rules, change the assigned value to max_weight. The second command assigns the new value of 10 to max_weight. Summarizing, Bill designed and implemented a feature that allows the program to both have and eat its cake, to attack problems in which the retention of a so-called messy formula was required (early in the study) and yet not drown before the assignment was completed.

Bill came through again in a totally different context, the following. In the mid-1980s, I suggested that we at Argonne have access to a new strategy, namely, the *hot list strategy* [15]. You may have witnessed many, many times in a textbook on mathematics the phenomenon in which some assumption for the theorem in question is frequently cited. (A glance at the literature reveals that, in various proofs in logic, researchers often do this. Branden Fitelson pointed this out, noting he sometimes assigned the value 8 to the "heat" parameter. C. A. Meredith, in effect, used the hot list strategy.) The value assigned to heat denotes how much so-to-speak recursion is being asked for. Members of the hot list are used to *complete* applications of inference rules and not to *initiate* applications. The use of the hot list strategy thus enables an automated reasoning program to briefly consider a newly retained conclusion whose complexity might otherwise prevent its use for perhaps many CPU-hours. For example, if the textbook is proving that commutativity holds for rings in which the cube of x is x for all x, that property, $xxx = x$, may be used in many of the steps of the proof. With this theorem, you might well include $xxx = x$ in the hot list and assign the value of 1, or greater, to heat. I suggested for paramodulation—and this particularization is crucial to this story—that one of our programs offer to the user access to the hot list strategy. The researcher would choose, at the beginning, one or more items to be placed in the "hot list", a list that is often consulted in making additional deductions. The hot list strategy was shortly thereafter added, by Overbeek and Lusk, to the repertoire of ITP.

Years passed. Then I said to Bill it would be nice to have OTTER offer the hot list strategy, of course, in the context of paramodulation. Not too long after our discussion, Bill called me and asked that I test, in his so-called presence, his implementation of the strategy. However—and here again you witness his inventiveness—he informed me that he had implemented the strategy not just for paramodulation but for whatever inference rules were in use. Among such rules was hyperresolution. I was, at the time, conducting various studies in the use of condensed detachment (of course, relying on hyperresolution), an inference rule frequently used in the study of some logical calculus. The following clause captures that rule when studying, say, two-valued sentential or (classical propositional) calculus. (For OTTER, "-" denotes logical **not** and "|" denotes logical **or**.)

```
% condensed detachment
-P(i(x,y)) | -P(x) | P(y).
```

Immediately, I made a run to test Bill's version of the hot list strategy, in the context of deducing one 3-axiom system from another. I had at the time a 22-step proof. Astounding: With the hot list strategy, OTTER found a 21-step proof, a proof I had been seeking for a long time.

Again, I note that Bill had generalized what I had in mind in the mid-1980s by implementing the strategy for all inference rules in use. Further, he had added substantially to my idea by implementing the means to, in effect, apply the strategy recursively; you simply assign a value to heat that is 2 or greater, depending on how much recursion you wish. Even more, he implemented an incarnation that, if you chose to use it, had the program adjoin new elements to the hot list during a run—the dynamic hot list strategy.

If you wonder whether researchers outside Argonne have found Bill's various generalizations useful, I mention as one example Fitelson, who used the strategy heavily in various incarnations.

5   Group Theory

I do not know what motivated Bill, but he completed research in group theory that had been begun by the logician John Kalman, who studied the area in a manner quite different from what you might be familiar with. Kalman's study relied on condensed detachment and on the Sheffer stroke, denoted by the function d, captured with the following clause and the use of hyperresolution.

```
-P(d(x,y)) | -P(x) | P(y).
```

Technically, Kalman focused on the right-group calculus. (For the researcher who enjoys relationships, I note that equivalential calculus corresponds to Boolean groups, the R-calculus and the L-calculus to Abelian groups, and the right-group calculus to ordinary groups.) Kalman proved the following axiom system for the right-group calculus.

```
P(d(z,d(z,d(d(x,d(y,y)),x)))).                     % R1
P(d(u,d(u,d(d(z,y),d(d(z,x),d(y,x)))))).           % R2
P(d(v,d(v,d(d(u,d(z,y)),d(u,d(d(z,x),d(y,x))))))). % R3
P(d(d(d(u,d(v,y)),d(z,d(v,x))),d(u,d(z,d(y,x))))). % R4
P(d(d(v,d(z,d(u,d(y,x)))),d(d(v,d(x,u)),d(d(z,d(x,y)),x))))  % R5
```

As for interpretation, $d(x, y)$ can be thought of as the Sheffer stroke (the **nand** of x and y, and, when preceded by the predicate P, the formula is equivalent to the identity. The theorems of the right-group calculus are a proper subset of those of the R-calculus, which in turn are a proper subset of the theorems of equivalential calculus.

Bill initiated a study of the Kalman 5-axiom system—out of simple curiosity, perhaps. His study produced charming results. In particular, Bill proved that each of the second, third, and fourth of Kalman's five axioms provides a complete axiomatization (by itself) for the calculus [6]. With his model-generation program MACE [4], you can prove that the first of Kalman's axioms is too weak to serve as a single axiom; you can find a 3-element model to yield this result. Again, I offer you a challenge, an open question, actually. Indeed, the status of the fifth remains undetermined—at least, such was the case in late 2003. (Bill's study was conducted, I believe, in 2001.)

Now, if you wonder about Bill and so-called standard group theory, yes, he did study aspects of that area of abstract algebra. Indeed, relying (I am almost certain) on his method for generating thousands of promising candidates, he sought (I believe in 1991) single axioms of Boolean groups, groups of exponent 2. A group has exponent 2 if and only if the square of every element x is the identity e. He was successful.

Upon learning of his achievement, I suggested he seek single axioms for groups of exponent 3, groups in which the cube of every element x is the identity e. Again, he succeeded, presenting to me four interesting single axioms. One important aspect of research is that it leads to further discoveries. Bill's certainly did. I used one of his four single axioms for groups of exponent 3 to find single axioms for groups of exponent 5, 7, 9, ..., 19. To permit you, if you choose, to attempt to produce the pattern that generalizes to all odd exponents, I give you single axioms for exponents 7 and 9.

```
(f(x,f(x,f(x,f(f(x,f(x,f(x,f(f(x,y),z)))),f(e,f(z,f(z,f(z,
   f(z,f(z,z)))))))))) = y).
```

```
(f(x,f(x,f(x,f(x,f(f(x,f(x,f(x,f(x,f(f(x,y),z)))))),f(e,
  f(z,f(z,f(z,f(z,f(z,f(z,f(z,z)))))))))))))) = y).
```

If you were to seek a proof that the given equation for exponent 7 is in fact a single axiom for the variety of groups of exponent 7, you could seek four proofs, for each of the following given in negated form.

```
(f(f(a,b),c) != f(a,f(b,c))) | $ANS(assoc).
(f(a,f(a,f(a,f(a,f(a,f(a,f(a,a)))))) != e) | $ANS(exp7).
(f(e,a) != a) | $ANS(lid).
(f(a,e) != a) | $ANS(rid).
```

My citing of "all odd exponents" is appropriate; indeed, Ken Kunen, with his student Joan Hart, proved [1] that my generalization through exponent 19 continues for all odd n with n greater than or equal to 3. Bill, not to be outdone so-to-speak, produced his own generalization for 3, 5, 7, ..., a set of single axioms in which the identity e is not explicitly present. (Tarski noted without proof that no single axiom exists in which product, inverse, and identity are all explicitly present; Neumann supplied a proof.) For the curious, and for an example of OTTER's going where no researcher has gone before, its occasional application of an inference rule to a set of hypotheses one or more of which is most complex has led to breakthroughs such as a detailed proof focusing on groups of exponent 19, a proof an unaided researcher would have found most burdensome to produce in view of relying on equations with more than 700 symbols.

6   Other Areas of Abstract Algebra

In the early part of this twenty-first century, Bill collaborated in studies of abstract algebra with Robert Veroff, Padmanahban, and (for a little while) his student Michael Rose. These collaborations proved indeed profitable, as the following evidences.

I never asked Bill about his choice of variety to study. For example, did he deliberately study a variety and then study subvarieties? For a nice example, commutative groups form a subvariety of groups. Did he begin by explicitly considering the following chain of algebraic varieties: Boolean algebra (BA), modular ortholattices (MOL), orthomodular lattices (OML), ortholattices (OL), complemented lattices (CL), lattice theory (LT), and quasilattice theory (QLT)? If you begin with an equational basis for quasilattice theory in terms of meet and join, with the addition of axioms, (in steps) you obtain bases for LT, CL, OL, OML, MOL, then BA. Mainly he and Veroff did find single axioms for Boolean algebra in terms of the Sheffer stroke; but I leave that topic for another paper. And, as cited, he made other contributions to Boolean algebra, most notably (from the world of long-standing open questions in mathematics) the splendid result focusing on Robbins algebra.

From Bill's many successes in algebra, I shall highlight a theorem from quasilattices and some results from lattice theory. The challenge offered by a theorem (denoted by Bill as QLT-3) in quasilattices was strikingly different from the search for new single axioms. Specifically, only model-theoretic proofs of the theorem existed—before OTTER, and Bill, entered the game to search for the missing axiomatic proof. The theorem asserts that the following self-dual equation can be used to specify distributivity, where "v" denotes join (union) and "^" denotes meet (intersection).

$$(((x ~\hat{}~ y) ~v~ z) ~\hat{}~ y) ~v~ (z ~\hat{}~ x) = (((x ~v~ y) ~\hat{}~ z) ~v~ y) ~\hat{}~ (z ~v~ x)$$

The first proof OTTER discovered has length 183. Access to the 183-step proof in turn prompted the search for yet another missing proof, still axiomatic, but simpler. With the various methodologies, a proof of length 108 was completed.

As for lattice theory, Bill did indeed make a monumental search. Bill's goal was not just some single axiom for lattice theory; after all, R. McKenzie had already devised a method that produces single axioms. The use of that method typically produces gigantic (in length) single axioms. Bill sought a single axiom of (undefined) "reasonable" length. Through a variety of techniques that keyed on the cited algorithm but incorporated the assistance of OTTER, he began with a single axiom of more than 1,000,000 symbols and eventually found a 79-symbol single axiom. The nature of his approach guaranteed that the result was sufficient, a theorem; no proof was needed. But, after a gap in time, Bill decided upon a new approach that would filter candidates, yielding equations that were promising. The goal was a far shorter single axiom.

Among the candidates, after one year, he found the following promising 29-symbol equation.

```
(((y v x)^x) v (((z^ (x v x)) v (u^x))^v))^ (w v ((v6 v x)^
(x v v7))) = x.
```

Success would be his when and if a proof of some basis could be found. One of the standard bases (axiom systems) for lattice theory consists of the following set of six equations.

```
y ^ x = x ^ y
(x ^ y) ^ z = x ^ (y ^ z)
y v x = x v y
(x v y) v z = x v (y v z)
x ^ (x v y) = x
x v (x ^ y) = x
```

Bill, however, preferred a 4-basis as he sought single axioms. The (nonstandard) 4-basis he chose as target was the following.

```
y v (x^ (y^z)) = y
y^ (x v (y v z)) = y
((x^y) v (y^z)) v y = y
((x v y)^ (y v z))^y = y
```

And he won [9].

But still he was not finished; indeed, as I learned, Bill had a 135-step proof that the cited equation sufficed to axiomatize lattice theory. He called me and asked if I would seek a shorter proof—a proof, I concluded from knowing him so well, that would be far shorter in length than 135 applications of paramodulation. After not too long, I did find what I thought he would like, and called and told him so.

He asked about its length. I told him the proof was of length 50, which brought from Bill a surprising response. "Can I buy it?" "No", I said; "you can have it". And he published it, from what I know. (For the curious or for one who might decide to seek a shorter proof, the following might be of interest. Specifically, in July 2007, I returned to the study of that single axiom and found a 42-step proof.)

Bill also found a second 29-letter single axiom, the following.

```
(((y v x) ^ x) v (((z ^ (x v x)) v (u ^ x)) ^ v)) ^ (((w v x) ^
(v6 v x)) v v7) = x.
```

From what I know, open questions still exist concerning Bill's two single axioms for lattice theory. In particular, is there a shorter single axiom, shorter than length 29? Are there other single axioms of length 29? What is the shortest single axiom for LT in terms of meet and join?

Bill contributed to other areas of mathematics, in geometry, with Padmanabhan, Rose, and Veroff, by finding single axioms for OL and OML [10].

7 Introducing Bill McCune

Who was Bill, really? Yes, of course, he was a fine researcher, attacking and answering various open questions from different fields of mathematics and logic. He designed, from my perspective, the most powerful automated reasoning program, OTTER, a program that I still use today, though it was designed and implemented mainly in 1988. But, what about the so-called nonprofessional side of Bill?

He played the piano; however, having never heard him play, I know not how well nor what type of music he played. He cooked, rather fancy sometimes; for example, he made his own mayonnaise. He immensely enjoyed eating; indeed, we shared many lunches, especially Thai food and Chinese food—not Americanized. He had a golden retriever he was deeply fond of. In the winter, they would go for long walks, the dog emerging from cold water with icicles on its coat. As evidence of Bill's deep ethical concerns, when the dog was diagnosed with cancer and Bill was given the choice of treatment or putting the dog down, he chose the latter. Yes, he did not want the dog to experience pain, and, instead, Bill made the supreme sacrifice of giving up his friend.

Had you worked at Argonne National Laboratory with Bill, sharing research experiments with him, you still would have known little about him. I would not say he was shy. Rather, his personal life was kept to himself. I did learn, after many, many years, that he was delighted with blueberries, wandering the trails in Maine, picking different varieties of wild blueberry. Did he enjoy dessert? Well, I had told him about my refrigerator, that it contained twenty-eight pints of ice cream. One late afternoon, Bill drove me home. Upon arriving, he brought up my claim of having twenty-eight pints of ice cream and expressed strong doubt about its accuracy. At my invitation, we entered my apartment, and I beckoned him to the refrigerator, indicating that he should investigate, which he did. After counting out loud, reaching the figure I had cited, he expressed amusement and surprise—and asked if he could try some. Of course, I told him to help himself. And he did, sampling, in one dish, three types of that frozen concoction.

Bill, as I said, was kind. Upon learning of my interests and also learning of how I worked, he wrote special programs for me. For example, he wrote a "subtract" program that takes as input two files and produces as output the set-theoretic difference. Another program he wrote for me interrogates an output file (from OTTER) containing numerous proofs, many of the same theorem, and returns in another file the shortest proof for each of the theorems proved in the experiment. Also, to enable the researcher to run without intervention a series of experiments, Bill wrote otter-loop and super-loop.

And, as many of you might know, Bill was aware of the chore some experienced when confronted with the array of choices OTTER offers. Perhaps because of his knowledge, he added to OTTER the "autonomous mode", a mode that removes from the user the need to make choices. In that mode, his program still proved to be of great assistance, often finding the proof(s) being sought.

Then there is the example of Bill's kindness combined with his thoroughness and professionalism. In particular, Kalman was writing a most detailed book about OTTER. That book promised to provide at the most formal level essentially most of what you would wish to know about OTTER. Before it was completed, however, Kalman notified me that he could not complete it because of a serious illness, one that eventually took his life. I promised him it would get finished. Indeed, I knew, or was almost certain, that I could count on Bill. And he did come through. After my informing him of the situation, Bill completed the book, enabling it to be delivered into Kalman's hands. A magnanimous gesture!

His sense of humor? The question is not whether Bill had one, but rather in what way it was expressed. Sometimes you gain insight into a person's view of life by having some information about that person's enjoyment of humor. I can say that, if only occasionally, some of my remarks did cause Bill to laugh heartily, to explode thunderously with enjoyment. For a different side of him, was he making a joke in one sense when he added to OTTER the command set(very_verbose)? That command has the program return copious, copious detail that enables you to, if you wish, check each inference, each application of demodulation, and such.

So long, Bill; you were unique; we do miss you. Mathematics, logic, computer science, and, even more, automated reasoning are each indebted to you, forever.

References

1. Hart, J., Kunen, K.: Single axioms for odd exponent groups. J. Automated Reasoning 14, 383–412 (1995)
2. Kolata, G.: With major math proof, brute computers show flash of reasoning power. New York Times (December 10, 1996)
3. Lusk, E., McCune, W., Overbeek, R.: ITP at Argonne National Laboratory. In: Siekmann, J.H. (ed.) CADE 1986. LNCS, vol. 230, pp. 697–698. Springer, Heidelberg (1986)
4. McCune, W.: Mace4 reference manual and guide. Technical Memorandum ANL/MCS-TM-264 (August 2003)
5. McCune, W.: Otter 1.0 users' guide. Tech. report ANL-88/44, Argonne National Laboratory (January 1989)
6. McCune, W.: Single axioms for the left group and right group calculi. Notre Dame J. Formal Logic 34(1), 132–139 (1993)
7. McCune, W.: Solution of the Robbins problem. J. Automated Reasoning 19(3), 263–276 (1997)
8. McCune, W., Padmanabhan, R.: Automated Deduction in Equational Logic and Cubic Curves. LNCS, vol. 1095. Springer, Heidelberg (1996)
9. McCune, W., Padmanabhan, R.: Single identities for lattice theory and for weakly associative lattices. Algebra Universalis 36(4), 436–449 (1996)
10. McCune, W., Padmanabhan, R., Rose, M.A., Veroff, R.: Automated discovery of single axioms for ortholattices. Algebra Universalis 52, 541–549 (2005)
11. McCune, W., Wos, L.: The absence and the presence of fixed point combinators. Theoretical Computer Science 87, 221–228 (1991)
12. Smullyan, R.: To mock a mockingbird, and other logic puzzles: Including an amazing adventure in combinatory logic. Knopf (1985)
13. Wos, L.: The kernel strategy and its use for the study of combinatory logic. J. Automated Reasoning 10(3), 287–343 (1993)
14. Wos, L., Pieper, G.W.: Automated reasoning and the discovery of missing and elegant proofs. Rinton Press (2003)
15. Wos, L., Pieper, G.W.: The hot list strategy. J. Automated Reasoning 22(1), 1–44 (1999)

The Strategy Challenge in SMT Solving

Leonardo de Moura[1] and Grant Olney Passmore[2,3]

[1] Microsoft Research, Redmond
leonardo@microsoft.com
[2] Clare Hall, University of Cambridge
[3] LFCS, University of Edinburgh
grant.passmore@cl.cam.ac.uk

Abstract. High-performance SMT solvers contain many tightly integrated, hand-crafted heuristic combinations of algorithmic proof methods. While these heuristic combinations tend to be highly tuned for
known classes of problems, they may easily perform badly on classes
of problems not anticipated by solver developers. This issue is becoming
increasingly pressing as SMT solvers begin to gain the attention of practitioners in diverse areas of science and engineering. We present a challenge to the SMT community: to develop methods through which users
can exert strategic control over core heuristic aspects of SMT solvers.
We present evidence that the adaptation of ideas of strategy prevalent
both within the Argonne and LCF theorem proving paradigms can go a
long way towards realizing this goal.

Prologue. Bill McCune, Kindness and Strategy, by Grant Passmore

I would like to tell a short story about Bill, of how I met him, and one way his
work and kindness impacted my life.

I was an undergraduate at the University of Texas at Austin, where in Autumn
2004 I was lucky enough to take an automated reasoning course with Bob Boyer.
One of our three main texts for the course (the other two being Robinson's 1965
JACM article on resolution and part of Gödel's 1931 Incompleteness paper in
which he defines his primitive recursive proof checker) was the wonderful book[1] of
Larry Wos and Gail Pieper on Bill's OTTER theorem prover. I was mesmerized
by OTTER's power and flexibility, and seduced by the playful way their book
taught us to apply it to real problems.

At the start of the Summer that followed, I began an internship at a company,
National Instruments in Austin, centered around the application of theorem
provers to LabVIEW/G, a concurrent, graphical programming language. Given
my exposure to them, OTTER and the Boyer-Moore prover ACL2 were the
first tools I applied to the job. Soon after I began, I contacted Bill and asked
his advice. He not only gave me help and direction, but invited me to come to
Argonne National Laboratory and give a talk. The first talk I ever gave on my
own research was at Argonne at Bill's invitation.

---

[1] A Fascinating Country in the World of Computing: Your Guide to Automated Reasoning by Larry Wos and Gail W. Pieper, World Scientific, 2000.

M.P. Bonacina and M.E. Stickel (Eds.): McCune Festschrift, LNAI 7788, pp. 15–44, 2013.

I can still remember the excitement I felt at flying into Chicago, knowing that I would soon get to meet and brainstorm with Bill and Larry, true heroes of mine. At some point during my visit, Bill took notice of how interested I was in one of Larry and Gail's new monographs. It was a book[2] on *proof strategy*, the study of how one might navigate (as Dolph Ulrich writes in his beautiful Foreword) the "unimaginably vast" sea of deductions, "a space in which all proofs solving a particular problem at hand might well be as unreachable as the farthest stars in the most distant galaxies." Bill also must have noticed that the price of the book was very steep for me as an undergraduate, that it would take me some time to save up to afford a copy. A few weeks later, imagine my surprise when I received a package in the mail consisting of the book, a present from Bill. I have it with me in my office to this day. I am certain his kindness played a role in my going into automated reasoning.

The work that Leo and I present in this paper has been hugely influenced by the work of Bill, Larry and the Argonne group. Their relentless championing of the importance of strategy in resolution theorem proving has made the solution of so many otherwise impossible problems a reality. Our dream with this paper is to translate their important lesson and conviction into a form suitable for a comparatively young branch of automated reasoning known as SMT.

1   Introduction

SMT (Satisfiability Modulo Theories) solvers are a powerful class of automated theorem provers which have in recent years seen much academic and industrial uptake [19]. They draw on some of the most fundamental discoveries of computer science and symbolic logic. They combine the Boolean Satisfiability problem with the decision problem for concrete domains such as arithmetical theories of the integers, rationals and reals, and theories of data structures fundamental in computing such as finite lists, arrays, and bit-vectors. Research in SMT involves, in an essential way, decision problems, completeness and incompleteness of logical theories and complexity theory.

The standard account of modern SMT solver architecture is given by the so-called *DPLL(T) scheme* [35]. DPLL(T) is a theoretical framework, a rule-based formalism describing, abstractly, how satellite *theory solvers* (T-solvers) for decidable theories such as linear integer arithmetic, arrays and bit-vectors are to be integrated together with DPLL-based SAT solving. Decision procedures (complete T-solvers) for individual theories are combined by the DPLL(T) scheme in such a way that guarantees their combination is a complete decision procedure as well. Because of this, one might get the impression that *heuristics* are not involved in SMT. However, this is not so: heuristics play a vital role in high-performance SMT, a role which is all too rarely discussed or championed.

By design, DPLL(T) abstracts away from many practical implementation issues. High-performance SMT solvers contain many tightly integrated,

---

[2] Automated Reasoning and the Discovery of Missing and Elegant Proofs by Larry Wos and Gail W. Pieper, Rinton Press, 2003.

hand-crafted heuristic combinations of algorithmic proof methods which fall out-side the scope of DPLL(T). We shall discuss many examples of such heuristics in this paper, with a focus on our tools **RAHD** [36] and **Z3** [18]. To mention but one class of examples, consider formula preprocessing. This is a vital, heav-ily heuristic component of modern SMT proof search which occurs outside the purview of DPLL(T). We know of no two SMT solvers which handle formula preprocessing in exactly the same manner. We also know of no tool whose doc-umentation fully describes the heuristics it uses in formula preprocessing, let alone gives end-users principled methods to control these heuristics.

While the heuristic components of SMT solvers tend to be highly tuned for known classes of problems (e.g., SMT-LIB [4] benchmarks), they may easily per-form very badly on new classes of problems not anticipated by solver developers. This issue is becoming increasingly pressing as SMT solvers begin to gain the attention of practitioners in diverse areas of science and engineering. In many cases, changes to the prover heuristics can make a tremendous difference in the success of SMT solving within new problem domains. Classically, however, much of the control of these heuristics has been outside the reach[3] of solver end-users[4]. We would like much more control to be placed in the hands of end-users, and for this to be done in a principled way.

We present a challenge to the SMT community:

The Strategy Challenge

> To build theoretical and practical tools allowing users to exert strategic control over core heuristic aspects of high-performance SMT solvers.

In this way, high-performance SMT solvers may be tailored to specific prob-lem domains, especially ones very different from those normally considered. We present evidence, through research undertaken with the tools **RAHD** and **Z3**, that the adaptation of a few basic ideas of strategy prevalent both within the Argonne and LCF theorem proving paradigms can go a long way towards real-izing this goal. In the process, we solidify some foundations for strategy in the context of SMT and pose a number of questions and open problems.

---

[3] Some SMT solvers such as **CVC3** [5], **MathSAT** [10] and **Z3** [18] expose a vast collection of parameters to control certain behavioral aspects of their core proof procedures. We view these parameters as a primitive way of exerting strategic control over the heuristic aspects of high-performance SMT solvers. As the use of SMT solvers continues to grow and diversify, the number of these options has steadily risen in most solvers. For instance, the number of such options in **Z3** has risen from 40 (v1.0) to 240 (v2.0) to 284 (v3.0). Many of these options have been requested by end-users. Among end-users, there seems to be a wide-spread wish for more methods to exert strategic control over the prover's reasoning.

[4] We use the term *end-user* to mean a user of an SMT solver who does not contribute to the essential development of such a solver. End-users regularly embed SMT solvers into their own tools, making SMT solvers a subsidiary theorem proving engine for larger, specialised verification tool-chains.

1.1   Caveat Emptor: What This Paper Is and Is Not

We find it prudent at the outset to make clear what this paper is and is not. Let us first enumerate a few things we shall not do:

- We shall not give a new *theoretical framework* or *rule-based formalism* capturing the semantics of heuristic proof strategies, as is done, for instance, in the influential STRATEGO work on term rewriting strategies [30].
- We shall not prove any theorems about the algebraic structure underlying the collection of heuristic proof strategies, as is done, for instance, in the insightful work on the *proof monad* for interactive proof assistants [27].
- We shall not propose a concrete syntax for heuristic proof strategies in the context of SMT, as, for instance, an extension of the SMT-LIB standard [4].

We shall not attempt any of the (worthy) goals mentioned above because we believe to do so would be premature. It is simply too early to accomplish them in a compelling way. For example, before a standard concrete syntax is proposed for heuristic SMT proof strategies, many different instances of strategy in SMT must be explicated and analyzed, in many different tools, from many contrasting points of view. Before any theorems are proved about the collection of heuristic SMT proof strategies, we must have a firm grasp of their scope and limitations. And so on. Instead, our goals with this Strategy Challenge are much more modest:

- To bring awareness to the crucial role heuristics play in high-performance SMT, and to encourage researchers in the field to be more explicit as to the heuristic strategies involved in their solvers.
- To convince SMT solver developers that providing flexible methods (i.e., a *strategy language*) for end-users to exert fine-grained control over heuristic aspects of their solvers is an important, timely and worthy undertaking.
- To show how the adaptation of some ideas of strategy prevalent both within the Argonne and LCF theorem proving paradigms can go a long way towards realizing these goals.
- To stress that from a scientific point of view, the explication of the actual heuristic strategies used in high-performance SMT solvers is absolutely crucial for enabling the reproducibility of results presented in publications. For instance, if a paper reports on a new decision procedure, including experimental findings, and these experiments rely upon an implementation of the described decision method incorporating some heuristic strategies, then these heuristics should be communicated as well.

Throughout this paper, we shall present many examples in (variants of) the concrete syntaxes we have developed for expressing heuristic proof strategies in our own tools. These are given to lend credence to the practical value of this challenge, not as a proposal for a standard strategy language.

§

As we shall see, our initial approach for meeting this Strategy Challenge is based on an SMT-specific adaptation of the ideas of *tactics* and *tacticals* as found in LCF-style [33,39] proof assistants. This adaptation has a strong relationship with the approach taken by the Argonne group on *theorem proving toolkits*, work that began with the Argonne systems NIUTP1 - NIUTP7 in the 1970s and early 1980s, and has continued through to modern day with Bill McCune's OPS (OTTER Parts Store) in the early 1990s, the foundation of EQP and its parallel version Peers [31,9], and LADR (Library for Automated Deduction Research), the foundation of McCune's powerful Prover9 and MACE4 tools [32]. The way in which we adapt ideas of tactics and tacticals to SMT results in notions of strategy which, though borrowing heavily from both of these sources, are quite distinct from those found in LCF-style proof assistants and Argonne-style theorem provers.

Finally, let us assure the reader that we are *not* proposing a tactic-based approach for implementing "low-level" aspects of SMT solvers such as unit propagation, nor for implementing core reasoning engines such as a SAT solver or Simplex. This would hardly be practical for high-performance SMT. Again, our goals are more modest: We are proposing only the use of this strategy machinery for facilitating the orchestration of "big" reasoning engines, that is, for prescribing heuristic combinations of procedures such as Gaussian Elimination, Gröbner bases, CNF encoders, SAT solvers, Fourier-Motzkin quantifer elimination and the like. In this way, "big" symbolic reasoning steps will be represented as tactics heuristically composable using a language of tacticals.

While we do not propose a particular concrete strategy language for SMT solvers, we will present some key features we believe a reasonable SMT strategy language should have. These features have been implemented in our tools, and examples are presented. One important requirement will be that a strategy language support methods for conditionally invoking different reasoning engines based upon features of the formula being analyzed. We shall exploit this ability heavily. In some cases, when high-level "big" engines contain within themselves heavily heuristic components made up of combinations of other "big" reasoning engines, we also propose that these components be modularized and made replaceable by user-specified heuristic proof strategies given as parameters to the high-level reasoning engine. These ideas will be explained in detail in Sec. 3, where we give the basis of what we call *big-step strategies* in SMT, and in Sec. 4, where we observe that certain strategically parameterized reasoning engines can be realized as tacticals.

1.2   Overview

In Sec. 2 we begin with the question *What is strategy?* and explore some possible answers by recalling important notions of strategy in the history of automated theorem proving. In Sec. 3, we give the foundations of big-step strategies in SMT. In Sec. 4, we observe that there is a natural view of some reasoning engines as tacticals, and we present a few examples of this in the context of **RAHD** and **Z3**. In Sec. 5, we show some promising results of implementations of many of these strategy ideas within the two tools. Finally, we conclude in Sec. 6 with a look towards the future.

2   Strategy in Mechanized Proof

There exists a rich history of ideas of *strategy* in the context of mechanized proof. In this section, we work to give a high-level, incomplete survey of the many roles strategy has played in automated proof. In the process, we shall keep an eye towards why many of the same issues which motivated members of the non-SMT-based automated reasoning community to develop powerful methods for user-controllable proof search strategies also apply, compellingly, to the case of SMT.

2.1   What is Strategy?

Before discussing strategy any further, we should certainly attempt to define it. *What is strategy, after all?* Even restricted to the realm of mechanized proof, this question is terribly difficult to answer. There are so many aspects of strategy pervasive in modern proof search methods, and there seems to be no obvious precise delineations of their boundaries. Where does one, for instance, draw the line between the "strategic enhancement" of an existing search algorithm and the advent of a new search algorithm altogether?

Despite the difficulties fundamental to defining precisely what *strategy* is, many researchers in automated reasoning have proposed various approaches to incorporating strategy into automated proof. Some approaches have been quite foundational in character, shedding light on the nature of strategy in particular contexts. For instance, in the context of term rewriting, the ideas found within the STRATEGO system have given rise to much fruit, both theoretical and applied, and elucidated deep connections between term rewriting strategy and concurrency theory [30]. In the context of proof strategies in interactive theorem provers, the recent work of Kirchner-Muñoz has given heterogeneous proof strategies found in proof assistants like PVS a firm algebraic foundation using the category-theoretic notion of a monad [27].

Though a general definition of *what strategy is* seems beyond our present faculties[5], we may still make some progress by describing a few of its most salient aspects. In particular, the following two statements seem a reasonable place to begin:

1. There is a natural view of automated theorem proving as being an exercise in combinatorial search.
2. With this view in mind, then *strategy* may be something like *adaptations of general search mechanisms which reduce the search space by tailoring its exploration to a particular class of problems.*

---

[5] Much deep work outside the scope of this paper has been done on notions of strategy in automated reasoning, especially in the context of first-order theorem proving. See, e.g., the articles [7,8,28,41].

We are most interested in these adaptations when end-users of theorem proving tools may be given methods to control them.

To gain a more concrete understanding of the importance of strategy in automated theorem proving, it is helpful to consider some key examples of its use. In working to gather and present some of these examples, we have found the vast number of compelling uses of strategy to be quite staggering. There are so many examples, in fact, that it seems fair to say that much of the history of automated reasoning can be interpreted as a series of *strategic advances*[6]. As automated reasoning is such a broad field, let us begin by focusing on the use of strategy in one of its particularly telling strands, the history of mechanized proof search in first-order predicate calculus (FOL).

§

When the field of first-order proof search began and core search algorithms were first being isolated, most interesting strategic advancements were so profound that we today consider them to be the advent of genuinely "new" theorem proving methods. For example, both the Davis-Putnam procedure and Robinson's resolution can be seen to be strategic advancements for the general problem of first-order proof search based on Herbrand's Theorem. But, compared to their predecessors, the changes these strategic enhancements brought forth were of such a revolutionary nature that we consider them to be of quite a different kind than the strategies we want to make user-controllable in the context of SMT.

Once resolution became a dominant focus of first-order theorem proving, however, then more nuanced notions of strategy began to take hold, with each of them using resolution as their foundation. Many of the most lasting ideas in this line of research were developed by the Argonne group. These ideas, including the set of support, term weighting, the given-clause algorithm (and, e.g., its use of pick-given ratio), hot lists and hints, did something very interesting: They provided a fixed, general algorithmic search framework upon which end-users could exert some of their own strategic control by prescribing restrictions to guide the general method. Moreover, beginning in the 1970s, the Argonne group promoted the idea of *theorem proving toolkits*, libraries of high-performance reasoning engines one could use to build customized theorem provers. This idea has influenced very much the approach we propose for strategy in SMT.

Let us now turn, in more detail, to uses of strategy in the history of mechanized first-order proof. Following this, we shall then consider some aspects of strategy

---

[6] As a very helpful reviewer of this paper pointed out to us, it is interesting to note that the boundary between "strategy" and "inference system" is a flexible one which has been constantly evolving in theorem proving. Take for instance ordered resolution: it is presently considered an inference system, but at its inception the notion of selecting literals based on an ordering could have been seen as a strategy. There are many examples of ideas which were born as "strategy" and then became full-fledged "inference systems" as they were formalized.

in the context of LCF-style interactive proof assistants. Ideas of strategy from both of these histories will play into our work.

2.2   Strategy in Automated Theorem Proving over FOL

One cannot attempt to interpret the history of mechanized proof in first-order logic without bearing in mind the following fact: Over two decades before the first computer experiments with first-order proof search were ever performed, the undecidability of FOL was established. This result was well-known to the logicians who began our field. Thankfully, this seemingly negative result was tempered with a more optimistic one: the fact that FOL is *semi-decidable*. This allowed programs such as the British Museum Algorithm (and quickly others of a more sophisticated nature) to be imagined which, in principle, will always find a proof of a conjecture C over an axiomatic theory T if C is in fact true in all models of T.

Early in the field, and long before the advent of algorithmic complexity theory in any modern sense, obviously infeasible approaches like the British Museum [34] were recognized as such. Speaking of the earliest (1950s) research in computer mechanized proof, Davis writes in "The Early History of Automated Deduction" [14]:

> [...] it was all too obvious that an attempt to generate a proof of something non-trivial by beginning with the axioms of some logical system and systematically applying the rules of inference in all possible directions was sure to lead to a gigantic combinatorial explosion.

Thus, though a semi-complete theorem proving method was known (and such a method is in a sense "best possible" from the perspective of computability theory), its search strategy was seen as utterly hopeless. In its place, other search strategies were sought in order to make the theorem proving effort more tractable. This point of view was articulated at least as early as 1958 by Hao Wang, who writes in [45]:

> Even though one could illustrate how much more effective partial strategies can be if we had only a very dreadful general algorithm, it would appear desirable to postpone such considerations till we encounter a more realistic case where there is no general algorithm or no efficient general algorithm, e.g., in the whole predicate calculus or in number theory. As the interest is presumably in seeing how well a particular procedure can enable us to prove theorems on a machine, it would seem preferable to spend more effort on choosing the more efficient methods rather than on enunciating more or less familiar generalities.

At the famous 1957 five week Summer Institute for Symbolic Logic held at Cornell University, the logician Abraham Robinson[7] gave a remarkably influential

---

[7] It is useful to note that this Abraham Robinson, the model theorist and inventor of non-standard analysis, is not the same person as John Alan Robinson who would some 8 years later invent the proof search method of first-order resolution.

short talk [43] in which he singled out Skolem functions and Herbrand's Theorem as potentially useful tools for general-purpose first-order theorem provers [14]. Aspects of this suggestion were taken up by many very quickly, notably Gilmore [22], Davis and Putnam [16], and eventually J.A. Robinson [44]. Let us examine, from the perspective of strategy, a few of the main developments in this exceedingly influential strand.

As noted by Davis [14], the first Herbrand-based theorem provers for FOL employed completely *unguided* search of the Herbrand universe. There was initially neither a top-level conversion to a normal form such as CNF nor a systematic use of Skolem functions. When these first Herbrand-based methods were applied, through for instance the important early implementation by Gilmore [22], they proved unsuccessful for all but the simplest of theorems. Contemporaneously, Prawitz observed the same phenomena through his early work on a prover based on a modified semantic tableaux [14]. The lesson was clear: unguided search, even when based on deep results in mathematical logic, is not a viable approach. Again, new *strategies* were sought for controlling search space exploration.

The flurry of theorem proving breakthroughs in the early 1960s led to a wealth of new search strategies (and new notions of strategy) which still form the foundation for much of our field today.

First, based on shortcomings they observed in Gilmore's unguided exploration of the Herband universe, in particular the reliance of his method upon a DNF conversion for (what we now call) SAT solving, Davis and Putnam devised a new Herbrand universe exploration strategy which systematically applied Skolemization to eliminate existential quantifiers and used a CNF input formula representation as the basis to introduce a number of enhanced techniques for recognising the unsatisfiability of ground instances [16]. In the process, they spent much effort on enhancing the tractable recognition of ground unsatisfiability, which they believed at the time to be the biggest practical hurdle in Herbrand-based methods [14]. When further enhanced by Davis, Logemann and Loveland, this propositional component of Davis and Putnam's general first-order search strategy gave us what we now call DPLL, the foundation of modern SAT solving [15]. Nevertheless, once implementations were undertaken and experiments performed, the power of their first-order method was still found completely unsatisfactory. As Davis states [14],

> Although testing for satisfiability was performed very efficiently, it soon became clear that no very interesting results could be obtained without first devising a method for avoiding the generation of spurious elements of the Herbrand universe.

In the same year, Prawitz devised an "on-demand" method by which the generation of unnecessary terms in Herbrand expansions could be avoided, at the cost of sequences of expensive DNF conversions [42]. Though these DNF conversions precluded the practicality of Prawitz's method, his new approach made clear the potential utility of unification in Herbrand-based proof search, and Davis soon proposed [13,14] that effort be put towards

> ... a new kind of procedure which seeks to combine the virtues of the
> Prawitz procedure and those of the Davis Putnam procedure.

Two years later, Robinson published his discovery of such a method: Resolution, a single, easily mechanizable inference rule (refutationally) complete for FOL. This new method soon became a dominant high-level strategy for first-order proof search [44]. That resolution was a revolutionary improvement over previous methods is without question. But, practitioners soon discovered that the game was by no means won. Even with resolution in hand, the act of proof search was utterly infeasible for the vast majority of nontrivial problems without the introduction of some techniques for guiding the generation of resolvents. It is here that a new class of strategies was born.

In the 1960s, the group centered around Larry Wos at Argonne National Laboratory contributed many fundamental developments to first-order proving. At Argonne, Robinson made his discovery of resolution, first communicated in a privately circulated technical report in 1964, and then published in his influential JACM paper the year that followed. Almost immediately, Wos and his colleagues championed the importance of user-controllable strategies during resolution proof search and began developing methods for their facilitation. At least two important papers in this line were published by the end of 1965: "The Unit Preference Strategy in Theorem Proving" [47] and "Efficiency and Completeness of the Set of Support Strategy in Theorem Proving" [48]. The first gave rise to a strategy that the resolution prover would execute without any influence from the user. The second, however, introduced a strategy of a different kind: a method by which end-users could exert strategic control over the proof search without changing its underlying high-level method or impinging upon its completeness.

In resolution, the idea of set of support is to partition the CNF representation of the negated input formula into two sets of clauses, a satisfiable set A and another set S, and then to restrict the generation of clauses to those who can trace their lineage back to at least one member of S. Importantly, the choice of this division of the input clauses between A and S may be chosen strategically by the end-user. This general approach to heuristic integration — parameterizing a fixed proof search method by user-definable data — has proven enormously influential. Other methods in this class include clause weighting, hot lists, pick-given ratio, and many others.

Beginning in the 1970s, the Argonne group made an important methodological decision. This was the introduction of *theorem proving toolkits*. As Lusk describes [29],

> The notion of a toolkit with which one could build theorem provers was
> introduced at this time and became another theme for later Argonne development. In this case the approach was used to build a series of systems
> incorporating ever more complex variations on the closure algorithm
> without changing the underlying data structures and inference functions.
> The ultimate system (NIUTP7) provided a set of user-definable theorem-
> proving "environments," each running a version of the closure algorithm

with different controlling parameters, and a meta-language for controlling their interactions. There was enough control, and there were enough built-in predicates, that it became possible to "program" the theorem prover to perform a number of symbolic computation tasks. With these systems, Winker and Wos began the systematic attack on open problems [...].

Finally, we shall end our discussion of strategy in first-order provers with a few high-level observations. Given that the whole endeavor is undecidable, researchers in first-order theorem proving recognized very early that strategy must play an indispensible role in actually finding proofs. In the beginning of the field, new strategies were often so different from their predecessors that we consider them to be genuinely new methods of proof search altogether. But, once a general method such as resolution had taken hold as a dominant foundation, then strategies were sought for allowing users to control specific aspects of this fixed foundation. Abstractly, this was accomplished by providing a resolution proof search loop which accepts strategic data to be given as user-specifiable parameters.

Let us turn our attention now to another important contributor to ideas of strategy in computer-assisted proof, the LCF-style of interactive proof assistants.

2.3 Strategy in LCF-Style Interactive Proof Assistants

In the field of interactive proof assistants, strategy appears in many forms. The fact that humans contribute much more to the proof development process in proof assistants than in fully automatic provers gives rise to ideas of strategy quite distinct from those encountered in our discussion of first-order provers. As above, we must limit our discussion to a very small segment of the field. Let us in this section discuss one key exemplar of strategy in proof assistants, the approach of LCF.

Strategy in the LCF Approach. The original LCF was an interactive proof checking system designed by Robin Milner at Stanford in 1972. This system, so-named for its mechanization of Scott's Logic of Computable Functions, provided a proof checker with the following high-level functionality [23]:

> Proofs [were] conducted by declaring a main goal (a formula in Scott's logic) and then splitting it into subgoals using a fixed set of subgoaling commands (such as induction to generate the basis and step). Subgoals [were] either solved using a simplifier or split into simpler subgoals until they could be solved directly.

Soon after the birth of Stanford LCF, Milner moved to Edinburgh and built a team to work on its successor. A number of shortcomings had been observed in the original system. In particular, Stanford LCF embodied only one high-level proof strategy: 'backwards' proof, working from goals to subgoals. Moreover, even within a backwards proof, Stanford LCF had only a fixed set of proof construction commands which could not be easily extended [23]. Finding principled techniques to free the system from these strategic shackles became a driving motivation behind the design of Edinburgh LCF.

To address these problems, Milner devised a number of ingenious solutions which still today form the design foundation for many widely-used proof assistants. Fundamentally, Edinburgh LCF took the point of view of treating proofs as computation. New theorems were to be computed from previously established theorems by a fixed set of theorem constructors. To ensure that this computation was always done in a correct way, Milner designed an abstract datatype thm whose predefined values were instances of axioms and whose constructors were inference rules [23]. The idea was that strict type-checking would guarantee soundness by making sure that values of type thm were always actual theorems. The strictly-typed programming language ML (the Meta Language of Edinburgh LCF) was then designed to facilitate programming *strategies* for constructing values of type thm. It was a remarkable achievement. The strategic shackles of Stanford LCF had most certainly been relinquished, but much difficult work remained in order to make this low-level approach to proof construction practical. Many core strategies, both for facilitating high-level proof methods like backwards proof, as well as for implementing proof methods such as simplifiers and decision procedures needed to be built before it would be generally useful to end-users.

As mentioned, with the bare foundation of a type thm and the meta language ML, the system did not directly support backwards proof. To remedy this, Milner introduced *tactics* and *tacticals*. The idea of backwards proof is that one begins with a goal, reduces it to simpler subgoals, and in the process forms a proof tree. When any of these subgoals have been made simple enough to be discharged, then a branch in the proof tree can be closed. A tactic reduces a goal to a set of subgoals such that if every subgoal holds then the goal also holds. If the set of unproved subgoals is empty, then the tactic has proved the goal. A tactic not only reduces a goal to subgoals, but it also returns a proof construction function to justify its action. *Tacticals* are combinators that treat tactics as data, and are used to construct more complex tactics from simpler ones. Gordon summarizes nicely [23]:

> By encoding the logic as an abstract type, Edinburgh LCF directly supported forward proof. The design goal was to implement goal directed proof tools by programs in ML. To make ML convenient for this, the language was made functional so that subgoaling strategies could be represented as functions (called "tactics" by Milner) and operations for combining strategies could be programmed as higher-order functions taking strategies as arguments and returning them as results (called "tacticals"). It was anticipated that strategies might fail (e.g. by being applied to inappropriate goals) so an exception handling mechanism was included in ML.

Since the time of Edinburgh LCF (and its soon-developed successor, Cambridge LCF), the technology within LCF-style proof assistant has grown considerably. However, the underlying design principles centered around user-definable proof strategies have remained more-or-less the same, and are found today in tools like Isabelle [40], HOL [24], Coq [6], MetaPRL [25] and Matita [3]. For each of these tools, immense work has gone into developing powerful tactics which embody particular proof strategies. Many of them, such as the proof producing real closed field quantifier elimination tactic in HOL-Light, are tactics embodying complete decision procedures for certain logical theories. Others, such as the implementation by Boulton of a tactic based on Boyer-Moore induction heuristics in HOL, are powerful, incomplete heuristic strategies working over undecidable theories.

There are many things to learn from the success of the LCF paradigm. One lesson relevant to our work is the following: By "opening up" strategic aspects of the proof effort and providing principled, sound programming methods through which users may write their own proof strategies, it has been possible for enormously diverse ecosystems of powerful proof strategies to be developed, contributed and shared by communities of users of LCF-style proof assistants. As these ecosystems grow, the theorem proving tools become stronger, and the realizable practical verification efforts scale up significantly more than they would if these user-specifiable strategic enhancements were not possible.

3   Foundations for Big-Step Strategies

In this section, we propose a methodology for orchestrating reasoning engines where "big" symbolic reasoning steps are represented as functions known as *tactics*, and tactics are composed using combinators known as *tacticals*. We define notions of goals, tactics and tacticals in the context of SMT. Our definitions diverge from the ones found in LCF for the following main reasons:

- in SMT, we are not only interested in proofs of goals, but also in counterexamples (models yielding satisfying instances), and
- we want to support over and under-approximations when defining strategies.

Goals. The SMT decision problem consists of deciding whether a set of formulas S is satifisfiable or not modulo some background theory. We say each one of the formulas in S is an *assumption*. This observation suggests that a goal might be simply a collection of formulas. For practical reasons, a goal will also have a collection of attributes. Later, we describe some of the attributes we use in our systems. Thus, a goal is a pair comprised of a sequence of formulas and a sequence of attributes. We use sequences instead of sets because the ordering is relevant for some operations in our framework. For example, we define operations to "split" the *first* clause occurring in a goal. We say a goal G is *satisfisfiable* iff the conjunction of the formulas occurring in G is satisfiable. Using ML-like syntax, we define:

$$goal \quad = formula\ sequence \times attribute\ sequence$$

We say a goal is *trivially satisfiable* if the formula sequence is empty, and it is *trivially unsatisfiable* if the formula sequence contains the formula *false*. We say a goal is *basic* if it is trivially satisfiable or unsatisfiable.

Tactics. In our approach, when a tactic is applied to some goal G, four different outcomes are possible:

- The tactic succeeds in showing G to be satisfiable,
- The tactic succeeds in showing G to be unsatisfiable,
- The tactic produces a sequence of subgoals,
- The tactic fails.

A tactic returns a *model* when it shows G to be satisfiable, and a *proof* when it shows G to be unsatisfiable. A model is a sequence of assignments of symbols to values. Such symbols may be constant symbols, function symbols, etc. Values may be booleans, integers, algebraic reals, bit-vectors, lambda terms, etc. An empty model is an empty sequence of assignments. A proof of unsatisfiability may be a full certificate that can be independently checked by a proof checker, or it may be just a subset of the goals (also known as an unsat core) that were used to demonstrate the unsatisfiability. In this paper, we intentionally do not specify how proofs and models are represented; the framework we describe is general enough to accommodate many different decisions in this regard. In fact, **RAHD** and **Z3** use different representations.

When reducing a goal G to a sequence of subgoals $G_1, \ldots, G_n$, we face the problems of proof and model conversion. A *proof converter* constructs a proof of unsatisfiability for G using the proofs of unsatisfiability for all subgoals $G_1, \ldots, G_n$. Analogously, a *model converter* constructs a model for G using a model for some subgoal G_i. In the type declarations below, we use *trt* to stand for "tactic return type."

$$
\begin{aligned}
proofconv\ &=\ proof\ sequence \to proof \\
modelconv\ &=\ model \times nat \to model \\
trt\ &=\ \mathsf{sat}\ model \\
&\mid\ \mathsf{unsat}\ proof \\
&\mid\ \mathsf{unknown}\ goal\ sequence \times modelconv \times proofconv \\
&\mid\ \mathsf{fail} \\
tactic\ &=\ goal \to trt
\end{aligned}
$$

The second parameter of a model converter is a natural number used to communicate to the model converter which subgoal was shown to be satisfiable.

Let us gain some intuition about tactics and tacticals in the context of SMT with a few simple examples.

The tactic elim eliminates constants whenever the given goal contains equations of the form $a = t$, where a is a constant and t is a term not containing a.

For example, suppose elim is applied to a goal containing the following sequence comprised of three formulas:

$$[\ a = b + 1,\ (a < 0 \lor a > 0),\ b > 3\ ]$$

The result will be $\mathsf{unknown}(s, mc, pc)$, where s is a sequence containing the single subgoal:

$$[\ (b + 1 < 0 \lor b + 1 > 0),\ b > 3\ ]$$

The model converter mc is a function s.t. when given a model M for the subgoal above, mc will construct a new model M' equal to M except that the interpretation of a in M' ($M'(a)$) is equal to the interpretation of b in M plus one (i.e., $M(b) + 1$). Similarly, the proof converter pc is a function s.t. given a proof of unsatisfiability for the subgoal will construct a proof of unsatisfiability for the original goal using the fact that $(b + 1 < 0 \lor b + 1 > 0)$ follows from $a = b + 1$ and $(a < 0 \lor a > 0)$.

The tactic split-or splits a disjunction of the form $p_1 \lor \ldots \lor p_n$ into cases and then returns n subgoals. If the disjunction to be split is not specified, the tactic splits the first disjunction occurring in the input goal. For example, given the goal G comprised of the following sequence of formulas:

$$[\ a = b + 1,\ (a < 0 \lor a > 0),\ b > 3\ ]$$

split-or G returns $\mathsf{unknown}([G_1, G_2], mc, pc)$, where G_1 and G_2 are the subgoals comprised of the following two formula sequences respectively:

$$[\ a = b + 1,\ a < 0,\ b > 3\ ]$$

$$[\ a = b + 1,\ a > 0,\ b > 3\ ]$$

The model converter mc is just the identity function, since any model for G_1 or G_2 is also a model for G. The proof converter pc just combines the proofs of unsatisfiability for G_1 and G_2 in a straighforward way. If G does not contain a disjunction, then split-or just returns the input goal unmodified. Another option would be to fail.

RAHD and **Z3** come equipped with several built-in tactics. It is beyond the scope of this paper to document all of them. Nevertheless, let us list some of them for illustrative purposes:

- simplify: Apply simplification rules such as constant folding (e.g., $x + 0 \rightsquigarrow x$).
- nnf: Put the formula sequence in negation normal form.
- cnf: Put the formula sequence in conjunctive normal form.
- tseitin: Put the formula sequence in conjunctive normal form, but use fresh Boolean constants and predicates for avoiding exponential blowup. The model converter produced by this tactic "erases" these fresh constants and predicates introduced by it.
- lift-if: Lift term if-then-else's into formula if-then-else's
- bitblast: Reduce bitvector terms into propositional logic.

- gb: Let E be the set of arithmetic equalities in a goal G, gb replaces E with the Gröbner basis induced by E.
- vts: Perform virtual term substitution.
- propagate-bounds: Perform bound propagation using inference rules such as $x < 1 \land y < x$ implies $y < 1$.
- propagate-values: Perform value propagation using equalities of the form $t = a$ where a is a numeral.
- split-ineqs: Split inequalities such as $t \leq 0$ into $t = 0 \lor t < 0$.
- som: Put all polynomials in sum of monomials form.
- cad: Apply cylindrical algebraic decomposition.

Tacticals. It is our hope[8] that tactics will be made available in the APIs of next generation SMT solvers. Developers of interactive and automated reasoning systems will be able to combine these tactics using their favorite programming language. Like in LCF, it is useful to provide a set of combinators (tacticals) that are used to combine built-in tactics into more complicated ones. The main advantage of using tacticals is that the resulting tactic is guaranteed to be correct, that is, it is sound if the used building blocks are sound, it connects the model converters and proof converters appropriately, and there is no need to keep track of which subgoals were already proved to be unsatisfiable. We propose the following basic tacticals:

then : $(tactic \times tactic) \to tactic$
   then(t_1, t_2) applies t_1 to the given goal and t_2 to every subgoal produced by t_1. The resulting tactic fails if t_1 fails when applied to the goal, or if t_2 does when applied to any of the resulting subgoals.

then* : $(tactic \times tactic\ sequence) \to tactic$
   then*$(t_1, [t_{2_1}, ..., t_{2_n}])$ applies t_1 to the given goal, producing subgoals $g_1, ..., g_m$. If $n \neq m$, the tactic fails. Otherwise, it applies, in parallel, t_{2_i} to every goal g_i. The resultant set of subgoals is the union of all subgoals produced by t_{2_i}'s. The resulting tactic fails if t_1 fails when applied to the goal, or if t_{2_i} does when applied to goal g_i. The resultant tactic also fails if the number of subgoals produced by t_1 is not n.

orelse : $(tactic \times tactic) \to tactic$
   orelse(t_1, t_2) first applies t_1 to the given goal, if it fails then returns the result of t_2 applied to the given goal.

par : $(tactic \times tactic) \to tactic$
   par(t_1, t_2) executes t_1 and t_2 in parallel to the given goal. The result is the one produced by the first tactic to terminate its execution. After one tactic terminates, the execution of the other one is terminated. The resulting tactic fails only if t_1 and t_2 fails when applied to the goal.

---

[8] In fact, **Z3** 4.0 is now available with all of the strategy machinery described in this paper. It uses the strategy language internally and publishes a strategy API. Bindings of the strategy API are also available within Python. This Python **Z3** strategy interface can be experimented with on the web at `http://rise4fun.com/Z3Py`.

repeat : *tactic → tactic*
> Keep applying the given tactic until no subgoal is modified by it. repeat(*t*) fails if *t* fails.

repeatupto : *tactic × nat → tactic*
> Keep applying the given tactic until no subgoal is modified by it, or the maximum number of iterations is reached. repeatupto(*t*) fails if *t* fails.

tryfor : *tactic × milliseconds → tactic*
> tryfor(*t, k*) returns the value computed by tactic *t* applied to the given goal if this value is computed within *k* milliseconds, otherwise it fails. The resulting tactic also fails if *t* fails when applied to the goal.

The tactic skip never fails and just returns the input goal unaltered, it is the unit for then: then(skip, *t*) = then(*t*, skip) = *t*; and fail always fails, and is the unit for orelse: orelse(fail, *t*) = orelse(*t*, fail) = *t*. Note that then, orelse and par are associative. From now on, in order to simplify the presentation of examples, we write t_1 ; t_2 to denote then(t_1, t_2), and $t_1 \mid t_2$ to denote orelse(t_1, t_2).

Formula Measures. Several SMT solvers use hard-coded strategies that perform different reasoning techniques depending on structural features of the formula being analyzed. For example, **Yices** [21] checks whether a formula is in the difference logic fragment or not. A formula is in the difference logic fragment if all atoms are of the form $x - y \bowtie k$, where x and y are uninterpreted constants, k a numeral, and $\bowtie$ is in $\{\leq, \geq, =\}$. If the formula is in the difference logic fragment, **Yices** checks if the number of inequalities divided by the number of uninterpreted constants is smaller than a threshold k. If this is the case, it uses the Simplex algorithm for processing the arithmetic atoms. Otherwise, it uses an algorithm based on the Floyd-Warshall all-pairs shortest distance algorithm. We call such structural features *formula measures*. This type of ad hoc heuristic strategy based upon formula measures is very common.

We use formula measures to create Boolean expressions that are evaluated over goals. The built-in tactic check : *cond → tactic* fails if the given goal does not satisfy the condition *cond*, otherwise it just returns the input goal. Many numeric and Boolean measures are available in **RAHD** and **Z3**. Here is an incomplete list for illustrative purposes:

bw: Sum total bit-width of all rational coefficients of polynomials.
diff: True if the formula is in the difference logic fragment.
linear: True if all polynomials are linear.
dim: Number of uninterpreted constants (of sort real or int).
atoms: Number of atoms.
degree: Maximal total multivariate degree of polynomials.
size: Total formula size.

Nontrivial conditions can be defined using the built-in measures, arithmetic operators (e.g., $+$, $-$, $\times$, $/$, $<$, $=$) and Boolean operators (e.g., $\wedge$, $\vee$, $\neg$). For example, the **Yices** strategy described above can be encoded as:

$$(\mathsf{check}(\neg\mathsf{diff} \vee \frac{\mathsf{atom}}{\mathsf{dim}} < k) \, ; \, \mathsf{simplex}) \mid \mathsf{floydwarshall}$$

Now, we define the combinators if and when based on the combinators and tactics defined so far.

$$\mathsf{if}(c, \ t_1, \ t_2) = (\mathsf{check}(c)\,;\, t_1)\,|\,t_2$$
$$\mathsf{when}(c, \ t) = \mathsf{if}(c, \ t, \ \mathsf{skip})$$

These are often helpful in the construction of strategies based upon formula measures.

Under and over-approximations. Under and over-approximation steps are commonly used in SMT solvers. An *under-approximation step* consists of reducing a set of formulas S to a set S' such that if S' is satisfiable, then so is S, but if S' is unsatisfiable, then nothing can be said about S. For example, any *strengthening* step that obtains S' by adding to S new formulas not deducible from S is an under-approximation.

A more concrete example is found in many SMT solvers for nonlinear integer arithmetic, where lower and upper bounds are added for every uninterpreted constant of sort int, and the resulting set of formulas is then reduced to SAT. Under-approximations are also used in finite model finders for first-order logic formulas, where the universe is assumed to be finite, and the first-order formula is then reduced into SAT. Analogously, an *over-approximation step* consists of reducing a set of formulas S into a set S' such that if S' is unsatisfiable, then so is S, but if S' is satisfiable, then nothing can be said about S. For example, any *weakening* step that removes formulas from S is an over-approximation. Boolean abstraction is another example used in many interactive theorem provers and SMT solvers. This comprises replacing every theory atom with a fresh propositional variable. Clearly, if the resulting set of formulas is unsatisfiable then so is the original set. Of course, given a set of formulas S, arbitrarily applying under and over-approximation steps result in set of formulas S' that cannot be used to establish the satisfiability nor the unsatisfiability of S. To prevent under and over-approximation steps from being incorrectly applied, we associate a precision attribute with every goal. A precision marker is an element of the set {prec, under, over}. A tactic that applies an under (over) approximation fails if the precision attribute of the input goal is over (under).

4   Parametric Reasoning Engines as Tacticals

Some reasoning engines utilize other engines as subroutines. It is natural to view these higher-level reasoning engines as tacticals. Given a subsidiary engine (a tactic given to the higher-level engine as a parameter), these tacticals produce a new tactic. Let us describe two examples of such parametric engines.

SMT Solvers. We observe three main phases in state-of-the-art SMT solvers: *preprocessing, search,* and *final check.*

Preprocessing. During preprocessing, also known as presolving, several simplifications and rewriting steps are applied. The main goal is to put the problem in a form that is suitable for solving. Another objective is to simplify the problem, eliminate uninterpreted constants, unconstrained terms, and redundancies. Some solvers may also apply reduction techniques such as bit-blasting where bit-vector terms are reduced to propositional logic. Another commonly used reduction technique is Ackermannization [2,12] where uninterpreted function symbols are eliminated at the expense of introducing fresh constants and additional constraints.

Search. During the search step, modern SMT solvers combine efficient SAT solving with "cheap" theory propagation techniques. Usually, this combination is an incomplete procedure. For example, consider problems containing arithmetic expressions. Most solvers ignore integrality and nonlinear constraints during the search step. These solvers will only propagate Boolean and linear constraints, and check whether there is a rational assignment that satisfies them. We say the solver is *postponing* the application of "expensive" and complete procedures to the final check step. Solvers, such as **Z3**, only process nonlinear constraints during final check. The word "final" is misleading since it may be executed many times for a given problem. For example, consider the following nonlinear problem comprised of three assumptions (over $\mathbb{R}$):

$$[\, x = 1, \; y \geq x + 1, \; (y \times y < 1 \vee y < 3 \vee y \times y > x + 3) \,]$$

In the preprocessing step, a solver may decide to eliminate x using Gaussian elimination obtaining:

$$[\, y \geq 2, \; (y \times y < 1 \vee y < 3 \vee y \times y > 4) \,]$$

During the search step, the solver performs only Boolean propagation and cheap theory propagation such as $y \geq 2$ implies $\neg(y < 2)$. Nonlinear monomials, such as $y \times y$, are treated as fresh uninterpreted constants. Thus, the incomplete solver used during the search may find the candidate assigment $y = 2$ and $y \times y = 0$. This assignment satisfies the atoms $y \geq 2$ and $y \times y < 1$, and all Boolean and linear constraints.

Final check. During final check, a complete procedure for nonlinear real arithmetic is used to decide $[\, y \geq 2, \; y \times y < 1 \,]$. The complete procedure finds it to be unsatisfiable, and the solver backtracks and learns the lemma $(\neg y \geq 2 \vee y \times y < 1)$. The search step resumes, and finds a new assignment that satisfies $[\, y \geq 2, \; y \times y > 4 \,]$. The final check step is invoked again, and this time it finds the constraints to be satisfiable and the SMT solver terminates. The procedure above can be encoded as tactic of the form:

preprocess ; smt(finalcheck)

where preprocess is a tactic corresponding to the preprocessing step, and finalcheck is another tactic corresponding to the final check step, and smt is a tactical. The smt tactical uses a potentially expensive finalcheck tactic to complement an incomplete and fast procedure based on SAT solving and cheap theory propagation.

Abstract Partial Cylindrical Algebraic Decomposition (AP-CAD). AP-CAD [36,38] is an extension of the well-known real closed field (RCF) quantifier elimination procedure *partial cylindrical algebraic decomposition* (P-CAD). In AP-CAD, arbitrary (sound but possibly incomplete) ∃-RCF decision procedures can be given as parameters and used to "short-circuit" certain expensive computations performed during CAD construction. The ∃-RCF decision procedures may be used to reduce the expense of the different phases of the P-CAD procedure. The key idea is to use some *fast, sound* and *incomplete* procedure P to improve the performance of a *complete* but potentially very expensive procedure. The procedure P may be the combination of different techniques based on interval constraint propagation, rewriting, Gröbner basis computation, to cite a few. These combinations may be tailored as needed for different application domains. These observations suggest that P is a tactic, and AP-CAD is tactical that given P returns a tactic that implements a complete ∃-RCF decision procedure.

We now illustrate the flexibility of our approach using the following simple strategy for nonlinear arithmetic:

$$\text{simplify}; \text{gaussian}; (\text{modelfinder} \mid \text{smt}(\text{apcad}(\text{icp})))$$

The preprocessing step is comprised of two steps. First, simple rewriting rules such as constant folding and gaussian elimination are applied. Then, a model finder for nonlinear arithmetic based on SAT [49] is used. If it fails, smt is invoked using AP-CAD (apcad) in the final check step. Finally, AP-CAD uses interval constraint propagation (icp) to speedup the AP-CAD procedure.

5   Strategies in Action

We demonstrate the practical value of our approach by describing successful strategies used in **RAHD** and **Z3**. We also provide evidence that the overhead due to the use of tactics and tacticals is insignificant, and the gains in performance substantial.

5.1   Z3 QF_LIA Strategy

SMT-LIB [4] is a repository of SMT benchmark problems. The benchmarks are divided in different divisions. The QF_LIA division consists of linear integer arithmetic benchmarks. These benchmarks come from different application domains such as scheduling, hardware verification, software analysis and bounded-model checking. The structural characteristics of these problems are quite diverse. Some of them contain a rich Boolean structure, and others are just the conjunction of linear equalities and inequalities. Several software analysis benchmark make extensive use of if-then-else terms that need to be eliminated during a preprocessing step. A substantial subset of the benchmarks are unsatisfiable even when integrality constraints are ignored, and can be solved using a procedure for linear real arithmetic, such as Simplex. We say a QF_LIA benchmark is *bounded* if every uninterpreted constant a has a lower ($k \leq a$) and upper bound ($a \leq k$), where k is a numeral. A bounded benchmark is said to be 0-1 (or *pseudo-boolean*) if the lower (upper) bound of every uninterpreted constant is 0 (1). Moreover, some of the problems in QF_LIA become bounded after interval constraint propagation is applied.

Z3 3.0 won the QF_LIA division in the 2011 SMT competition[9] (SMT-COMP'11). The strategy used by **Z3** can be summarized by the following tactic:

preamble ; (mf | pb | bounded | smt)

where the preamble, mf, pb and bounded tactics are defined as

$$\begin{aligned}
\text{preamble} &= \text{simplify} ; \text{propagate-values} ; \text{ctx-simplify} ; \text{lift-if} ; \text{gaussian} ; \text{simplify} \\
\text{mf} &= \text{check(is-ilp)} ; \text{propagate-bounds} ; \\
&\quad \begin{pmatrix} \text{tryfor(mip, 5000)} & | \text{ tryfor(smt-no-cut(100), 2000)} \; | \\ \text{(add-bounds}(-16, 15)\text{)} ; \text{smt)} \; | \text{ tryfor(smt-no-cut(200), 5000)} \; | \\ \text{(add-bounds}(-32, 31)\text{)} ; \text{smt)} \; | \text{ mip} \end{pmatrix} \\
\text{pb} &= \text{check(is-pb)} ; \text{pb2bv} ; \text{bv2sat} ; \text{sat} \\
\text{bounded} &= \text{check(bounded)} ; \begin{pmatrix} \text{tryfor(smt-no-cut(200), 5000)} & | \\ \text{tryfor(smt-no-cut-no-relevancy(200), 5000)} \; | \\ \text{tryfor(smt-no-cut(300), 15000)} \end{pmatrix}
\end{aligned}$$

The tactic smt is based on the **Yices** approach for linear integer arithmetic. The tactic ctx-simplify performs contextual simplification using rules such as:

$$(a \neq t \vee F[a]) \; \rightsquigarrow \; (a \neq t \vee F[t])$$

The tactic mip implements a solver for mixed integer programming. It can only process conjunctions of linear equations and inequalities. The tactic fails if the input goal contains other Boolean connectives. The tactic smt-no-cut(seed) is a variation of the **Yices** approach in which Gomory cuts are not used. The parameter seed is a seed for the pseudo-random number generator. It is used to randomize the search. The tactic smt-no-cut-no-relevancy(seed) is yet another variation where "don't care" propagation is disabled [17]. The tactic pb2bv converts a pseudo-boolean formula into a bit-vector formula. It fails if the input goal is not pseudo-boolean. Similarly, the tactic bv2sat bitblasts bit-vector terms into propositional logic. The tactic sat implements a SAT solver. Finally, the tactic add-bounds(lower, upper) performs an under-approximation by adding lower and upper bounds to all uninterpreted integer constants. The idea is to guarantee that the branch-and-bound procedure used in smt and mip terminates. The tactic mf is essentially looking for models where all integer variables are assigned to

---

[9] http://www.smtcomp.org

small values. The tactic pb is a specialized 0-1 (Pseudo-Boolean) solver. It fails if the problem is not 0-1.

To demonstrate the benefits of our approach we run all QF_LIA benchmarks using the following variations of the strategy above:

pre	= preamble; smt
pre+pb	= preamble; (pb | smt)
pre+bounded	= preamble; (bounded | smt)
pre+mf	= preamble; (mf | smt)
combined	= preamble; (mf | pb | bounded | smt)

All experiments were conducted on an Intel Quad-Xeon (E54xx) processor, with individual runs limited to 2GB of memory and 600 seconds. The results of our experimental evaluation are presented in Table 1. The rows are associated with the individual benchmark families from QF_LIA division, and columns separate different strategies. For each benchmark family we write the number of benchmarks that each strategy failed to solve within the time limit, and the cumulative time for the solved benchmarks.

Table 1. Detailed Experimental Results

benchmark family	smt		pre		pre+pb		pre+bounded		pre+mf		combined	
	failed	time (s)	failed	time (s)	failed	time (s)	failed	time (s)	failed	time (s)	failed	time (s)
Averest (19)	0	4.0	0	5.9	0	6.0	0	5.9	0	5.9	0	5.9
bofill sched (652)	1	1530.7	1	1208.3	1	1191.5	1	1205.4	1	1206.0	1	1205.9
calypto (41)	1	2.0	1	7.7	1	8.0	1	7.8	1	7.9	1	7.8
CAV 2009 (600)	190	1315.3	190	1339.3	190	1329.7	190	1342.7	1	8309.5	1	8208.1
check (5)	0	0.1	0	0.1	0	0.1	0	0.1	0	0.1	0	0.1
CIRC (51)	17	188.4	17	239.8	17	238.2	17	336.1	17	221.5	8	158.66
convert (319)	206	1350.5	190	3060.6	190	3025.9	0	112.7	190	3030.2	0	112.5
cut lemmas (100)	48	2504.0	48	2532.4	48	2509.2	48	2543.4	27	3783.9	27	3709.0
dillig (251)	68	1212.0	68	1237.6	68	1226.9	68	1242.5	3	2677.8	3	2763.9
mathsat (121)	0	171.4	0	150.2	0	149.9	0	151.1	0	150.9	0	150.2
miplib2003 (16)	5	53.8	5	57.7	5	424.4	5	109.5	5	58.8	5	430.5
nec-smt (2780)	147	224149.0	8	59977.4	8	59968.3	8	59929.3	8	60042.1	8	60032.9
pb2010 (81)	43	90.3	43	96.2	25	2581.2	43	146.3	43	96.2	25	2583.1
pidgeons (19)	0	0.3	0	0.4	0	0.4	0	0.3	0	0.3	0	0.3
prime-cone (37)	13	9.6	13	9.5	13	9.5	13	9.7	0	11.0	0	11.0
rings (294)	48	4994.4	46	5973.7	46	6016.2	48	9690.0	46	6024.6	48	9548.2
rings pre (294)	57	441.5	54	1288.7	54	1261.9	54	1260.9	54	1274.7	54	1261.5
RTCL (2)	0	0.1	0	0.1	0	0.1	0	0.1	0	0.1	0	0.1
slacks (251)	135	1132.9	136	550.0	136	545.9	136	550.8	53	8969.3	53	8803.9
total (5938)	978	239153.0	819	77737.4	801	80495.2	631	78646.5	449	95872.4	234	98995.2

Overall, the combined strategy is the most effective one, solving the most problems. It fails only on 234 out of 5938 benchmarks. In constrast, the basic smt strategy fails on 978 benchmarks. The results also show which tactics are effective in particular benchmark families. The tactics ctx-simplify and lift-if are particularly effective on the NEC software verification benchmarks (sf nec-smt). The pseudo-boolean strategy reduces by half the number of failures in the industrial pseudo-boolean benchmarks coming from the 2010 pseudo-boolean competition. The convert software verification benchmarks become trivial when Gomory cuts are disabled by the tactic bounded. Finally, the model finder tactic mf is very effective on crafted benchmark families such as CAV 2009, cut lemmas, dillig, prime-cone, and slacks.

Figure 1 contains scatter plots comparing the strategies described above. Each point on the plots represents a benchmark. The plots are in log scale. Points below (above) the diagonal are benchmarks where the strategy on y-axis (x-axis) is faster than the strategy on the x-axis (y-axis). Note that in some cases, the combined strategy has a negative impact, but it overall solves more problems.

We observed several benefits when using tactics and tacticals in **Z3**. First, it is straighforward to create complex strategies using different solvers and techniques. The different solvers can be implemented and maintained independently of each other. The overhead of using tactics and tacticals is insignificant. We can provide custom strategies to **Z3** users in different application domains. Finally, the number of SMT-LIB problems that could be solved by **Z3** increased dramatically. **Z3** 2.19 uses only the (default) smt tactic, and fails to solve 978 (out of 5938) QF_LIA benchmarks with a 10 minutes timeout. In contrast, **Z3** 3.0 fails in only 234 benchmarks. In **Z3** 4.0, tactic and tacticals are available in the programmatic API and SMT 2.0 frontend.

5.2   Calculemus RAHD Strategies

The calculemus **RAHD** strategies[10] combine simplification procedures, interval constraint propagation, Gröbner basis computation, non-strict inequality splitting, DNF conversion, OpenCAD and CAD. OpenCAD is a variation of the CAD procedure that can be applied to problems containing $\wedge\vee$ combinations of polynomial strict inequalities. OpenCAD takes advantage of the topological structure of the solution set of these formulas (such solution sets are always open subsets of $\mathbb{R}^n$) to yield a proof procedure substantially faster than general CAD. This speed-up is caused by (i) the use rational numbers instead of irrational real algebraic numbers to represent CAD sample points, and (ii) the use of an efficient projection operator which avoids the costly computation of polynomial subresultants.

The key idea of the calculemus strategy is to split non-strict inequalities ($p \leq 0$) appearing in a conjunctive formula F into ($p < 0 \vee p = 0$), resulting in two sub-problems $F_<$ and $F_=$. The branch $F_<$ containing the strict inequality is then closer to being able to be processed using OpenCAD, while the branch $F_=$ containing the equality has an enriched equational structure which is then used, via Gröbner basis computation, to inject equational information into the polynomials appearing in the strict inequalities in $F_=$. If the ideal generated by the equations in the branch $F_=$ is rich enough and the original formula is unsatisfiable, then this unsatisfiability of $F_=$ may be recognized by applying

---

[10] Detailed descriptions of these strategies may be found in Passmore's PhD thesis [36].

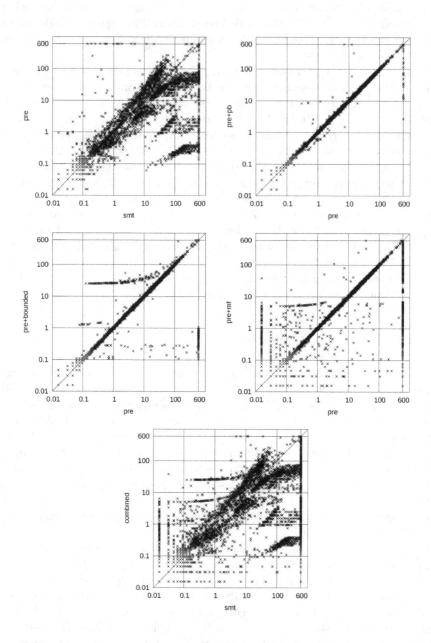

Fig. 1. smt, pre, pre+pb, pre+bounded, pre+mf and combined strategies

OpenCAD only to the resulting strict inequational fragment of $F_=$ after this Gröbner basis reduction has been performed.

In this section, we consider the basic calculemus strategy calc-0, and two refinements: calc-1 and calc-2. These refinements use formula measures to control inequality splitting. Moreover, interval constraint propagation is used to close goals before further splitting is performed.

Table 2. The three **RAHD calculemus** proof strategies compared with QEPCAD-B and Redlog on twenty-four problems

benchmark	dimension	degree	calc-0 time (s)	calc-1 time (s)	calc-2 time (s)	qepcad-b time (s)	redlog/rlqe time (s)	redlog/rlcad time (s)
P0	5	4	0.9	1.6	1.7	416.4	40.4	>600.0
P1	6	4	1.7	3.1	3.4	>600.0	>600.0	>600.0
P2	5	4	1.3	2.4	2.6	>600.0	>600.0	>600.0
P3	5	4	1.5	2.5	2.7	>600.0	>600.0	>600.0
P4	5	4	1.1	2.0	2.7	>600.0	>600.0	>600.0
P5	14	2	0.3	0.3	0.3	>600.0	97.4	>600.0
P6	11	5	147.4	<0.1	<0.1	>600.0	<0.1	<0.1
P7	8	2	<0.1	<0.1	<0.1	<0.1	<0.1	<0.1
P8	7	32	4.5	0.1	<0.1	8.4	<0.1	>600.0
P9	7	16	4.5	0.2	<0.1	0.3	<0.1	6.7
P10	7	12	100.7	20.8	8.9	>600.0	>600.0	>600.0
P11	6	2	1.6	0.5	0.5	<0.1	<0.1	<0.1
P12	5	3	0.8	0.3	0.4	<0.1	<0.1	<0.1
P13	4	10	3.8	3.9	4.0	>600.0	>600.0	>600.0
P14	2	2	4.5	1.7	<0.1	<0.1	>600.0	>600.0
P15	4	3	0.2	0.2	0.1	<0.1	<0.1	<0.1
P16	4	2	10.0	2.2	2.1	<0.1	<0.1	<0.1
P17	4	2	0.6	0.6	0.7	0.3	<0.1	0.6
P18	4	2	1.3	1.3	1.3	<0.1	<0.1	<0.1
P19	3	6	3.3	1.7	2.1	<0.1	<0.1	0.7
P20	3	4	1.2	0.7	0.7	<0.1	<0.1	0.3
P21	3	2	<0.1	<0.1	<0.1	<0.1	<0.1	<0.1
P22	2	4	<0.1	<0.1	<0.1	<0.1	<0.1	<0.1
P23	2	2	<0.1	<0.1	<0.1	<0.1	<0.1	<0.1

Table 2 shows the performance of the calculemus **RAHD** strategies on the twenty-four benchmarks considered in [37] and compares this performance to that of QEPCAD-B [11] and two quantifier elimination procedures available in Reduce/Redlog [1]:

- Rlqe, which is an enhanced implementation by Dolzmann and Sturm of Weispfenning's quadratic virtual term substitution (VTS) [46], and
- Rlcad, which is an implementation by Seidl, Dolzmann and Sturm of Collins-Hong's partial CAD [20].

Comparing these calculemus strategies with each other and with other tools (QEPCAD-B, Redlog/rlqe, Redlog/rlcad) is interesting in this setting, as in addition to having its own implementation of many real algebraic decision methods, **RAHD** also provides access to these other tools as tactics. For instance, the OpenCAD tactic used in the calculemus strategies is actually executed by **RAHD** invoking QEPCAD-B in a special mode which only lifts over full-dimensional cells. Thus, the comparison given in **Table 2** is indeed a comparison between six **RAHD** strategies: the three calculemus strategies, and three simple strategies which only invoke QEPCAD-B, Redlog/Rlqe and Redlog/Rlcad, respectively.

For each benchmark we write the dimension and maximal total multivariate degree of polynomials, and the total runtime for each strategy and solver. Experiments were performed on a 2 x 2.4 GHz Quad-Core Intel Xeon PowerMac with 10GB of 1066 MHz DDR3 RAM.

For brevity, let us only compare calc-0 with the QEPCAD-B and Redlog procedures. With this restriction, the results of these experiments can be broadly summarized as follows:

- The calc-0 strategy is able to solve a number of high-dimension, high-degree problems that QEPCAD-B, Redlog/Rlqe, and Redlog/Rlcad are not. It is interesting that while the calc-0 strategy involves an exponential blow-up in its reliance on inequality splitting followed by a DNF normalisation, for many benchmarks the increase in complexity caused by this blow-up is overshadowed by the decrease in complexity of the CAD-related computations this process induces.
- Redlog/Rlqe is able to solve a number of high-dimension, high-degree benchmarks that QEPCAD-B and Redlog/Rlcad are not.
- Redlog/Rlqe is able to solve a number of benchmarks significantly faster than the calc-0 strategy, Redlog/Rlcad, and QEPCAD-B.
- For the benchmarks QEPCAD-B is able to solve directly, using QEPCAD-B directly tends to be much faster than using the calc-0 strategy.

Overall, the final refinement, calc-2, substantially improves upon the strategy calc-0 on benchmarks P6, P8, P10, P11, P12, P14, P16, P19 and P20, often by many orders of magnitude. On benchmarks P0, P1, P2, P3, P4, calculemus-2 is slower than calc-0 by roughly a factor of two. Strategies calc-1 and calc-2 are roughly equal for most benchmarks, except for P1 and P19 where calc-2 is slightly ($\cong$ 10-20%) slower, and P10 and P14 where calc-2 is substantially ($\cong$ 2-25x) faster.

6 Conclusion

We have demonstrated the practical value of heuristic proof strategies within the context of our **RAHD** and **Z3** tools. We have illustrated that not only is a strategy-language based approach practical in the context of high-performance solvers, it is also desirable. A key take-away message is the following: In difficult (i.e., infeasible or undecidable) theorem proving domains, the situation with heuristic proof strategies is rarely "one size fits all." Instead, given a class of problems to solve, it is often the case that one heuristic combination of reasoning engines is far more suited to the task than another. SMT solver developers cannot anticipate all classes of problems end-users will wish to analyze. By virtue of this, heuristic components of high-performance solvers will never be sufficient in general when they are beyond end-users' control. Without providing end-users mechanisms to control and modify the heuristic components of their solvers, solver developers are inhibiting their chances of success.

Beyond the situation with end-users, let us also make the following anecdotal remarks as solver developers. By introducing a strategy language foundation into our solvers, we have found our productivity radically enhanced, especially when faced with the goal of solving new classes of problems. The strategy language framework allows us to easily modify and experiment with variations of our solving heuristics. Before we had such strategy language machinery in place, with its principled handling of goals, models and proofs, this type of experimentation with new heuristics was cumbersome and error-prone.

We have proposed a Strategy Challenge to the SMT community: To build theoretical and practical tools allowing users to exert strategic control over core heuristic aspects of high-performance SMT solvers. We discussed some of the rich history of ideas of strategy in the context of mechanized proof, and presented an SMT-oriented approach for orchestrating reasoning engines, where "big" symbolic reasoning steps are represented as tactics, and these tactics are composed using combinators known as tacticals. We demonstrated the practical value of this approach by describing a few examples of how tactics and tacticals have been successfully used in our **RAHD** and **Z3** tools.

There are several directions for future work. First, we believe that many other authors of SMT solvers should take up this Strategy Challenge, and much experimentation should be done — from many different points of view and domains of application — before a standard strategy language for SMT can be proposed. When the time is right, we believe that the existence of a strategy standard (extending, for instance, the SMT-LIB standard) and the development and study of theoretical frameworks for SMT strategies could give rise to much progress in the practical efficacy of automated reasoning tools.

Second, we would like to understand how one might *efficiently* exert "small step" strategic control over reasoning engines. Abstract proof procedures, such as Abstract DPLL [35], DPLL(T) [35] and cutsat [26], represent a proof procedure as a set of transition rules. In these cases, a strategy comprises a *recipe* for applying these "small" step rules. Actual implementations of these abstract procedures use carefully chosen efficient data-structures that depend on the pre-selected strategy. It is not clear to us how to specify a strategy for these abstract procedures so that an efficient implementation can be automatically generated. Another topic for future investigation is to explore different variations of the LCF approach, such as the ones used by the interactive theorem provers Isabelle [40], HOL [24], Coq [6], MetaPRL [25] and Matita [3].

Acknowledgements. We thank Rob Arthan, Jeremy Avigad, Maria Paola Bonacina, Bob Boyer, Barbara Galletly, Paul B. Jackson, Joost Joosten and Larry Paulson for helpful comments on a draft of this paper. We thank the referees for their many useful corrections and suggestions. Grant Passmore was supported during this research by the UK Engineering and Physical Sciences Research Council [grant numbers EP/I011005/1 and EP/I010335/1].

References

1. Dolzmann, T.S.A.: Redlog User Manual - Edition 2.0. MIP-9905 (1999)
2. Ackermann, W.: Solvable cases of the decision problem. Studies in Logic and the Foundation of Mathematics (1954)
3. Asperti, A., Ricciotti, W., Sacerdoti Coen, C., Tassi, E.: The Matita Interactive Theorem Prover. In: Bjørner, N., Sofronie-Stokkermans, V. (eds.) CADE 2011. LNCS (LNAI), vol. 6803, pp. 64–69. Springer, Heidelberg (2011)
4. Barrett, C., Stump, A., Tinelli, C.: The Satisfiability Modulo Theories Library, SMT-LIB (2010), http://www.SMT-LIB.org
5. Barrett, C.W., Tinelli, C.: CVC3. In: Damm, W., Hermanns, H. (eds.) CAV 2007. LNCS, vol. 4590, pp. 298–302. Springer, Heidelberg (2007)
6. Bertot, Y., Castéran, P.: Interactive Theorem Proving and Program Development. Coq'Art: The Calculus of Inductive Constructions. Texts in Theoretical Computer Science. Springer (2004)
7. Bonacina, M.P.: A Taxonomy of Theorem-Proving Strategies. In: Veloso, M.M., Wooldridge, M.J. (eds.) Artificial Intelligence Today. LNCS (LNAI), vol. 1600, pp. 43–84. Springer, Heidelberg (1999)
8. Bonacina, M.P., Hsiang, J.: On the modeling of search in theorem proving–towards a theory of strategy analysis. Inf. Comput. 147(2), 171–208 (1998)
9. Bonacina, M.P., McCune, W.W.: Distributed Theorem Proving by Peers. In: Bundy, A. (ed.) CADE 1994. LNCS, vol. 814, pp. 841–845. Springer, Heidelberg (1994)
10. Bozzano, M., Bruttomesso, R., Cimatti, A., Junttila, T., Rossum, P., Schulz, S., Sebastiani, R.: MathSAT: Tight Integration of SAT and Mathematical Decision Procedures. J. Autom. Reason. 35(1-3), 265–293 (2005)
11. Brown, C.W.: QEPCAD-B: A System for Computing with Semi-algebraic Sets via Cylindrical Algebraic Decomposition. SIGSAM Bull. 38, 23–24 (2004)
12. Bruttomesso, R., Cimatti, A., Franzén, A., Griggio, A., Santuari, A., Sebastiani, R.: To Ackermann-ize or Not to Ackermann-ize? On Efficiently Handling Uninterpreted Function Symbols in $SMT(\mathcal{EUF} \cup \mathcal{T})$. In: Hermann, M., Voronkov, A. (eds.) LPAR 2006. LNCS (LNAI), vol. 4246, pp. 557–571. Springer, Heidelberg (2006)
13. Davis, M.: Eliminating the Irrelevant from Mechanical Proofs. In: Proc. Symp. Applied Math., vol. XV, pp. 15–30 (1963)
14. Davis, M.: The early history of automated deduction. In: Robinson, J.A., Voronkov, A. (eds.) Handbook of Automated Reasoning, pp. 3–15. Elsevier and MIT Press (2001)
15. Davis, M., Logemann, G., Loveland, D.: A machine program for theorem-proving. Commun. ACM 5, 394–397 (1962)
16. Davis, M., Putnam, H.: A computing procedure for quantification theory. J. ACM 7, 201–215 (1960)
17. de Moura, L., Bjørner, N.: Relevancy propagation. Technical Report MSR-TR-2007-140, Microsoft Research (2007)
18. de Moura, L., Bjørner, N.: Z3: An Efficient SMT Solver. In: Ramakrishnan, C.R., Rehof, J. (eds.) TACAS 2008. LNCS, vol. 4963, pp. 337–340. Springer, Heidelberg (2008)

19. de Moura, L.M., Bjørner, N.: Satisfiability modulo theories: introduction and applications. Commun. ACM 54(9), 69–77 (2011)
20. Dolzmann, A., Seidl, A., Sturm, T.: Efficient Projection Orders for CAD. In: ISSAC 2004: Proceedings of the 2004 International Symposium on Symbolic and Algebraic Computation, pp. 111–118. ACM (2004)
21. Dutertre, B., de Moura, L.: A Fast Linear-Arithmetic Solver for DPLL(T). In: Ball, T., Jones, R.B. (eds.) CAV 2006. LNCS, vol. 4144, pp. 81–94. Springer, Heidelberg (2006)
22. Gilmore, P.C.: A Proof Method for Quantification Theory: its Justification and Realization. IBM J. Res. Dev. 4, 28–35 (1960)
23. Gordon, M.: From LCF to HOL: a short history, pp. 169–185. MIT Press, Cambridge (2000)
24. Gordon, M.J.C., Melham, T.F.: Introduction to HOL: a theorem-proving environment for higher-order logic. Cambridge University Press (1993)
25. Hickey, J.J.: The MetaPRL Logical Programming Environment. PhD thesis, Cornell University, Ithaca, NY (January 2001)
26. Jovanović, D., de Moura, L.: Cutting to the Chase Solving Linear Integer Arithmetic. In: Bjørner, N., Sofronie-Stokkermans, V. (eds.) CADE 2011. LNCS, vol. 6803, pp. 338–353. Springer, Heidelberg (2011)
27. Kirchner, F., Muñoz, C.: The proof monad. Journal of Logic and Algebraic Programming 79(3-5), 264–277 (2010)
28. Kowalski, R.: Search Strategies for Theorem Proving. Machine Intelligence 5, 181–201 (1969)
29. Lusk, E.L.: Controlling Redundancy in Large Search Spaces: Argonne-Style Theorem Proving Through the Years. In: Voronkov, A. (ed.) LPAR 1992. LNCS, vol. 624, pp. 96–106. Springer, Heidelberg (1992)
30. Luttik, B., Visser, E.: Specification of rewriting strategies. In: Sellink, M.P.A. (ed.) 2nd International Workshop on the Theory and Practice of Algebraic Specifications (ASF+SDF 1997), Electronic Workshops in Computing. Springer, Berlin (1997)
31. McCune, W.: Solution of the Robbins Problem. J. Autom. Reason. 19(3), 263–276 (1997)
32. McCune, W.: Prover9 and Mace4 (2005-2010), http://www.cs.unm.edu/mccune/prover9/
33. Milner, R.: Logic for computable functions: description of a machine implementation. Technical Report STAN-CS-72-288, Stanford University (1972)
34. Newell, A., Shaw, J.C., Simon, H.A.: Elements of a Theory of Human Problem Solving. Psychological Review 65, 151–166 (1958)
35. Nieuwenhuis, R., Oliveras, A., Tinelli, C.: Solving SAT and SAT Modulo Theories: From an abstract Davis–Putnam–Logemann–Loveland procedure to DPLL(T). J. ACM 53(6), 937–977 (2006)
36. Passmore, G.O.: Combined Decision Procedures for Nonlinear Arithmetics, Real and Complex. PhD thesis, University of Edinburgh (2011)
37. Passmore, G.O., Jackson, P.B.: Combined Decision Techniques for the Existential Theory of the Reals. In: Carette, J., Dixon, L., Coen, C.S., Watt, S.M. (eds.) Calculemus/MKM 2009. LNCS (LNAI), vol. 5625, pp. 122–137. Springer, Heidelberg (2009)
38. Passmore, G.O., Jackson, P.B.: Abstract Partial Cylindrical Algebraic Decomposition I: The Lifting Phase. In: Cooper, S.B., Dawar, A., Löwe, B. (eds.) CiE 2012. LNCS, vol. 7318, pp. 560–570. Springer, Heidelberg (2012)
39. Paulson, L.: Logic and Computation: Interactive Proof with Cambdrige LCF, vol. 2. Cambridge University Press (1987)

40. Paulson, L.: Isabelle: The next 700 theorem provers. In: Logic and Computer Science, pp. 361–386. Academic Press (1990)
41. Plaisted, D.A.: The Search Efficiency of Theorem Proving Strategies. In: Bundy, A. (ed.) CADE 1994. LNCS, vol. 814, pp. 57–71. Springer, Heidelberg (1994)
42. Prawitz, D.: An Improved Proof Procedure. Theoria 26(2), 102–139 (1960)
43. Robinson, A.: Short Lecture. Summer Institute for Symbolic Logic, Cornell University (1957)
44. Robinson, J.A.: A machine-oriented logic based on the resolution principle. J. ACM 12, 23–41 (1965)
45. Wang, H.: Toward mechanical mathematics. IBM J. Res. Dev. 4, 2–22 (1960)
46. Weispfenning, V.: Quantifier Elimination for Real Algebra - the Quadratic Case and Beyond. Appl. Algebra Eng. Commun. Comput. 8(2), 85–101 (1997)
47. Wos, L., Carson, D., Robinson, G.: The unit preference strategy in theorem proving. In: Proceedings of the Fall Joint Computer Conference, AFIPS 1964, Part I, October 27-29, pp. 615–621. ACM, New York (1964)
48. Wos, L., Robinson, G.A., Carson, D.F.: Efficiency and completeness of the set of support strategy in theorem proving. J. ACM 12, 536–541 (1965)
49. Zankl, H., Middeldorp, A.: Satisfiability of Non-linear (Ir)rational Arithmetic. In: Clarke, E.M., Voronkov, A. (eds.) LPAR-16. LNCS (LNAI), vol. 6355, pp. 481–500. Springer, Heidelberg (2010)

Simple and Efficient Clause Subsumption
with Feature Vector Indexing

Stephan Schulz

Institut für Informatik, Technische Universität München,
D-80290 München, Germany
`schulz@eprover.org`

Abstract. This paper describes *feature vector indexing*, a new, non-perfect indexing method for clause subsumption. It is suitable for both forward (i.e., finding a subsuming clause in a set) and backward (finding all subsumed clauses in a set) subsumption. Moreover, it is easy to implement, but still yields excellent performance in practice. As an added benefit, by restricting the selection of features used in the index, our technique immediately adapts to indexing modulo arbitrary AC theories with only minor loss of efficiency. Alternatively, the feature selection can be restricted to result in *set subsumption*. Feature vector indexing has been implemented in our equational theorem prover E, and has enabled us to integrate new simplification techniques making heavy use of subsumption. We experimentally compare the performance of the prover for a number of strategies using feature vector indexing and conventional sequential subsumption.

Keywords: First-order theorem proving, indexing, subsumption.

1 Introduction

First-order theorem proving is one of the core areas of automated deduction. In this field, saturating theorem provers currently show a significant lead compared to systems based on other paradigms, such as top-down reasoning or instance-based methods. One of the reasons for this lead is the compatibility of saturating calculi with a large number of redundancy elimination techniques, as e.g. tautology deletion, rewriting, and *clause subsumption*. Subsumption allows us to discard a clause (i.e., exclude it from further proof search) if a (in a suitable sense) more general clause exists. In many cases, subsumption can eliminate between 50% and 95% of all clauses under consideration, with a corresponding decrease in the size of the search state.

Subsumption of multi-literal clauses is an NP-complete problem [6]. If some attention is paid to the implementation, the worst case is rarely (if ever) encountered in practice, and single clause-clause subsumption tests rarely form a critical bottleneck. However, the sheer number of possible subsumption relations to test for means that a prover can spend a significant amount of time in subsumption-related code. Even in the case of our prover E [17,19], which,

M.P. Bonacina and M.E. Stickel (Eds.): McCune Festschrift, LNAI 7788, pp. 45–67, 2013.

because of its DISCOUNT loop proof procedure, minimizes the use of subsumption, frequently between 10% and 20% of all time was spent on subsumption, with much higher values observed occasionally. The cost of subsumption systematically increases if other simplification techniques based on subsumption are implemented.

In a saturating prover, we are most often interested in subsumption relations between whole sets of clauses and a single clause. In *forward subsumption* (FS), we want to know if *any* clause from a set subsumes a given clause. In *backward subsumption* (BS), we want to find all clauses in a set that are subsumed by a given clause. This observation can be used to speed up subsumption, by using *indexing techniques* that return only candidates suitable for a given subsumption relation from a set of clauses, thus reducing the number of explicit subsumption tests necessary. A *perfect index* will return exactly the necessary clauses, whereas a *non-perfect* index should return a superset of candidates for which the desired relationship has to be verified.

Term indexing techniques have been used in theorem provers for some time now (see [10] for first implementations in Otter or [3,4,22] for increasingly up-to-date overviews). However, lifting term indexing to clause indexing is not trivial because the associative and commutative properties of the disjunction, and the symmetry of the equality predicate, are hard to handle. In many cases, (perfect) term indexing is used only to retrieve subsumption candidates, i.e., to implement non-perfect clause indexing (see e.g. [26]). Moreover, often two different indices are used for forward- and backward subsumption, as e.g. in the very advanced indexing schemes currently implemented in Vampire [15]. Similar problems affect term indexing modulo associativity or commutativity. McCune's EQP, for example, performs unit-subsumption, but disables all indexing as soon as some symbols are declared AC or C [9].

We suggest a new indexing technique based on *subsumption-compatible* numeric clause features. It is much easier to implement than known techniques, and the same, relatively compact data structure can be used for both forward- and backward subsumption. We have implemented the new technique for E 0.8, and, in more polished and configurable ways, for later versions, with excellent results.

In this paper, describe the new technique. We also discuss how it has been integrated into E, and how it also serves to speed up *contextual literal cutting*, a subsumption-based simplification technique that has given another boost to E. We present the results of various experiments to support our claims.

2   Preliminaries

We are primarily interested in first-order formulae in clause normal form in this paper. We assume the following notations and conventions. Let F be a finite set of function symbols. We write $f/n \in F$ to denote f as a function symbol with arity n. Functions symbols are written as lower case letters. We usually employ a, b, c for function symbols with arity 0 (constants), and f, g, h for other function symbols. Let V be an enumerable set of variable symbols. We use upper case letters, usually X, Y, Z to denote variables. The set of all *terms* over F and V, $Term(F, V)$, is defined as the smallest set fulfilling the following conditions:

1. $X \in Term(F, V)$ for all $X \in V$
2. $f/n \in F$, $s_1, \ldots, s_n \in Term(F, V)$ implies $f(s_1, \ldots, s_n) \in Term(F, V)$

We typically omit the parentheses from constant terms, as for example in the expression $f(g(X), a) \in Term(F, V)$.

An (equational) *atom*[1] is an unordered pair of terms, written as $s \simeq t$. A *literal* is either an atom, or a negated atom, written as $s \not\simeq t$. We define a negation operator on literals as $\overline{s \simeq t} = s \not\simeq t$ and $\overline{s \not\simeq t} = s \simeq t$. If we want to write about arbitrary literals without specifying polarity, we use $s \dot\simeq t$, or, in less precise way, $l, l_1, l_2, \ldots$. Note that $\simeq$ is commutative in this notation.

A *clause* is a multiset of literals, interpreted as an implicitly universally quantified disjunction, and usually written as $l_1 \vee l_2 \ldots \vee l_n$. Please note that in this notation, the $\vee$ operator is associative and commutative (but not idempotent). The empty clause is written as $\square$, and the set of all clauses as $Clauses(F, V)$. A *formula* in clause normal form is a set of clauses, interpreted as a conjunction.

A *substitution* is a mapping $\sigma : V \to Term(F, V)$ with the property that $Dom(\sigma) = \{X \in V \mid \sigma(X) \neq X\}$ is finite. It is extended to a function on terms, atoms, literals and clauses in the obvious way.

A *match* from a term (atom, literal, clause) s to another term (atom, literal, clause) t is a substitution σ such that $\sigma(s) \equiv t$, where $\equiv$ on terms denotes syntactic identity and is lifted to atoms, literal, clauses in the obvious way, using the unordered pair and multiset definitions.

3 Subsumption

If we consider a (multi)set of clauses not all of the clauses necessarily contribute to the meaning of it. Often, some clauses are *redundant*. Some clauses do not add any new constraints on the possible models of a formula, because they are already implied by other clauses. Depending on the mechanism of reasoning employed, we can delete some of these clauses, thus reducing the size of the formula (and hence the difficulty of finding a proof). In the case of current saturating calculi, *subsumption* is a technique that allows us to syntactically identify certain clauses that are implied by another clause, and can usually be discarded without loss of completeness. We can specify the (multiset) subsumption rule as a deleting simplification rule (i.e., the clauses in the precondition are *replaced* by the clauses in the conclusion) as follows:

(CS) $\dfrac{\sigma(C) \vee \sigma(R) \qquad C}{C}$ where σ is a substitution, C and R are arbitrary (partial) clauses

---

[1] For our current discussion, the non-equational case is a simple special case and can be handled by encoding non-equational atoms as equalities with a reserved constant $\$true$. We will still write non-equational literals in the conventional manner, i.e., $p(a)$ instead of $p(a) \simeq \$true$.

In other words, a clause C' is subsumed by another clause C if there is an instance $\sigma(C)$ that is a sub-multiset of C'.

This version of subsumption is used by most modern saturation procedures. It is particularly useful in reducing search effort, since it allows us to discard larger clauses in favor of smaller clauses. Smaller clauses typically have fewer inference positions and generate fewer and smaller successor clauses.

Individual clause-clause subsumption relations are determined by trying to find a simultaneous match from all literals in the potentially subsuming clause to corresponding literals in the potentially subsumed clause. This is usually implemented by a backtracking search over permutations of literals in the potentially subsumed clause (and in the equational case, permutations of terms in equational literals).

Most of the techniques used to speed up subsumption try to detect failures early by testing necessary conditions. Those include compatibility of certain clause measures (discussed in more detail below) and existence of individually matched literals in the potentially subsumed clause for each literal in the potentially subsuming clause. Additionally, in many cases certain permutations of literals can be eliminated by partially ordering literals in a clause with a suitable ordering.

However, while individual subsumption attempts are reasonably cheap in practice, the number of potential subsumption relations to test for in saturation procedures is very high. Using a straightforward implementation of subsumption, we have measured up to 100,000,000 calls to the subsumption subroutine of our prover E in just 5 minutes on a 300 MHz SUN Ultra-60 for some proof tasks. Thus, the overall cost of subsumption is significant.

3.1 Subsumption Variants

In addition to standard multiset subsumption, there are a number of other subsumption variants and related techniques.

The definition of *set subsumption* is identical to that of multiset subsumption, except in that clauses are viewed as sets of literals (i.e. a single literal occurs at most once in a given clause). This allows for a slightly stronger subsumption relation: $p(X) \vee p(Y)$ can subsume $p(a)$ with set subsumption, but not with multiset subsumption. Set subsumption can e.g. be used in preprocessing. However, for most saturation-based calculi (especially those for which factorization is an explicit inference rule), the fact that a clause can subsume some of its factors causes loss of completeness.

Subsumption modulo AC is a stronger version of multiset or set subsumption, where we do not require that the instantiated subsuming clause is a subset of the subsumed clause, but only that it is equal to a subset modulo a specified theory for associative and commutative function symbols. For example, if f is commutative, then $p(f(a, X))$ subsumes $p(f(b, a)) \vee q(a)$. This variant of subsumption is useful for systems that reason modulo AC, as e.g. SNARK [25].

Equality subsumption (also known as *functional subsumption*) allows an equational unit clause to potentially subsume another clause with an equational literal implied by the potential subsumer. It can be described by the following simplification rule:

$$(ES) \quad \frac{s \simeq t \quad u[p \leftarrow \sigma(s)] \simeq u[p \leftarrow \sigma(t)] \vee R}{s \simeq t}$$

It is typically only applied if $s \simeq t$ cannot be used for rewriting, i.e. if $\sigma(s) \simeq \sigma(t)$ cannot be oriented). This rule is implemented by E and a number of other provers, including at least the completion-based systems Waldmeister [8] and DISCOUNT [2], as well as in EQP [11].

Finally, a simplification rule that has been popularized by implementation in SPASS [28] and Vampire [14], and is sometimes called *subsumption resolution* or *clausal simplification*, combines resolution and subsumption to cut a literal out of a clause. In the context of a modern superposition calculus, we believe the rule can be better described as *contextual literal cutting*:

$$(CLC) \quad \frac{\sigma(C) \vee \sigma(R) \vee \sigma(l) \qquad C \vee \bar{l}}{\sigma(C) \vee \sigma(R) \qquad C \vee \bar{l}} \qquad \text{where } \bar{l} \text{ is the negation of } l \text{ and } \sigma \text{ is a substitution}$$

It can be implemented via a normal subsumption engine (by negating each individual literal in turn, and then testing for subsumption) and is implemented thus at least in E and Vampire. Depending on how and when this rule is applied, it can increase the number of required subsumption tests by many orders of magnitude.

3.2 Saturation Procedures and Clause Set Subsumption

Most modern saturating theorem provers use a variant of the *given clause algorithm* that was popularized by Bill McCune's *Otter*. This algorithm maintains two sets of clauses, the *processed clauses* P and the *unprocessed clauses* U. At the start, all clauses are unprocessed. The algorithm repeatedly picks one clause from U, and performs all generating inferences using this *given clause* and the clauses in P as premises. It then moves the given clause to P and adds all newly generated clauses to U.

The original *Otter loop* performs simplification and subsumption between all clauses, both processed an unprocessed. A variant, originally implemented in DISCOUNT, restricts simplification to only allow processed clauses as side premises in simplification. Figure 1 shows a sketch of the main proof procedure of our prover E, an implementation of the DISCOUNT loop for full superposition.

Please observe that subsumption appears in exactly two different places and exactly two different roles in this procedure: First, we test if the *given clause* g is subsumed by *any* clause in P. In other words, we want to know if a single clause is subsumed by any clause from a set. This is usually called *forward subsumption*.

If the given clause is not redundant, we next want to find *all* clauses in P that are subsumed by g. Again, we have an operation between a single clause and a whole set, in this case called *backward subsumption*.

It is obvious that we can implement forward and backward subsumption naively by sequentially testing each clause from P against g. This implementation was

---

Prover state: $U \cup P$
U contains *unprocessed* clauses, P contains *processed* clauses.
Initially, all clauses are in U, P is empty.
The *given clause* is denoted by g.

---

while $U \neq \{\}$
   $g = $ delete_best(U)
   $g = $ simplify(g, P)
   if $g == \Box$
      SUCCESS, Proof found
   if g is not subsumed by any clause in P (or otherwise redundant w.r.t. P)
      $P = P \backslash \{c \in P \mid c$ subsumed by (or otherwise redundant w.r.t.) $g\}$
      $T = \{c \in P \mid c$ can be simplified with $g\}$
      $P = (P \backslash T) \cup \{g\}$
      $T = T \cup $ generate(g, P)
      foreach $c \in T$
         $c = $ cheap_simplify(c, P)
         if c is not trivial
            $U = U \cup \{c\}$
SUCCESS, original U is satisfiable

Remarks: delete_best(U) finds and extracts the clause with the best heuristic evaluation from U. generate(g, P) performs all generating inferences using g as one premise, and clauses from P as additional premises. simplify(c, S) applies all simplification inferences in which the main (simplified) premise is c and all the other premises are clauses from S. This typically includes full rewriting and (CLC). cheap_simplify(c, S) works similarly, but only applies inference rules with a particularly efficient implementation, often including rewriting with orientable units, but usually not (CLC).
Similarly, in this context, a clause is *trivial*, if it can be shown redundant with simple, local syntactic checks. If we test for redundancy, we also apply more complex and non-local techniques.

Fig. 1. Saturation procedure of E

used, for instance, in early versions of SPASS [28], and was used in E up to version 0.71. However, this does not make use of the fact that we are interested in subsumption relations between individual clauses and usually only slowly changing *clause sets*. The idea behind clause indexing is to *preprocess* the clause set so that subsumption queries can be answered more efficiently than by sequential search.

4   Feature Vector Indexing

Indexing for subsumption is used by a number of provers. Most existing implementations (e.g. [26,27,10]) use a variant of *discrimination tree indexing* on terms to build a index for forward subsumption, often for non-perfect indexing.

Indexing for backward subsumption is less frequent, and usually based on a variant of path indexing. We will now present a new and much simpler technique suitable for both forward and backward subsumption.

Our technique is based on the compilation of necessary conditions on numeric clause features. Essentially, a clause is represented by a vector of feature values, and subsumption candidates are identified by comparisons of feature vectors. Feature vectors for clause sets are compiled into a *trie* data structure to quickly identify candidate sets.

4.1   Subsumption-Compatible Clause Features

A *(numeric) clause feature function* (or just *feature*) is a function mapping clauses to natural numbers, $f : Clauses(F, V) \rightarrow \mathbf{N}$. We call f *compatible with subsumption* if $f(C) \leq f(C')$ whenever C subsumes C'. In other words, if f is a subsumption-compatible clause feature, then $f(C) \leq f(C')$ is a necessary condition for the subsumption of C' by C. Unless we specify a particular subsumption variant, we assume multiset subsumption.

We will define a number of clause features now, all of which are compatible with multiset subsumption, and many of which are compatible with other subsumption variants.

Let C be a clause. We denote the sub-multiset of positive literals in C by C^+, and similarly the sub-multiset of negative literals by C^-. Please note that both C^+ and C^- are clauses as well. $|C|$ is the number of literals in C. $|C|_f$ is the number of occurrences of the symbol f in C, e.g. $|p(a, b) \vee f(a, a) \not\simeq a|_a = 4$.

Let t be a term, and let f/n be a function symbol. We define $d_f(t)$ as follows:

$$d_f(t) = \begin{cases} 0 & \text{if } f \text{ does not occur in } t \\ \max\{1, d_f(t_1) + 1, \ldots, d_f(t_n) + 1\} & \text{if } t \equiv f(t_1, \ldots, t_n) \\ \max\{d_f(t_1) + 1, \ldots, d_f(t_m) + 1\} & \text{if } t \equiv g(t_1, \ldots, t_m), g/m \neq f, \\ & \quad f \text{ occurs in } t \end{cases}$$

Intuitively, $d_f(t)$ is the depth of the deepest occurrence of f in t (or 0). The function is continued to atoms, literals and clauses as follows:

$$d_f(s \simeq t) = \max\{d_f(s), d_f(t)\}$$
$$d_f(s \not\simeq t) = d_f(s \simeq t)$$
$$d_f(l_1 \vee \ldots \vee l_k) = \max\{d_f(l_1), \ldots, d_f(l_k)\}$$

If we assume the standard representation of terms given above, the following theorems hold:

Theorem 1. *The feature functions defined by the following expressions are compatible with multiset subsumption, subsumption modulo AC, and equality subsumption:* $|C^+|$, $|C^-|$, $|C^+|_f$ *(for all f)*, $|C^-|_f$ *(for all f)*.

The argument is essentially always the same: instantiation can only add new symbols, and a superset (super-multiset) or superstructure always contains at least as many symbols as the subset or substructure.

Theorem 2. *The feature functions defined by the following expressions are compatible with multiset subsumption, set subsumption, and equality subsumption:* $d_f(C^+)$ *(for all f)*, $d_f(C^-)$ *(for all f)*.

The argument is similar: Instantiation can only introduce function symbols at new positions, never take them away at an existing depth.

If AC terms are represented in flattened form, symbol-counting features need to correct for the effects of this representation to be subsumption-compatible. On the other hand, if all terms are fully flattened, then the depth-based features also become compatible with AC-subsumption.

A final theorem allows us to combine different feature functions while maintaining compatibility with different subsumption types:

Theorem 3. *If any two feature functions f_1, f_2 are compatible with one of the listed subsumption types, then any linear combination of the two with non-negative coefficients is also compatible with that subsumption type. That is, $f(C) = af_1(C) + bf_2(C)$ with $a, b \in \mathbb{R}^+$ is also a compatible feature function.*

Many provers already use the criterion that a subsuming clause cannot have more function symbols that the subsumed one. In our notation, this can be described by the requirement that $\sum_{f \in F} |C|_f \leq \sum_{f \in F} |C'|_f$. This will, on average, already decide about half of all subsumption attempts. However, by looking at and combining more fine-grained criteria, we can do a lot better.

4.2   Clause Feature Vectors and Candidate Sets

Let π_n^i be the projection function for the ith element of a vector with n elements. A *clause feature vector function* is a function $F : Clauses(F, V) \to \mathbf{N}^n$. We call F subsumption-compatible (for a given subsumption type) if $\pi_n^i \circ F$ is a subsumption compatible feature for each $i \in \{1, \dots, n\}$. In other words, a subsumption compatible feature vector function combines a number of subsumption compatible feature functions. We will now assume that F is a subsumption-compatible feature vector function. If $F(C) = v$, we call v the feature vector of C.

We define a partial ordering $\leq_s$ on vectors by $v \leq_s v'$ iff $\pi_n^i(v) \leq \pi_n^i(v')$ for all $i \in \{1, \dots, n\}$. By definition of the feature vector, if C subsumes C', then $F(C) \leq_s F(C')$. This allows us to succinctly identify the candidate sets of clauses for forward subsumption and backward subsumption. Let C be a clause and P be a clause set. Then

$$candFS_F(P, C) = \{c \in P | F(c) \leq_s F(C)\}$$

is a superset of all clauses in P that subsume C and

$$candBS_F(P, C) = \{c \in P \mid F(C) \leq_s F(c)\}$$

is a superset of all clauses in P that are subsumed by C.

As our experiments show, if a reasonable number of clause features are used in the clause feature vector, these supersets are usually fairly small. Restricting subsumption attempts to members of these candidate sets reduces the number of attempts often by several orders of magnitude.

4.3   Index Data Structure

Whereas it is possible to store complete feature vectors with every clause in a set, this approach is rather inefficient in terms of memory consumption, and still requires the full comparison of all feature vectors. If, on the other hand, we compile feature vectors into a *trie*-like data structure, with all clauses sharing a vector stored at the corresponding leaf, large parts of the vectors are shared, and candidate sets can be computed much more efficiently.

Assume a (finite) set P of clauses with associated feature vectors $F(P)$ of length n. A *clause feature vector index* for P and F is a tree of uniform depth n (i.e., each path from the root to a leaf has length n). It can be recursively constructed as follows: If n is equal to 0, the tree consists of just a leaf node, which we associate with all clauses in P. Otherwise, let $D = \{\pi_n^1(F(C)) \mid C \in P\}$, let $P_i = \{C \mid \pi_n^1(F(C)) = i\}$ for $i \in D$ (the set of all clauses for which the first feature has a given value i), and let $F' = \langle \pi_n^2, \ldots, \pi_n^n \rangle \circ F$ (shortening the original feature vectors by the first element). Then the index consist of a root node with sub-trees T_i, such that each T_i is an index for P_i and F'. Inserting and deleting is linear in the number of features and independent of the number of elements in the index.

As an example, consider F defined by $F(C) = \langle |C^+|_a, |C^+|_f, |C^-|_b| \rangle$, the clauses $C1 = p(a) \vee p(f(a)), C2 = p(a) \vee \neg p(b), C3 = \neg p(a) \vee p(b), C4 = p(X) \vee p(f(f(b)))\}$, and the set of clauses $P = \{C1, C2, C3, C4\}$. The feature vectors are as follows: $F(C1) = \langle 2, 1, 0 \rangle$, $F(C2) = \langle 1, 0, 1 \rangle$, $F(C3) = \langle 0, 0, 0 \rangle$, $F(C4) = \langle 0, 2, 0 \rangle$. Figure 2 shows the resulting index.

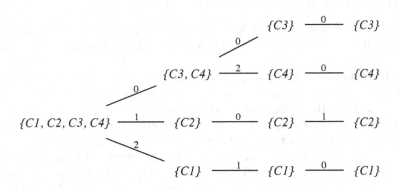

Fig. 2. Example of a Clause Feature Vector Index. For illustration purposes, each node is annotated with the clauses compatible with the features leading to that node. The actual implementation only stores clauses at the leaf nodes.

4.4   Forward Subsumption

For forward subsumption, we do not need to compute the full candidate set $candFS_F(P,C)$. Instead, we can just enumerate the elements and stop as soon as a subsuming clause is found. Assume a clause set P, a feature function F with feature vector length n, and an index I. We denote by $I[v]$ the sub-tree of I associated with value v. The clause to be subsumed is C. Figure 3a) shows the algorithm for indexed subsumption.

Note that it is trivial to return the subsuming clause (if any), instead of just a Boolean value. We traverse the sub-trees in order of increasing feature values, so that (statistically) smaller clauses with a higher chance of subsuming get tested first.

The subsumption test in the leaves of the tree is implemented by sequential search. In particular, finding the candidate sets and applying the actual subsumption test are clearly separated, i.e., it is trivially possible to use any subsumption concept as long as F is compatible with it.

4.5   Backward Subsumption

The algorithm for backward subsumption is quite similar, except that we traverse nodes with feature values greater than or equal to that of the subsuming clause, and that we cannot terminate the search early, since we have to find (and return) *all* subsumed clauses. We use the same conventions as above. Additionally, $mv(I)$ is the largest feature value associated with any sub-tree in I. Figure 3b) shows the algorithm.

4.6   Optimizing the Index Data Structure

Each leaf in the feature vector index corresponds to a given feature vector. If we ignore the internal structure of the trie, and the order of features in the vector, we can associate each leaf with an unordered set of tuples $(f, f(C))$ of individual feature functions and corresponding feature value. It is easy to see that any order of features in the feature vector will generate the same number of leaves, and that each leaf is either compatible with a given set of feature function/feature value tuples, or not. Thus, at least for a complete search as in the backward subsumption algorithm, we always have to visit the same number of leaves.

However, we can certainly minimize the internal number of nodes in the trie, and thus the total number of nodes. Consider for a simple example feature vectors with two features f_1, f_2, where f_1 yields the same value for all clauses from a set P, whereas f_2 perfectly separates the set into n individual clauses. If we test f_1 first, our tree has just one internal node (plus the root). Traversing all leaves touches $n+2$ nodes (counting the root). If on the other hand we evaluate the more informative f_2 first, we will immediately split the tree into n internal nodes, each of which has just one leaf as the successor. Thus, to traverse all leaves we would touch $2n+1$ nodes, or, for a reasonably sized n, nearly twice as many nodes.

(a) Forward subsumption

// Note that d is ranging over the length of $F(C)$
function search_subsuming(I, d, C)
  if I is a leaf node then
    if a clause in I subsumes C
      return true
    else
      return false
  else
    for $i \in \{0, \ldots, \pi_n^d(F(C))\}$
      if search_subsuming($I[i]$, $d + 1$, C)
        return true
  return false

function is_subsumed(I, C) // Return true if clause in I subsumes C
  return search_subsuming(I, 1, C)

(b) Backward subsumption

function search_subsumed(I, d, C)
  if I is a leaf node then
    return $\{C' \in I \mid C'$ subsumed by $C\}$
  else
    res = {}
    for $i \in \{\pi_n^d(F(C)), \ldots, mv(I)\}$
      res = res $\cup$ search_subsumed($I[i]$, $d + 1$, C)
    return res

function find_subsumed(I, C) // Return clauses in I subsumed by C
  return search_subsumed(I, 1, C)

Fig. 3. Forward and backward subsumption with feature vector indexing

This example easily generalizes to longer vectors. In general, we want the least informative features first in a feature vector, so that as many initial paths as possible can be shared. This is somewhat surprising, since for most exclusion tests it is desirable to have the most informative features first, so that impossible candidates are excluded early. Of course, if we have totally uninformative features, we can just as well drop them completely, thus shrinking the tree depth.

Unless we want to deal with the complexity and computational cost of dynamically adapting the index, we have to determine the feature vector function before we start building the index, i.e., in practice before the proof search starts. We can estimate the informativeness of a given feature by looking at the distribution of its values in the initial clause set, and assume that this is typical for the later clauses.

For best results, we could view application of a feature function to a clause as a probability experiment and the results on the initial clause set as a sample. We could then sort features by increasing estimated entropy[2] [23] or even conditional entropy. However, we decided to use a much simpler estimator first, namely the range of the feature value over the initial clause set. We have implemented three different mappings: *Direct mapping*, where the place of a feature in the vector is determined by the internal representation of function symbols used by the system (i.e. the first function symbol in the signature is responsible for the first 2 or 4 features, the second for the next, and so on), *permuted*, where features are sorted by feature value range, and *optimized permuted*, where additionally features with no estimated usefulness (i.e., features which evaluate to the same value for all initial clauses) are dropped.

Our experimental results show that both permuted and optimized permuted feature vectors perform much better than direct mapped ones, with optimized permuted ones being best if we allow only a few features, whereas plain permuted ones gain if we allow more features. Generally, we can decrease the number of nodes in an index by about 50% using permuted feature vectors. We explain this behaviour by noting that the degree of informativeness is generally estimated correctly, but the prediction whether a feature will be useful at all is less precise. We have especially observed the situation where only a single negative literal occurs in the initial clause set (e.g. all unit-equational proof problems with a single goal), and hence all features restricted to negative literals have an initial range of zero, although a large and varied set of negative literals is generated during the proof search.

5   Implementation Notes

We have implemented clause feature vector indexing in our prover E, using essentially simple versions of standard trie algorithms for inserting and deleting feature vectors (and hence clauses), and the algorithms described in section 4.4 and 4.5 for forward and backward subsumption. In E's implementation of the given-clause algorithm, unprocessed clauses are passive, i.e. they don't impose a computational burden once normalized. Hence we are using subsumption only between the set of processed clauses P and the given clause g and vice versa, not on the full set of unprocessed clauses. However, we have also implemented contextual literal cutting using the index. It can be optionally applied either to the newly generated clauses during simplification (using clauses from P for cutting) or between g and P, in both directions.

Feature vector indexing is used for forward and backward non-unit multiset subsumption, all versions of contextual literal cutting, (unit) equality backward subsumption, and backward simplify-reflect (equational unit cutting, see [17]) inferences. Forward equality subsumption and forward simplify-reflect have been implemented using discrimination tree indexing (on maximal terms in the unit

---

[2] The *entropy* of a probability experiment is the expected information gain from it, or, in other words, the expected cost of predicting the outcome. In our case, a feature with higher entropy splits the clause set into more (or more evenly distributed) parts. See e.g. [16] or, for a more comprehensive view, [5].

clause used) since early versions of E. Paramodulation and and backwards rewriting are implemented using fingerprint indexing[21].

The existing multiset subsumption code, used both for conventional subsumption and to check indexed candidates for actual subsumption, already is fairly optimized. It uses a number of simple criteria to quickly determine unsuitable candidates, including tests based on literal- and symbol count, and trying to match individual literals in the potential subsumer onto literals in the potentially subsumed clause. Only if all these tests succeed do we start the recursive permutation of terms and literals to find a common match.

The feature vector index is implemented in a fairly straightforward way, using a recursive data structure. Note that all our features in practice yield small integers. Originally, we had implemented the mapping from a feature value to the associated sub-tree via a dynamic array. However, the current implementation uses the *IntMap* data structure, a self-optimizing data structure that reorganizes itself as either a dynamic array or a *splay tree* [24], depending on the fraction of elements used.

Clauses in a leaf node are stored in a simple set data structure (which is implemented throughout E as a splay tree using pointers as keys). Empty subtrees are deleted eagerly.

It may be interesting to note that the first (and working) version of the indexing scheme took only about three (part-time) days to implement and integrate from scratch. It took approximately 7 more days to arrive at the current (production-quality) version that allows for a large number of different clause feature vector functions to be used and applies the index to many different operations, and about half a day to change the implementation to IntMaps. Compared to other indexing techniques, feature vector indexing seems to be easy to implement and easy to integrate into existing systems.

6   Experimental Results

We used all untyped first-order problems from TPTP 5.2.0 for the experimental evaluation. There are 15356 problems, about evenly split between 7712 clause normal form problems, and 7674 full first-order problems. The problems were not modified in any way. Tests were run on the University of Miami *Pegasus cluster*, under the Linux 2.6.18-164.el5 SMP Kernel in 64 bit mode. Each node of the cluster is equipped with 8 Intel Xeon cores, running at 2.6 GHz, and 16 GB of RAM.[3] Test runs were done with a CPU time limit of 300 seconds per job, a memory limit of 512 MB per job, and with 8 jobs scheduled per node. Detailed results of these and additional test runs, including an archive of the source package of the prover version, are available at http://www.eprover.eu/E-eu/FVIndexing.html.

---

[3] See [13]. Jobs were submitted on the "Small" queue, which schedules only to Intel Xeon systems.

6.1 Prover Instrumentation and Configuration

In first-order theorem proving, even small changes to the order of inferences can influence the course of the proof search significantly. Indexing affects both the order in which clauses are generated and the internal memory layout of the process. There is no guarantee that the system performs the same search with different subsumption implementations. To minimize this effect, E has been modified with an option that imposes a total ordering on newly generated clauses in each iteration of the main loop.[4] For the detailed quantitative analysis of the run times, we use only cases where the prover performs the same number of iterations of the main loop, and had the same number of processed and unprocessed clauses at termination time. These three indicators give a high likelihood that the proof searches followed very similar lines for the indexed and non-indexed case.

To gain more insights into the time spent in various parts of the prover, we have added a generic profiling mechanism to E by instrumenting the source code. The system can maintain an arbitrary number of profiling points. The system computes (at microsecond resolution) the difference between the time a profiled code segment is entered and left. The times for each profiled segment are summed over the life time of the process. Here, we use performance counters measuring the time spent for the whole proof search, non-unit clause-clause subsumption, set subsumption (i.e. forward- and backward subsumption involving a clause and a set of clauses), index maintenance, and feature vector computation.

As for all profiling solutions using portable high-level timing interfaces as defined e.g. in POSIX, the times measured are only statistically valid. Many functions have run times much shorter than the microsecond resolution of the UNIX system clock. However, over sufficiently many calls, the average values become increasingly more reliable. The same holds for the small variations invariably caused by small differences in process scheduling and even instruction scheduling in modern, pipelined multiple-issue processors. We performed tests on thousands of problems, with many thousands of calls to small profiled functions for each problem. Thus, noise effects largely average out.

---

[4] The exact options given to the prover were --definitional-cnf=24 --tstp-in --split-clauses=4 --split-reuse-defs --simul-paramod --forward-context-sr-aggressive --backward-context-sr --destructive-er-aggressive --destructive-er --prefer-initial-clauses -tKBO6 -winvfreqrank -c1 -Ginvfreqconjmax -F1 -s --delete-bad-limit=512000000 -WSelectMaxLComplexAvoidPosPred -H'(4*RelevanceLevelWeight2(SimulateSOS,0,2,1,2,100,100,100,400,1.5, 1.5,1), 3*ConjectureGeneralSymbolWeight(PreferNonGoals,200,100,200,50, 50,1,100,1.5,1.5,1),1*Clauseweight(PreferProcessed,1,1,1), 1*FIFOWeight(PreferProcessed))' --detsort-new --fvindex-maxfeatures=<XXX> --fvindex-featuretypes=<YYY> --subsumption-indexing=<ZZZ>.
   Tests were done with E 1.4-011 and E 1.4-012 (for the **BI** strategy), versions of E 1.4 which has been instrumented for profiling feature vector operations.

6.2   Feature Selection

The indexed version of the prover evaluated here uses a maximum feature vector length of around 150 elements. Features used in the vectors are $|C^+|$, $|C^-|$, $|C^+|_f$, $|C^-|_f$, $d_f(C^+)$ and $d_f(C^-)$ (for some function symbols f). Some feature vector functions use a "catch-all" feature that summarizes the value of a given feature type for all function symbols not represented by an individual feature.

The vector might be slightly shorter than 150 elements if only a few symbols occur in the input formula. The lengths can vary slightly because the feature vector functions are systematically generated from a set of function symbols, with either two or four features generated per symbol.

We used 4 different feature selection schemes:

- **AC** uses the AC-compatible features, $|C^+|$, $|C^-|$, $|C^+|_f$, $|C^-|_f$, with a catch-all feature implemented individually for positive and negative literals. The catch-all sums occurrences of all otherwise uncounted symbols.
- **DF** uses the set-subsumption compatible features, $d_f(C^+)$ and $d_f(C^-)$, with a catch-all feature implemented individually for positive and negative literals. The catch-all feature represents the maximum depths of any otherwise unrepresented function symbols.
- **AL** uses all the features used by **AC** or **DF**, including the 4 catch-all features.
- **BI** was suggested by Bill McCune[5] and used by him in Prover9. McCune noted that predicate symbols always occur at depth 1, and hence $d_f(C)$-type features add no information for predicate symbols. The following features are used:
  - $|C^+|$, $|C^-|$
  - $|C^+|_f$, $|C^-|_f$ for all symbols f in the signature
  - $d_f(C^+)$ and $d_f(C^-)$ for all proper function symbols (as opposed to predicate symbols) in the signature

  BI does not use a catch-all feature.

6.3   Feature Vector Optimization

We have implemented the two optimizations described in section 4.6. So each of the 4 different feature selection schemes can be combined with 3 different orderings of features in the vectors:

- **DRT** uses direct mapping: Features are ordered by some arbitrary order naturally generated by the implementation. In practice, different feature types are grouped together, and sorted by the index of the function symbol in the symbol table.
- **PRM** uses permuted feature vectors, with features ordered by their span on the axiom set.
- **OPT** uses the same order as PRM, but completely removes features with span 0 on the axiom from the vector.

---

[5] Personal communication.

We know from previous experiments that **PRM** is generally the strongest of these three [18] and hence performed most tests with this schema.

6.4 Basic Performance

Table 1 shows the performance of the prover over the whole first-order part of TPTP 5.2.0 with conventional subsumption and different versions of feature-vector indexing for subsumption and contextual literal cutting. Feature vector functions are named in the obvious way. **NONE** represents the prover using conventional sequential subsumption. The table is sorted by number of solutions.

Table 1. Basic performance different subsumption methods

Index	Number of solutions
NONE	8,471
DRT-DF	8,721
PRM-DF	8,769
DRT-AL	8,832
PRM-BI	8,858
DRT-AC	8,881
PRM-AL	8,897
OPT-AC	8,914
PRM-AC	8,922

The weakest feature vector indexing (DRT-DF) can solve 250 problems more than the system with conventional subsumption. The best feature vector index-ing (PRM-AC) can solve an extra 201 problems. We can see that in general, the use of the **AC** features is best, followed by **AL** and finally **DF**. Also, the previous result that **PRM** (marginally) outperforms **OPT** which outperforms **DRT** is confirmed.

It is somewhat surprising that **PRM-AL** outperforms **PRM-BI**. Since **BI** should capture the same information as **AL** in the regular features, the most likely reason for this is the lack of the catch-all feature in **BI**.

In all pairings of results, the larger set of solutions is very nearly a strict superset of a smaller set. As an example, there is only one problem solved by **NONE** that is not solved by **PRM-AC**, while there are 441 problems solved by **PRM-AC** but not by **NONE**.

Figure 4 is a scatter plot that illustrates the superiority of indexed (**PRM-AC**) over conventional (**NONE**) subsumption. Each cross represents the perfor-mance on a single problem. Marks below the diagonal show better performance for the indexed versions. Marks on the right border represent problems where the system with conventional subsumption timed out, marks on the upper border to problems where the indexed system timed out. It is obvious that, with very few exceptions, the indexed version is much superior in performance.

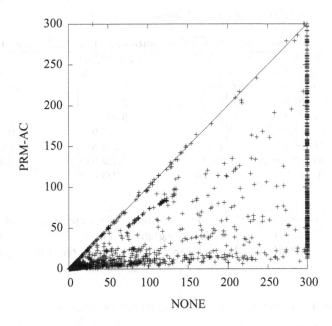

Fig. 4. Run times (in seconds) of **PRM-AC** over **NONE**

6.5   Profiling and Time Behavior

In the following we consider only problems solved by all subsumption strategies under consideration. Moreover, we consider only problems where the system has, with high likelihood, performed the same proof search. Filtering for these criteria, 8386 problems remain.

Table 2 shows where the indexed and non-indexed versions of the prover spend time on this set of problems. The order of strategies is the same as in Table 1. The columns contain the name of the index, total run-time on all 8386 problems, time spent in the computation of feature vectors, time spent on index maintenance (insert and delete), time spent directly in the clause-clause subsumption code, and time spent in forward- and backward subsumption, including the overhead of iterating through the sets or the index.

The first row shows that subsumption (including contextual literal cutting) has a drastic influence on total run time. For the non-indexed version, about 65% of total run time is spent in forward- and backwards subsumption operations, about 40% in individual non-unit clause-clause subsumption. Feature vector indexing drastically changes this, cutting the time for subsumption itself by a factor of more than 22 for **PRM-AC**. If we consider forward- and backward subsumption (including the sequential or indexed iteration over candidate clauses), we still see an improvement by a factor of more than 10.

Comparing the different indices, we can see that the run times in Table 2 are roughly, but not strictly, decreasing with the number of solutions in Table 1, indicating that the major reason for the improved performance is indeed the

Table 2. Time (in seconds) spend in various code segments

Name	Total time	FVs	Index	C/C subsumption	F&B subsumption
NONE	69,523.375	N/A	N/A	28,379.300	46,160.810
DRT-DF	34,059.599	81.250	51.460	6,848.240	17,548.180
PRM-DF	30,813.326	102.440	30.680	6,845.000	14,675.360
DRT-AL	31,301.364	75.820	86.410	2,051.890	13,322.010
PRM-BI	23,909.235	106.180	27.530	3,065.410	7,792.850
DRT-AC	25,360.085	81.490	72.590	1,217.920	9,017.900
PRM-AL	21,758.335	83.710	23.030	2,104.840	5,624.780
OPT-AC	20,963.105	100.580	35.930	1,423.630	4,550.500
PRM-AC	20,663.982	103.040	38.440	1,283.890	4,418.090

speed-up of subsumption. We can also see that index maintenance and feature vector computation are comparatively negligible. The cost for computing permuted vectors is slightly higher than for direct vectors, but this is more than compensated for by the smaller cost for index maintenance.

Table 3. Number of (non-unit) clause/clause subsumption attempts

Name	Non-unit subsumption calls	Recursive subsumption calls
NONE	120,934,644,536	14,142,932,647
DRT-DF	24,954,121,073	4,774,601,640
PRM-DF	25,897,859,434	4,851,875,239
DRT-AL	4,545,278,003	2,537,935,468
PRM-BI	8,832,113,671	4,189,693,241
DRT-AC	2,840,076,064	1,691,440,719
PRM-AL	5,795,826,572	2,936,641,523
OPT-AC	3,550,612,558	1,933,200,370
PRM-AC	3,254,649,092	1,846,439,309

The most dramatic improvement is in the time spent in actual subsumption code. The reason for this becomes obvious in Table 3. Column 2 shows the number of (non-unit) clause-clause subsumption attempts, column 3 the number of those that cannot be rejected by non-recursive tests and actually go into the exponential matching algorithm. The number of subsumption attempts drops by 97% of subsumption attempts when comparing **NONE** and **PRM-AC**. Even more dramatically, the number of recursive calls drops by nearly two full orders of magnitude. The reduction in subsumption attempts is also illustrated in the scatter plot of Figure 5.

Note the double logarithmic scale necessary to adequately display the large variation in numbers. The conventional version needs, over all problems, about 40 times more calls than the indexed version. For individual problems the improvement factor varies from 1 (for some trivial problems) to approximately 7500 e.g. for PUZ080+2, where the number of subsumption attempts drops from 69504083 with **NONE** to only 9284 with **PRM-AC**.

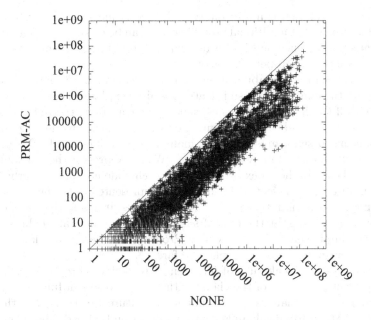

Fig. 5. Non-unit subsumption attempts of **PRM-AC** over **NONE**

6.6   Performance in Automatic Mode

The previous section listed results for a consistent search heuristic. However, much of the power of modern provers comes from an automatic selection of various search parameters in an auto-mode. We have also run E in its automatic mode with both **NONE**, **DRT-AC** and **PRM-AC**. The results are presented in Table 4.

Table 4. Performance in automatic mode

Index	Number of solutions
NONE	9,241
DRT-AC	9,782
PRM-AC	9,823

With E in automatic mode, the prover gains more than 600 solutions if **PRM-AC** is used.

7   Future Work

While we are very satisfied with the performance of our current implementation, there are a number of open research opportunities.

On the one side, we can still improve the technique itself. Mark Stickel has pointed out to us that additional useful features can be constructed not only from the greatest occurrence depth of a function symbol, but also from the smallest occurrence depth (if properly transformed).[6]

Recent theorem prover applications yield problems with very large signatures. In this case, the number of possible features is also very large. However, as Table 2 shows, for all the indexed strategies, time spent traversing the index is already longer than actual subsumption time. Several approaches to dealing with this are promising. If signatures are large, many feature values will be 0, since most symbols will not occur in any single clause. We can represent these default value feature implicitly in the index by annotating each node not only with the feature value, but also with the feature itself (usually represented as the numerical index describing its position in the vector). This would result in a fairly compact index even for large vectors, at the cost of slightly more complex algorithms. This is the version implemented by Korovin in iProver[7]. However, to our knowledge no systematic evaluation has taken place so far.

Secondly, we can make use of Theorem 3 to define more complex features that represent properties of the clause with respect to sets of function symbols. This requires only changes in the individual feature functions, not the other algorithms. An prototypical implementation of this in E already shows promising results, cutting total time for subsumption by another factor of two.

On a different track, while we developed feature vector indexing with the aim of finding a good solution for clause subsumption indexing, it can also be used to index terms for the retrieval relations *matches* and *is matched by*. While there are very good techniques for term indexing, most of them cannot handle AC symbols well. Feature vector indexing, on the other hand, handles associativity and commutativity easily. This might make an attractive choice for implementing forward and backward rewriting modulo AC.

8   Conclusion

Feature vector indexing has proved to be a simple, but effective answer to the subsumption problem for saturating first-order theorem provers. In our experiments, it is able to reduce the number of subsumption tests by, on average, about 97% compared to a naive sequential implementation, and thus reduces cost of subsumption in our prover to a level that makes it hard to measure using standard UNIX profiling tools. In addition to the direct benefit, this gain in efficiency has enabled us to implement otherwise relatively expensive subsumption-based simplification techniques (like contextual literal cutting), further improving overall performance of our prover.

---

[6] Personal communication.

[7] Personal communication.

Feature vector indexing has been successful not only in E and Prover9, but has also been implemented in the description logic reasoner KAON-2 [12], in the SMT-solver Z3 [1], and in the instance-generation based first-order prover iProver [7].

Afterword

A preliminary version of this paper was presented in 2004 at the *Workshop on Empirically Successful First-Order Reasoning (ESFOR)*, associated with IJCAR in Cork. This updated version of the paper describes some of the evolution of the implementation and adds a more detailed analysis of the costs and benefits of feature vector indexing.

ESFOR was the first of a series of workshops [20] with a renewed focus on practically useful systems and applications. I was one of the organizers, and we considered ourselves very lucky when Bill McCune joined the program committee. As it turned out, Bill was also one of the anonymous reviewers of my paper. Half a year later, he send me an email stating:

> A few weeks ago I tried your feature vector indexing in one of my provers. Powerful, easy to implement, elegant. The best new idea in indexing I've seen in many years.

It was one of the proudest moments of my life.

This informal style of cooperation was typical for Bill. He was always generous with credit, help, and advice, even when I first met him as a brand-new Ph.D. student in 1996. It is hard to accept that he is no longer around.

Acknowledgements. I thank the University of Miami's Center for Computational Science HPC team for making their cluster available for the extended experimental evaluation.

References

1. de Moura, L., Bjørner, N.S.: Engineering DPLL(T) + Saturation. In: Armando, A., Baumgartner, P., Dowek, G. (eds.) IJCAR 2008. LNCS (LNAI), vol. 5195, pp. 475–490. Springer, Heidelberg (2008)
2. Denzinger, J., Kronenburg, M., Schulz, S.: DISCOUNT: A Distributed and Learning Equational Prover. Journal of Automated Reasoning 18(2), 189–198 (1997); Special Issue on the CADE 13 ATP System Competition
3. Graf, P.: Term Indexing. LNCS, vol. 1053. Springer, Heidelberg (1996)
4. Graf, P., Fehrer, D.: Term Indexing. In: Bibel, W., Schmitt, P.H. (eds.) Automated Deduction — A Basis for Applications. Applied Logic Series, vol. 9(2), ch. 5, pp. 125–147. Kluwer Academic Publishers (1998)
5. Jaynes, E.T.: Probability Theory: The Logic of Science. Cambridge University Press (2003)

6. Kapur, D., Narendran, P.: NP-Completeness of the Set Unification and Matching Problems. In: Siekmann, J.H. (ed.) CADE 1986. LNCS, vol. 230, pp. 489–495. Springer, Heidelberg (1986)
7. Korovin, K.: iProver – An Instantiation-Based Theorem Prover for First-Order Logic (System Description). In: Armando, A., Baumgartner, P., Dowek, G. (eds.) IJCAR 2008. LNCS (LNAI), vol. 5195, pp. 292–298. Springer, Heidelberg (2008)
8. Löchner, B., Hillenbrand, T.: A Phytography of Waldmeister. Journal of AI Communications 15(2/3), 127–133 (2002)
9. McCune, W.W.: EQP 0.9 Users' Guide (1999), http://www.cs.unm.edu/~mccune/eqp/Manual.txt (accessed December 11, 2012)
10. McCune, W.W.: Experiments with Discrimination-Tree Indexing and Path Indexing for Term Retrieval. Journal of Automated Reasoning 9(2), 147–167 (1992)
11. McCune, W.W.: 33 Basic Test Problems: A Practical Evaluation of Some Paramodulation Strategies. In: Veroff, R. (ed.) Automated Reasoning and its Applications: Essays in Honor of Larry Wos, ch. 5, pp. 71–114. MIT Press (1997)
12. Motik, B., Sattler, U.: A Comparison of Reasoning Techniques for Querying Large Description Logic ABoxes. In: Hermann, M., Voronkov, A. (eds.) LPAR 2006. LNCS (LNAI), vol. 4246, pp. 227–241. Springer, Heidelberg (2006)
13. University of Miami Center for Computational Science. Pegasus - Introduction (2007-2011), http://ccs.miami.edu/?page_id=3749 (accessed December 09, 2012)
14. Riazanov, A., Voronkov, A.: The Design and Implementation of VAMPIRE. Journal of AI Communications 15(2/3), 91–110 (2002)
15. Riazanov, A., Voronkov, A.: Efficient Instance Retrieval with Standard and Relational Path Indexing. In: Baader, F. (ed.) CADE 2003. LNCS (LNAI), vol. 2741, pp. 380–396. Springer, Heidelberg (2003)
16. Schulz, S.: Information-Based Selection of Abstraction Levels. In: Russel, I., Kolen, J. (eds.) Proc. of the 14th FLAIRS, Key West, pp. 402–406. AAAI Press (2001)
17. Schulz, S.: E – A Brainiac Theorem Prover. Journal of AI Communications 15(2/3), 111–126 (2002)
18. Schulz, S.: Simple and Efficient Clause Subsumption with Feature Vector Indexing. In: Sutcliffe, G., Schulz, S., Tammet, T. (eds.) Proc. of the IJCAR-2004 Workshop on Empirically Successful First-Order Theorem Proving, Cork, Ireland (2004)
19. Schulz, S.: System Description: E 0.81. In: Basin, D., Rusinowitch, M. (eds.) IJCAR 2004. LNCS (LNAI), vol. 3097, pp. 223–228. Springer, Heidelberg (2004)
20. Schulz, S.: Empirically Sucessful Topics in Automated Deduction (2008), http://www.eprover.org/EVENTS/es_series.html
21. Schulz, S.: Fingerprint Indexing for Paramodulation and Rewriting. In: Gramlich, B., Miller, D., Sattler, U. (eds.) IJCAR 2012. LNCS, vol. 7364, pp. 477–483. Springer, Heidelberg (2012)
22. Sekar, R., Ramakrishnan, I.V., Voronkov, A.: Term Indexing. In: Robinson, A., Voronkov, A. (eds.) Handbook of Automated Reasoning, vol. II, ch. 26, pp. 1853–1961. Elsevier Science and MIT Press (2001)
23. Shannon, C.E., Weaver, W.: The Mathematical Theory of Communication. University of Illinois Press (1949)
24. Sleator, D.D., Tarjan, R.E.: Self-Adjusting Binary Search Trees. Journal of the ACM 32(3), 652–686 (1985)
25. Stickel, M.E.: SNARK - SRI's New Automated Reasoning Kit (2008), http://www.ai.sri.com/~stickel/snark.html (accessed October 04, 2009)

26. Tammet, T.: Towards Efficient Subsumption. In: Kirchner, C., Kirchner, H. (eds.) CADE 1998. LNCS (LNAI), vol. 1421, pp. 427–441. Springer, Heidelberg (1998)
27. Voronkov, A.: The Anatomy of Vampire: Implementing Bottom-Up Procedures with Code Trees. Journal of Automated Reasoning 15(2), 238–265 (1995)
28. Weidenbach, C.: SPASS: Combining Superposition, Sorts and Splitting. In: Robinson, A., Voronkov, A. (eds.) Handbook of Automated Reasoning, vol. II, ch. 27, pp. 1965–2013. Elsevier Science and MIT Press (2001)

Superposition for Bounded Domains

Thomas Hillenbrand and Christoph Weidenbach

Max-Planck-Institut für Informatik
Stuhlsatzenhausweg 85
D-66123 Saarbrücken
{hillen,weidenbach}@mpi-inf.mpg.de

Abstract. Reasoning about bounded domains in resolution calculi is often painful. For explicit and small domains and formulas with a few variables, grounding can be a successful approach. This approach was in particular shown to be effective by Bill McCune. For larger domains or larger formula sets with many variables, there is not much known. In particular, despite general decidability, superposition implementations that can meanwhile deal with large formula sets typically will not necessarily terminate. We start from the observation that lifting can be done more economically here: A variable does not stand anymore for every ground term, but just for the finitely many domain representatives. Thanks to this observation, the inference rules of superposition can drastically be restricted, and redundancy becomes effective. We present one calculus configuration which constitutes a decision procedure for satisfiability modulo the cardinality bound, and hence decides the Bernays-Schönfinkel class as a simple consequence. Finally, our approach also applies to bounded sorts in combination with arbitrary other, potentially infinite sorts in the framework of soft sorts. This frequent combination – which we recently explored in a combination of SPASS and Isabelle – is an important motivation of our study.

1 Introduction

Reasoning about bounded domains in resolution-style calculi is often painful. Despite general decidability, superposition implementations typically will not terminate. Bounded domain means that the domain size is bounded from above: In virtue of a clause like $x \simeq 1 \vee \ldots \vee x \simeq n$, any domain element equals one of some given n "digits", which need not be distinct.

Traditionally, attacking bounded domain problems has been done by so-called finite-domain model generators where the most prominent and influential one is MACE4 [25], developed by Bill McCune. In particular, MACE4 has been successfully applied by mathematicians for reasoning in algebraic structures. In addition to the development of MACE4, Bill McCune was also influential in the development of general-purpose automated theorem provers, in particular OTTER [27], which was later on replaced by PROVER9 [26]. The first version MACE2 of Bill McCune's MACE program encodes a finite domain problem into a SAT problem and then applies a SAT solver. The recent version MACE4 works directly on the finite domain first-order structure at the advantage of using first-order reductions, such as rewriting. So MACE4 is already a big step from MACE2 towards a

M.P. Bonacina and M.E. Stickel (Eds.): McCune Festschrift, LNAI 7788, pp. 68–100, 2013.

first-order logic reasoning procedure. Actually, Bill McCune was already thinking of integrating MACE4 and PROVER9 more closely and suggested "MACE4 *can be a valuable complement to* PROVER9, *looking for counterexamples before (or at the same time as) using* PROVER9 *to search for a proof.*" Although MACE4 and PROVER9 rely on the same coding infrastructure, they actually do not work together. Our contribution is a suggestion of a calculus to bridge the gap between classical first-order reasoning and finite-model-search reasoning in the style of MACE4.

The superposition model operator R (see page 74) serves as a kind of MACE4 component in our calculus. It builds a partial model assumption including function tables for all ground instances, due to the function definition clauses $f(\vec{x}) \simeq 1 \vee \ldots \vee f(\vec{x}) \simeq n$ added by our calculus. Then, as is customary for superposition calculi (Definition 3), inferences can be restricted to a minimal false clause with respect to R. We do not exploit this explicitly in the definition of our calculus, but it is part of the completeness approach. When our calculus terminates by saturation, R is a model of the clause set and R is explicitly given by unit rewrite rules in the saturation, see Section 4.4. So in fact, our calculus combines explicit model building in the style of MACE4 and first-order theorem proving in the style of PROVER9. An approach Bill McCune might have thought of already.

Our approach extends the search of finite-domain model generators, which search for suitable interpretations in domains of increasing order. In our approach such interpretations are implicitly constructed by a (partial) model assumption where the calculus itself operates in a superposition style manner on clauses with variables. The problem of (finite) model computation has gained renewed interest, as witnessed by various recent contributions. For example, new approaches via transformation into certain fragments of logic have been presented in [8] and [14]. A variant tailored to instantiation-based methods is given in [6]. The fruitful interplay of superposition and decision procedures is testified, for example, by [2] and [11].

We start from the observation that lifting can be done more economically here: A variable does not stand anymore for every ground term, but just for the finitely many digits (Sect. 3.1). Conversely, an inference only has to be considered if the range of the pertaining most general unifier does exclusively consist of variables and digits. Secondly, for any non-ground inference one can easily determine those instantiations that satisfy its ordering constraints. Thirdly, redundancy also considers digit instances only, such that stronger simplifications become possible in some situations, but compatibility with the corresponding notion of standard superposition is mostly preserved (Sect. 3.2). In order to obtain this, the above cardinality-bounding clause needs to be exchanged for its functional instances $f(\vec{x}) \simeq 1 \vee \ldots \vee f(\vec{x}) \simeq n$ in order not to lose completeness.

The lifting modification applies to the family of superposition calculi. Soundness and refutational completeness are preserved. We demonstrate this for a domain-specific calculus configuration in which non-Horn clauses are dealt with not by equality factoring, but by aggressive splitting. We give a termination result based on the detection of particular loops, and another one based on ordered rewriting with some instantiation (Sect. 4). Both decision procedures for satisfiability modulo the cardinality bound naturally also cope with the Bernays-Schönfinkel class (Sect. 4.5) as a special case. This solves yet another classical decidability problem by superposition. Finally, the lifting modification is also applicable to bounded sorts in combination with arbitrary other, potentially infinite sorts (Sect. 5) in the framework of dynamic sort theories. This frequent combination – think for example of finite enumeration types in programming languages, or any verification problem that involves a component with finite state space – is an important motivation of our study. Two application scenarios are discussed in Sect. 6. Tedious proofs have been omitted here, but can be found in a technical report [21].

Compared to instantiation-based methods for finite-domain problems with explicit instantiation such as MACE [25], PARADOX [13], FINDER [34], SEM [39], and related calculi such as hyper-tableaux [7], our calculus does not instantiate variables a priori, but exploits the boundedness of the domain on the level of non-ground clauses. In particular, this offers advantages if the problem has structure that can be employed by inference and reduction rules [32]. As a first simple example, not a single inference is possible between the two unit clauses $P(x_1, \ldots, x_k, x_1)$ and $\neg P(a, y_1, \ldots, y_{k-1}, b)$, but instantiation-based methods will generate more than n^k clauses for domain size n. In general, a superposition inference or simplification that involves variables simulates up to exponentially many ground steps. Likewise, proving one inference redundant may save an exponential amount of work. As a second example, consider an equation $f(x) \simeq x$ and an atom $P(f(g(x)))$, which standard rewriting would simplify to $P(g(x))$. After instantiation with digits this reduction is no longer possible, as any term $g(\ldots)$ is not a digit. For examples of this form, inferring and simplifying at the non-ground level has the potential to exponentially shorten proofs and model representations. In Sect. 6 we elaborate two real-world examples of this form.

Transformation-based methods [24,8] translate a given clause set into a form on which standard inference mechanisms like hyperresolution search for a model in a bottom-up way. This work is orthogonal to ours, because the problem is transformed, whereas we exploit the boundedness of the domain truly at the calculus level. However, neither the instantiation-based nor the transformation-based approach currently support the combination with general first-order theories, in contrast to our calculus.

Starting with a simple Bernays-Schönfinkel style setting, where all function symbols are constants, we prove that a cardinality-bounding clause $x \simeq 1 \vee \ldots \vee x \simeq n$ is not needed and can be dropped. Nevertheless, superposition is not a decision procedure for this class. It may generate arbitrary long clauses with an unbounded number of variables. Consider for example a clause expressing a confluence property, such as

$$\neg P(x,y) \vee \neg P(x,z) \vee P(y,z).$$

All occurrences of P literals are incomparable by the reduction ordering underlying superposition, so in particular superposition self inferences with this clause produce arbitrary long clauses with an unbounded number of variables. Our solution here is to extend superposition inferences by lazy instantiation with digits such that the literals triggering the inference become (strictly) greatest in the ordering. For the above example, a potential inference with a clause $\neg P(1,1)$ is not possible, because whatever digit is substituted for x after unifying $P(y,z)$ with $P(1,1)$ a negative literal will become greatest, assume the natural ordering on the digits $1 \prec 2 \dots \prec n$ (see Definition 3). Together with a splitting rule this style of reasoning basically already guarantees termination on this fragment. For a bounded fragment with non-constant function symbols, the situation gets more involved. In particular, on such a fragment we do want to perform rewriting as much as possible in the standard first-order style. We achieve this goal by exchanging the cardinality-bounding clause $x \simeq 1 \vee \dots \vee x \simeq n$ for its functional instances $f(\vec{x}) \simeq 1 \vee \dots \vee f(\vec{x}) \simeq n$. Still we limit unifiers in inferences to digits or variables, but support almost unlimited rewriting with arbitrary terms and matchers. Basically, these two ingredients lift our approach from the Bernays-Schönfinkel class to full first-order clause sets with a cardinality-bounding clause.

Finally, a sort discipline supports combination of finite domain sorts with others as it naturally occurs in real-world application. Think for example of a network model where the single bits 0 and 1 for building bit vectors representing network addresses must not be confused with other terms. For example, a cardinality bounding clause

$$\neg Bit(x) \vee x \simeq 0 \vee x \simeq 1.$$

should not be involved in any inference at any position with a clause talking about bit vectors, such as performing a logical and operation for IP-addresses:

$$IP(x_1 \circ y_1, \dots) \simeq ipand(IP(x_1, \dots), IP(y_1, \dots))$$

where $\circ$ represents bitwise "and". This property is supported by our bounded domain calculus introduced in Sect. 5.

2   Getting Started

For most logical notions and notations, we refer to [29]. In particular we work in a logic with built-in equality. We stipulate a single-sorted signature Σ that contains the constant symbols 1 through n, which we name *digits*, besides arbitrary other function symbols, possibly including constants. So equality is the only predicate symbol, but free predicate symbols will briefly be discussed in Sect. 4.5. Moving on, a set $\mathcal{V}$ provides an infinite supply of variables. For a term t we denote by $\mathrm{var}(t)$ the set of variables that occur in t; the set $\mathrm{var}(C)$ is defined correspondingly for every clause C. If σ is a substitution, then $\mathrm{dom}\,\sigma$ is the set of all variables for which $x\sigma \neq x$, $\mathrm{ran}\,\sigma$ is the image of $\mathrm{dom}\,\sigma$ under σ, and $\mathrm{cdom}\,\sigma$ is the set of variables occurring in $\mathrm{ran}\,\sigma$. We say that a substitution σ can be *refined into* a substitution ρ if $\rho = \sigma\tau$ for some substitution τ, and that σ is *more general than* ρ if it can be refined into ρ, but not vice versa. For simplicity of notation the equality symbol $\simeq$ is supposed to be symmetric. A literal $s \bowtie t$ is either an equation $s \simeq t$ or a disequation $s \not\simeq t$. A clause is a disjunction of literals; a Horn clause is a clause with at most one positive literal;

the empty clause is denoted by $\perp$. We assume that a reduction ordering $\succ$ is given which is total on ground terms. To every literal, we assign a *complexity* according to $s \simeq t \mapsto \{s, t\}$ and $s \not\simeq t \mapsto \{s, s, t, t\}$. Literals are compared in the multiset extension of $\succ$ on their complexities, and clauses in the two-fold multiset extension on the multisets of the respective literal complexities. If C is a clause and M a clause set, then $M^{\prec C}$ holds all elements of M smaller than C, and $\mathrm{gnd}(C)$ consists of all ground instances of C.

We study the theory $\mathcal{T}$ given by the formula

$$\forall x.\, x \simeq 1 \vee \ldots \vee x \simeq n$$

and will introduce a superposition-based calculus to tackle the $\mathcal{T}$-satisfiability of clause sets over Σ. Note that this also covers the case that the domain size is exactly n, since one can add clauses $i \not\simeq j$ for any distinct $i, j \in [1; n]$.

The calculus will be described by rule patterns of three different types in a fraction-like notation. Clauses occurring in the numerator are generally called *premises*, and in the denominator *conclusions*. As usually, premises are assumed to share no variables. Finite clause sequences $C_1, \ldots, C_m$ where $m \geq 0$ are abbreviated as $\vec{C}$, and $\bigwedge \vec{C}$ is the conjunction of all C_i. If C denotes a clause and M a clause set, then M, C is shorthand notation for $M \cup \{C\}$.

(i) *Inference rules:* $\qquad \mathcal{I} \dfrac{\vec{C}}{D} \qquad$ if *condition*

denotes any transition from a clause set $M, \vec{C}$ to $M, \vec{C}, D$ provided *condition* is fulfilled. Occasionally the rightmost of the premises is named *main premise*, and the remaining ones are the *side premises*.

(ii) *Reduction rules:* $\qquad \mathcal{R} \dfrac{C}{\vec{C'}} \vec{D} \qquad$ if *condition*

stands for any transition from a clause set $M, C, \vec{D}$ to a clause set $M, \vec{C'}, \vec{D}$ whenever *condition* holds. In essence, the clause C is replaced by the clauses $\vec{C'}$, the sequence of which may be empty.

(iii) *Split rules:* $\qquad \mathcal{S} \dfrac{C}{D \mid D'} \qquad$ if *condition*

describes any transition from a clause set M, C to the pair of clause sets $(M, C, D \mid M, C, D')$ constrained by *condition*. Note that the premise is part of each of the descending clause sets.

In the *condition* part of inference rules, frequently some terms, say s and t, are required to have a most general unifier σ. We stipulate that σ satisfies $\mathrm{dom}\,\sigma \cup \mathrm{cdom}\,\sigma \subseteq \mathrm{var}(s, t)$, which for syntactic unification is without loss of generality. Furthermore, occurrences of terms or literals may be restricted to maximal ones. In the former case this refers to the enclosing literal, and in the latter to the enclosing clause. Maximality means that no other occurrence is greater, and is strict if none is greater or equal. Correspondingly we will speak of greatest occurrences, which are greater than or equal to the remaining ones, and of strictly greater ones, that are greater than all the rest. There is no difference between being greatest or maximal in case the underlying ordering is total, as happens in the case of ground clauses and a reduction ordering total on ground terms.

An application of one of the above rules is called an *inference,* a *reduction* or a *split,* respectively. Given an inference with premises $\vec{C}$ and conclusion D, then an *instance* of this inference is every inference with premises $\vec{C}\sigma$ and conclusion $D\sigma$.

A *derivation* from a (not necessarily finite) clause set M with respect to a calculus specified that way is a finitely branching tree such that (i) the nodes are sets of clauses, (ii) the root is M, and (iii) if a node N has the immediate descendants $N_1, \ldots, N_k$, respectively, then there is a transition from N to $N_1, \ldots, N_k$ in the calculus. Infinite inputs could, for example, arise from the instantion of finite sets, or from the enumeration of some theory. A *complete path* $N_1, N_2, \ldots$ in a derivation tree starts from the root, ends in a leaf in case the path is finite, and has the *limit* $N_\infty = \bigcup_i \bigcap_{j \geq i} N_j$. Note that the term "complete" has been chosen simply to stress that the path, if finite, indeed reaches a leaf. Given a redundancy notion for inferences and clauses, a derivation is said to be *fair* if for every complete path $N_1, N_2, \ldots$ the following applies to the transitions from N_∞: (i) Every inference is redundant in some N_i, and (ii) in case a split rule is present in the calculus, then for every split, one of its conclusion is in some N_i or redundant with respect to it. A clause set M is *saturated* if (i) every inference with premises in M is redundant with respect to M, and (ii) for every split, one of its conclusion is in M or redundant with respect to it. In the context of full first-order logic without a bounded domain, fairness and completeness is more involved [16].

3   A Calculus for $\mathcal{T}$-unsatisfiability

3.1   Calculus Rules

Let us first recapitulate a standard variant of superposition. For the sake of simplicity, selection of negative equations is not taken into account yet.

Definition 1. The *standard superposition calculus* $\mathcal{S}$ consists of the rules

$$\text{Negative super-position} \quad \mathcal{I} \, \frac{C \vee \underline{s \simeq t} \quad \underline{u[s']} \not\simeq v \vee D}{(C \vee u[t] \not\simeq v \vee D)\sigma} \qquad \text{Equality resolution} \quad \mathcal{I} \, \frac{C \vee \underline{s \not\simeq s'}}{C\sigma}$$

$$\text{Positive super-position} \quad \mathcal{I} \, \frac{C \vee \underline{s \simeq t} \quad \underline{u[s']} \simeq v \vee D}{(C \vee u[t] \simeq v \vee D)\sigma} \qquad \text{Equality factoring} \quad \mathcal{I} \, \frac{C \vee \underline{s \simeq t} \vee s' \simeq u}{(C \vee t \not\simeq u \vee s \simeq u)\sigma}$$

subject to the *restrictions*

(i) $\sigma = \mathrm{mgu}(s, s')$, and $s' \notin \mathcal{V}$ in case of the superposition rules

(ii) under σ, the underlined occurrences are <u>maximal</u> or <u>strictly maximal</u>

(iii) the main premise is strictly maximal under σ

and augmented with a notion of *redundancy* with respect to a clause set M:

(a) for a clause C, if $\mathrm{gnd}(M)^{\prec C\sigma} \models C\sigma$ for every ground substitution σ

(b) for an inference, if for every ground instance with maximal premise $C\sigma$ and conclusion $D\sigma$ we have $\mathrm{gnd}(M)^{\prec C\sigma} \models D\sigma$

Let us remark that condition (iii) shows up, for example, in [4, Sect. 3], but not in all presentations of superposition. It excludes, for instance, the superposition of $s \simeq t$ into $s \simeq u$ if $s \succ t \succ u$, and may facilitate termination proofs.

The calculus $\mathcal{S}$ is sound and refutationally complete in the sense that $M \models \bot$ and $\bot \in M$ coincide for every saturated set M, and the limit of every fair derivation is saturated and equivalent to the input. The completeness proof relies on a model functor that associates with M a convergent ground rewrite system R. For saturated M free of $\bot$, the model is the quotient R^* of the free ground term algebra modulo the congruence generated by R. The rewrite system is given in terms of sequences $\mathrm{Gen}(C)$ and R_C which are defined by mutual recursion. For every ground clause C let $\mathrm{Gen}(C) = \{s \to t\}$ if (i) $C \equiv C' \vee s \simeq t \in \mathrm{gnd}(M)$, (ii) $R_C^* \not\models C$, (iii) $R_C^* \not\models t \simeq u$ for all literals $s \simeq u$ in C, (iv) s is $\overline{R_C}$-irreducible; and let $\mathrm{Gen}(C) = \{\}$ otherwise. Furthermore R_C is $\bigcup_{D \prec C} \mathrm{Gen}(D)$, and finally R is $\bigcup_D \mathrm{Gen}(D)$.

As already mentioned, the rationale of our calculus refinement is that a variable will just stand for the digits, not for every ground term. Let us introduce some notions to make this precise: For a substitution τ we say that it *numbers* if $\mathrm{ran}\,\tau \subseteq [1; n]$. Note that τ is more general than another numbering substitution τ if and only if $\tau \subset \tau'$, in the set-theoretic sense. So we say that τ *minimally numbers* with respect to a set of conditions if these are satisfied by τ and by no other numbering τ' more general than τ. Furthermore τ *ground numbers* a clause C if τ numbers and $C\tau$ is ground. The set of all ground instances of C under such substitutions is denoted by $\Omega(C)$, and its elements are called the *Ω-instances* of C.

Alas, if we apply the new interpretation of variables to the clause $x \simeq 1 \vee \ldots \vee x \simeq n$ which defines our theory, then each of its instances in terms of digits is a tautology. Hence we exchange $\mathcal{T}$ for the set $\mathcal{T}'$ of its functional instances, which consists of these clauses:

$$f(\vec{x}) \simeq 1 \vee \ldots \vee f(\vec{x}) \simeq n \qquad \text{for any } f \in \Sigma \setminus [1; n]$$

$\mathcal{T}'$ is weaker than $\mathcal{T}$ in the sense that the upper cardinality bound is only applied to function values; but it satisfies the same universal formulae. There is an increase in the initial number of clauses, but this will be outshined by the fact that no inferences with complex unifiers are necessary. Interestingly, within the Bernays-Schönfinkel class the set $\mathcal{T}'$ is empty, as we will demonstrate in Sect. 4.5.

The following equivalence makes our consideration precise:

Proposition 2. A clause set M is $\mathcal{T}$-satisfiable iff $\Omega(M \cup \mathcal{T}')$ is satisfiable.

How to exploit this proposition? Assume M is a non-ground clause set, and we are interested in its $\mathcal{T}$-satisfiability. In standard superposition, if N is a saturated presentation of $M \cup \mathcal{T}$, then $\mathrm{gnd}(N)$ is saturated as well. However, it is sufficient that $\Omega(N)$ is saturated, provided the saturation started from $M \cup \mathcal{T}'$. The benefit is that lifting can be made more economically: An inference only has to be considered if the range of the pertaining most general unifier consists of variables and digits only. Secondly, for any non-ground inference one can, via partial instantiation with digits, determine those instantiations that satisfy its ordering constraints. Thirdly, redundancy also considers digit instances only and becomes effective. We formulate the following refinement of standard superposition:

Definition 3. *Superposition for bounded domains* S_B *refines* S *as follows:*

Negative super-position $\quad\mathcal{I}\,\dfrac{C \vee \underset{=}{\underline{s \simeq t}} \quad \underline{u[s']} \not\simeq v \vee D}{(C \vee u[t] \not\simeq v \vee D)\sigma}$

Equality resolution $\quad\mathcal{I}\,\dfrac{C \vee \underline{s \not\simeq s'}}{C\sigma}$

Positive super-position $\quad\mathcal{I}\,\dfrac{C \vee \underset{=}{\underline{s \simeq t}} \quad \underline{u[s']} \simeq v \vee D}{(C \vee u[t] \simeq v \vee D)\sigma}$

Equality factoring $\quad\mathcal{I}\,\dfrac{C \vee \underline{s \simeq t} \vee \underline{s' \simeq u}}{(C \vee t \not\simeq u \vee s \simeq u)\sigma}$

under the *restrictions*

(i) $\sigma = \mathrm{mgu}(s, s')$, and $s' \notin \mathcal{V}$ in case of the superposition rules
(ii) $\mathrm{ran}\,\sigma \subseteq \mathcal{V} \cup [1; n]$
(iii) there is a minimally numbering substitution τ such that under $\sigma\tau$
  – the underlined occurrences are <u>greatest</u> or <u>strictly greatest</u>
  – the main premise is strictly greatest

where *redundancy* with respect to a clause set M is given

(a) for a clause C if $\Omega(M)^{\prec C\rho} \models C\rho$ for every ground numbering ρ
(b) for an inference with main premise C, most general unifier σ, minimally numbering substitution τ and conclusion D if for every ground numbering ρ we have $\Omega(M)^{\prec C\sigma\tau\rho} \models D\rho$.

Based on the notion of redundancy, *simplification,* in its general form, is making a clause redundant by adding (zero or more) entailed smaller clauses. Here it is already enough if these conditions hold on the Ω-instances.

$$\mathcal{R}\,\dfrac{C \quad \vec{D}}{\vec{C'}} \qquad \text{if} \quad \begin{array}{l} \cdot\ C \text{ is redundant w.r.t. } \vec{C'}, \vec{D} \\ \cdot\ \Omega(C, \vec{D}) \models \Omega(\bigwedge \vec{C'}) \\ \cdot\ \Omega(C) \succ \Omega(\vec{C'}) \end{array}$$

Testing the existence of a substitution τ in (iii) is effective for every decidable reduction ordering. Actually, instead of just testing, one could alternatively enumerate *all* such minimally numbering substitutions τ for which the mentioned maximality conditions hold, and for each τ add the inference conclusion instantiated by τ. The number of these substitutions is always finite, but it may become large. Just to give an example, equality resolution would become the following:

$$\mathcal{I}\,\dfrac{C \vee s \not\simeq s'}{C\sigma\tau} \quad \text{if} \quad \begin{array}{l} \cdot\ \sigma = \mathrm{mgu}(s, s') \text{ and } \mathrm{ran}\,\sigma \subseteq \mathcal{V} \cup [1; n] \\ \cdot\ \tau \text{ minimally numbers such that} \\ \quad s \not\simeq s' \text{ is greatest under } \sigma\tau \end{array}$$

Let us stress that condition (ii) – absence of complex unifiers – is easy to test and should exclude many of the inferences drawn in the standard calculus. For example, with the lexicographic path ordering [22] induced by the precedence $+ \succ s$, from the two clauses $(x+y)+z \simeq x+(y+z)$ and $u+s(v) \simeq s(u+v)$ one would normally obtain every $s^i(x + y) + z \simeq x + (s^i(y) + z)$. But since y needs to be bound to $s(v)$, not a single inference is drawn in the calculus $\mathcal{S}_B$.

We stipulate that from now on the smallest ground terms are the digits from $[1; n]$, say such that $n \succ \ldots \succ 1$. Then the calculus $\mathcal{S}_B$ is sound and refutationally complete in the sense that a clause set M is $\mathcal{T}$-unsatisfiable if and only if every fair $\mathcal{S}_B$-derivation from $M \cup \mathcal{T}'$ eventually produces the empty clause. Notably the minimality of the digits is indispensable: Assume that $\succ$ is the lexicographic path ordering induced by the precedence $n \succ \ldots \succ 1 \succ f \succ c$. Then from the unsatisfiable clause set $\{f(x) \simeq 1,\ 1 \simeq c,\ 1 \not\simeq f(c)\}$ nothing but the clause $f(c) \not\simeq c$ is inferable. When lifting ground-level inferences to the non-ground level, this minimality is needed to show that variable overlaps are non-critical; and indeed the variable overlap from $1 \simeq c$ into $f(x) \simeq 1$ would produce $f(c) \simeq 1$ and eventually lead to the empty clause.

3.2   Redundancy in $\mathcal{S}_B$ and in $\mathcal{S}$

In the calculus $\mathcal{S}_B$, redundancy on the general level is defined via redundancy of Ω-instances on the ground level, whereas in standard superposition one goes back to redundancy of all ground instances. Let us compare under which conditions a clause C is redundant with respect to a clause set M. In the calculus $\mathcal{S}_B$ we require $\Omega(M)^{\prec C\rho} \models C\rho$ for every ground numbering ρ. The condition in standard superposition is $\mathrm{gnd}(M)^{\prec C\sigma} \models C\sigma$ for every ground substitution σ. So for redundancy in the sense of $\mathcal{S}_B$ fewer instances need to be shown redundant, but on the other hand there are fewer premises for doing so. For example, $f(g(1)) \simeq 1$ is not redundant with respect to $f(x) \simeq 1$, since it is not entailed from $f(1) \simeq 1, \ldots, f(n) \simeq 1$. Fortunately, in $\mathcal{S}_B$-derivations the set M with respect to which redundancy is studied always contains the clauses of $\mathcal{T}'$, possibly simplified. Therefore we additionally have $g(1) \simeq 1 \vee \ldots \vee g(1) \simeq n$ at hand, with which $f(g(1)) \simeq 1$ does become redundant.

This subsection contains two results that generalize this observation. Firstly, if every digit instance $C\rho$ is entailed from smaller ground instances of M except some problematic ones, then C is redundant in the sense of $\mathcal{S}_B$. Secondly, if every $C\rho$ follows from arbitrary smaller ground instances, but C is not of a particular form, then C is also redundant. These results permit us to adapt concrete simplification techniques like rewriting or subsumption to our calculus. The subsection ends with a demonstration that $\mathcal{S}_B$ should not be mixed with the standard notion of redundancy.

We reserve the identifier f for non-digit function symbols, whereas i, j, k denote digits and $\vec{\imath}$ a vector thereof. For any term t, let $\mathrm{Dig}(t)$ denote the clause $t \simeq 1 \vee \ldots \vee t \simeq n$.

Given a clause C with ground substitution σ, we call the pair C, σ *problematic* if $x\sigma \equiv f(\vec{\imath})$ for some $x \in \mathrm{var}(C)$ and $C\sigma \preceq \mathrm{Dig}(f(\vec{\imath}))$. Otherwise the pair is called *unproblematic*. Furthermore, let $\mathrm{gnd}(C)$ denote the set of all ground instances $C\sigma$ for which C, σ is unproblematic, and let gnd extend to clause sets in the usual way. Here are two necessary and quite restrictive conditions for C, σ to be problematic: Firstly, some variable $x \in \mathrm{var}(C)$ may occur only in literals of the form $x \simeq i$ and $x \simeq y$. Secondly, the greatest literal of $C\sigma$ must have the form $f(\bar{\imath}) \simeq j$.

Additionally, a clause C is called *critical* if it has an Ω-instance $C\rho$ with greatest term $f(\bar{\imath})$ such that $C\rho \preceq \text{Dig}(f(\bar{\imath}))$. Otherwise C is called *noncritical*. Note that this notion refers to Ω-instances, whilst in a problematic pair C, σ, the second element is an arbitrary ground substitution,

Lemma 4. Consider a path in an $\mathcal{S}_B$-derivation from $M \cup \mathcal{T}'$ to N and a clause C. Then C is redundant with respect to N if one of the following conditions holds, where ρ ranges over all ground numbering substitutions:

   (i) $\mathring{\text{gnd}}(N)^{\prec C\rho} \models C\rho$ for all ρ,
   (ii) C is noncritical and $\text{gnd}(N)^{\prec C\rho} \models C\rho$ for all ρ.

The difference between our redundancy notion and the one of standard superposition may show up in practice: Assume $n = 2$ and some input M which via $\mathcal{S}_B$ eventually leads to the clause set $N = \{x \simeq 1, f(1) \simeq 2, f(2) \simeq 2, f(1) \not\simeq 1\}$. Now the clause $x \simeq 1$ has the ground instances $2 \simeq 1$ and $f(1) \simeq 1$ which make the second and the third clause redundant in the standard sense. Since $f(1) \simeq 1$ is not an Ω-instance of $x \simeq 1$, these clauses are not redundant in the sense of $\mathcal{S}_B$. Note also that $x \simeq 1, \{x \mapsto f(1)\}$ is problematic and that both $f(1) \simeq 2$ and $f(2) \simeq 2$ are critical, such that Lem. 4 does not apply.

Going further, the example shows that combining $\mathcal{S}_B$ with standard redundancy is problematic: If $f(1) \simeq 2$ and $f(2) \simeq 2$ were deleted from N, then the rest $\{x \simeq 1, f(1) \not\simeq 1\}$ would be $\mathcal{S}_B$-saturated, despite the apparent unsatisfiability. Summing up, refutational completeness would be lost. However, because of Lem. 4 only in rare cases is standard redundancy stronger than redundancy in the sense of $\mathcal{S}_B$.

Notably the opposite relation can be observed as well: Let $n = 2$, $C \equiv x \simeq y \vee f(1) \simeq y$ and $N = \{f(1) \simeq 1 \vee f(1) \simeq 2, f(2) \simeq 1 \vee f(2) \simeq 2, 1 \simeq 2, C\} \supseteq \mathcal{T}'$. The clause C is redundant in the sense of $\mathcal{S}_B$ because $C\rho$ is a tautology if $x\rho \equiv y\rho$, and because otherwise $C\rho$ is subsumed by $1 \simeq 2$. However C is not redundant in the standard sense: Consider the ground instance $C\sigma \equiv f(1) \simeq 1 \vee f(1) \simeq 1$. We obtain $\text{gnd}(N)^{\prec C\sigma} = \{1 \simeq 2, 1 \simeq 1 \vee f(1) \simeq 1, 2 \simeq 1 \vee f(1) \simeq 1\}$, which is equivalent to $\{1 \simeq 2\}$. Clearly, this does not entail $f(1) \simeq 1$. One cannot hold the exchange of $\mathcal{T}'$ for $\mathcal{T}$ responsible for this phenomenon, since it also occurs in case of $N' = \{x \simeq 1 \vee x \simeq 2, 1 \simeq 2, C\}$.

3.3 Application to Unit Rewriting

For a set E of unit equations, the *ordered rewrite relation* $\rightarrow_E$ is commonly defined as the smallest relation on terms such that $u[s\sigma] \rightarrow_E u[t\sigma]$ whenever $s \simeq t \in E$ and $s\sigma \succ t\sigma$, where σ is a substitution such that $s\sigma$ occurs as subterm u. If t contains variables that do not show up in s, as in $f(x) \simeq f(y)$, then one has to guess an instantiation of these in order to achieve decreasingness. However, in case a solution exists, then binding to the minimal constant works as well. So we stipulate that additionally $(\text{var}(t) \setminus \text{var}(s))\sigma \equiv \{1\}$ holds. As usual, the reflexive-transitive closure of $\rightarrow_E$ is denoted by $\rightarrow_E^*$.

We extend the ordered rewrite relation $\rightarrow_E$ from terms to clauses in the obvious way. As such, it is a simplification in the sense of our calculus only if the clause to be simplified is above the simplifying equation instances. For example,

$$f(3) \simeq 1 \;\rightarrow_{\{f(3)\simeq 2\}}\; 2 \simeq 1$$

is a rewrite step, but not a simplification, because the clause to be rewritten is smaller than the one used for rewriting. In order to capture this, let $\rightarrow_E^{\succ}$ denote the smallest relation on clauses such that $C[s\sigma] \rightarrow_E^{\succ} C[t\sigma]$ whenever $s \simeq t \in E$, $s\sigma \rightarrow_E t\sigma$ and $C\rho \succ (s \simeq t)\sigma\rho$ for all ground numbering ρ. A further condition is necessary to ensure that the rewritten clause is redundant according to Lem. 4: Rewriting $C \rightarrow_E^* D$ is called Ω-admissible for any noncritical C. If C is critical, however, then it contains literals of the shape $f(\vec{s}) \simeq t$ where t and every s_i is a digit or a variable, such that with a suitable ground numbering substitution ρ the term $f(\vec{s})\rho$ is the greatest of $C\rho$. Then $C \rightarrow_E^* D$ is Ω-admissible only if rewrite steps on the left-hand side of such literals $f(\vec{s}) \simeq t$ with equations $x \simeq i \in E$ or $x \simeq y \in E$ only take place below f. So the following is an instance of simplification in the calculus $\mathcal{C}$:

Ordered unit rewriting

$$\mathcal{R} \frac{C}{D} E \qquad \text{if} \quad \begin{array}{l} \cdot \; E \text{ is a set of unit equations} \\ \cdot \; C \rightarrow_E^{\succ} \circ \rightarrow_E^* D \\ \cdot \; C \rightarrow_E^* D \text{ is } \Omega\text{-admissible} \end{array}$$

4   Obtaining a Decision Procedure

4.1   Calculus Rules

Refutational completeness of the calculus $\mathcal{S}_B$ means that if M is $\mathcal{T}$-unsatisfiable, then in every fair derivation eventually the empty clause will show up, even for infinite M. If M is $\mathcal{T}$-satisfiable, however, then derivations without suitable simplification steps may become infinite. We will present a calculus configuration that enforces termination. To make this effective, naturally the input clause set M must be finite, as well as the signature Σ; and the ordering $\succ$ must be decidable.

Going back to standard superposition $\mathcal{S}$, without simplifications this calculus does not decide the satisfiability of finite ground clause sets: If $\succ$ is a lexicographic path ordering induced by the precedence $a \succ f \succ b$, then from the equations $f(a) \simeq a$ and $a \simeq f(b)$ one obtains an infinite series $f(f(b)) \simeq a$, $f(f(f(b))) \simeq a$, ... by positive superposition. However, if all clauses are units, then every inference conclusion makes its main premise redundant and hence can be turned into a simplification. The clause set decreases in the multiset extension of the clause ordering, which guarantees termination.

The satisfiability of finite ground Horn clause sets can be decided the same way if in every clause with negative literals, at least one of them shall be *selected*. This eager selection leads to a positive unit literal strategy [15], where the side premise of superposition inferences is always a positive unit clause. We denote this ground-level calculus variant by $\mathcal{G}$. Via splitting of non-Horn clauses into Horn clauses, decidability extends to the non-Horn case. In [20], we have encoded Sudoku puzzles as ground satisfiability problems for the SPASS theorem prover, which proceeding that way succeeded within a blink of an eye. Therefore, we decided to choose $\mathcal{G}$ as a basis for the formulation of a decision procedure.

The resulting calculus $\mathcal{C}$ for arbitrary clauses is an instance of $\mathcal{S}_B$ where in every Horn clause with negative literals at least one of them shall be selected. Besides, equality factoring is exchanged for an aggressive splitting rule. If a clause contains positive literals with shared variables, then the digit instances of this clause are split. The number of split conclusions can become large, but remains finite, as opposed to the general case without theory $\mathcal{T}$. In order to simplify the treatment, a clause should always be split at the same position. Hence we assume that for every non-Horn clause an arbitray partitioning into two subclauses is *designated* where each subclause has strictly fewer positive literals. Now, the calculus rules are the following:

Negative superposition

$$\mathcal{I} \frac{l \simeq r \quad s[l'] \not\simeq t \vee C}{(s[r] \not\simeq t \vee C)\sigma\tau} \quad \text{if}$$

- $l' \notin \mathscr{V}$ and $\sigma = \mathrm{mgu}(l, l')$
- $\mathrm{ran}\,\sigma \subseteq \mathscr{V} \cup [1; n]$
- τ minimally numbers such that l and s are strictly greatest under $\sigma\tau$
- $s \not\simeq t$ is selected
- C is Horn

Positive superposition

$$\mathcal{I} \frac{l \simeq r \quad s[l'] \simeq t}{(s[r] \simeq t)\sigma\tau} \quad \text{if}$$

- $l' \notin \mathscr{V}$ and $\sigma = \mathrm{mgu}(l, l')$
- $\mathrm{ran}\,\sigma \subseteq \mathscr{V} \cup [1; n]$
- τ minimally numbers such that l and s are strictly greatest under $\sigma\tau$ and $(l \simeq r)\sigma\tau \prec (s \simeq t)\sigma\tau$

Equality resolution

$$\mathcal{I} \frac{C \vee t \not\simeq t'}{C\sigma} \quad \text{if}$$

- $\sigma = \mathrm{mgu}(t, t')$
- $\mathrm{ran}\,\sigma \subseteq \mathscr{V} \cup [1; n]$
- $t \not\simeq t'$ is selected
- C is Horn

Split

$$\mathcal{S} \frac{C \vee s \simeq t \vee l \simeq r \vee D}{(C \vee s \simeq t)\tau \mid (l \simeq r \vee D)\tau} \quad \text{if}$$

- the partitioning is designated
- τ minimally numbers such that the conclusions share no variables

In the two superposition rules, applying the numbering substitution τ in the conclusion guarantees that the number of variables in the latter is not higher than in the main premise, which is exploited in one of our termination proofs. Conversely, if there is no increase in the number of variables before applying τ, then one can *avoid enumerating* all such substitutions and just add the *single* conclusion instance where τ is the identity.

4.2 Soundness and Refutational Completeness

Next we give a formulation of lifting, which is at the heart of our approach. We reduce completeness of our calculus, on the non-ground level, to that of calculus

$\mathcal{G}$ on the ground level. Therefore, no dedicated model functor will be necessary for proving $\mathcal{C}$ complete. For any clause set M, let $\widehat{M}$ denote its Ω-instances that are Horn clauses.

Proposition 5. If a clause set M is $\mathcal{C}$-saturated, then $\widehat{M}$ is $\mathcal{G}$-saturated.

Proof. We adapt the usual lifting arguments (see for example [29, p. 393]) to our calculus, inspecting $\mathcal{G}$-inferences with premises from $\widehat{M}$. If a clause $D \in \widehat{M}$ contains negative literals, then let the literal selection be inherited from one arbitrary $C \in M$ that instantiates into D.

- Ground positive superposition: Given two clauses $l \simeq r$ and $s \simeq t$ from M with ground numbering substitution ρ, consider the $\mathcal{G}$-inference with premises $l\rho \simeq r\rho$ and $s\rho[l\rho]_p \simeq t\rho$, and conclusion $s\rho[r\rho]_p \simeq t\rho$. The position p is not introduced by ρ because the range of ρ consists of digits only. This $\mathcal{G}$-inference corresponds to a variable overlap if $s|_p \equiv x \in \mathcal{V}$, and to a non-variable overlap otherwise.

  In the former case we have $x\rho \equiv l\rho$, such that $l\rho$ is a digit. Because $l\rho \succ r\rho$ and the digits are the smallest ground terms, the term $r\rho$ must be a digit as well. Let ρ' denote the substitution identical to ρ except that $x\rho' \equiv r\rho$. Then $(s \simeq t)\rho'$ is contained in $\Omega(M)$ and makes the inference redundant.

  Now we come to non-variable overlaps. Let $l' \equiv s|_p$, furthermore $\sigma = \mathrm{mgu}(l, l')$ with $\mathrm{dom}\,\sigma \subseteq \mathrm{var}(l, l')$, and $\rho = \sigma\sigma'$. Because ρ is a ground numbering substitution, we know that $x\rho$ is a digit for every $x \in \mathrm{dom}\,\sigma$. Given $\rho = \sigma\sigma'$, every $x\sigma$ is either a digit or a variable, because the range of σ contains only digits and variables.

  The substitution σ' numbers the clauses $s\sigma \simeq t\sigma$ and $l\sigma \simeq r\sigma$ in such a way that the literals $l\sigma$ and $s\sigma$ are greatest under σ', respectively, and that $(l \simeq r)\sigma\sigma' \prec (s \simeq t)\sigma\sigma'$. If τ is a more general such substitution, then it satisfies $\mathrm{dom}\,\tau \subseteq \mathrm{dom}\,\sigma'$ and $x\tau \equiv x\sigma'$ for every $x \in \mathrm{dom}\,\tau$, which implies $\tau \subseteq \sigma'$. There exists a $\subseteq$-minimal such τ because all descending $\subseteq$-chains are finite. Summing up: $l \simeq r,\ s[l'] \simeq t \vdash (s[r] \simeq t)\sigma\tau$ is a $\mathcal{C}$-inference with premises from M, and is redundant with respect to M because M is saturated. If $\sigma' = \tau\tau'$, then the inference instance under τ' is redundant with respect to $\Omega(M)$.

- Ground equality resolution: Consider a Horn clause $C \vee t \not\simeq t' \in M$ with ground numbering substitution ρ such that $C\rho \vee t\rho \not\simeq t'\rho \vdash C\rho$ is a $\mathcal{G}$-inference. We may assume that $t \not\simeq t'$ is selected in $C \vee t \not\simeq t'$. As usual, t and t' have a most general unifier σ, which specializes into ρ say via σ'. We obtain $\mathrm{ran}\,\sigma \subseteq \mathcal{V} \cup [1; n]$ like for ground positive superposition. So $C \vee t \not\simeq t' \vdash C\sigma$ is a $\mathcal{C}$-inference with premises from M; and its redundancy carries over to that of the above instance.

- Ground negative superposition: similar to ground positive superposition, but taking selectedness into account like for ground equality resolution.

□

The calculus $\mathcal{C}$ is sound and refutationally complete:

Lemma 6. For every clause set M, the following are equivalent:

(i) M is $\mathcal{T}$-satisfiable.

(ii) Every fair derivation from $M \cup \mathcal{T}'$ contains a complete path $N_1, N_2, \ldots$ such that the empty clause is not in N_∞.

Proof sketch. A series of propositions rewrites characterization (i) into (ii). First, the clause set M is $\mathcal{T}$-satisfiable iff in every derivation from $M \cup \mathcal{T}'$ there exists a complete path $N_1, N_2, \ldots$ such that every $\Omega(N_i)$ is satisfiable. Second, every $\Omega(N_i)$ in this path is satisfiable iff $\Omega(N_\infty)$ is. Third, we note that N_∞ is $\mathcal{C}$-saturated provided the derivation is fair. Fourth, $\Omega(N_\infty)$ and $\widehat{N_\infty}$ are equivalent because in case of saturated sets, split conclusions are redundant or contained. Fifth, $\widehat{N_\infty}$ is saturated with respect to $\mathcal{G}$ according to Prop. 5. Finally, since $\mathcal{G}$ is sound and complete, the satisfiability of $\widehat{N_\infty}$ is equivalent to $\bot \notin \widehat{N_\infty}$, which is the same as $\bot \notin N_\infty$. $\qquad\square$

4.3   Termination by Loop Detection

In this subsection, we will pinpoint where the non-terminating behaviour in $\mathcal{C}$-derivations arises from, and then look for a remedy. So we study here fair $\mathcal{C}$-derivations that start from some finite input $M \cup \mathcal{T}'$ and exclusively consist of inferences, splits and simplifications. In order to avoid trivial loops, no inference or split shall be repeated while the parent clauses persist. Furthermore, simplifications shall not increase the number of variables in a clause, a condition that inferences and splits satisfy:

Proposition 7. Inference and split conclusions do not have more variables than one of the premises.

Compared to standard superposition, the calculus $\mathcal{C}$ is far more restrictive: There are no inferences with complex unifiers; both inferences and splits do not increase the number of variables; and in each satisfiable path, every ground term can eventually be rewritten into a digit. A further observation is the following:

Proposition 8. Consider an inference or a split or a simplification

$$\mathcal{I}\,\frac{C_1 \;\ldots\; C_m}{D_1} \quad \text{or} \quad \mathcal{S}\,\frac{C_m}{D_1 \mid D_2} \quad \text{or} \quad \mathcal{R}\,\frac{C_m \quad \vec{C'}}{\vec{D}}$$

in the calculus $\mathcal{C}$. In each case we have $\Omega(C_m) \succ \Omega(D_i)$, for every i.

By König's lemma, a derivation is infinite if and only if it contains an infinite path. Such a path is only possible with infinitely many inference steps. The clauses that occur in a path can be arranged in a forest with the input clauses as root nodes, and with each inference conclusion or split clause or reduct being attached to its corresponding parent clause C_m of Prop. 8. Inductively it is clear that whenever a clause does not persist in a path of a derivation, but is generated again later, then the two occurrences produce distinct nodes. Therefore, a path in a derivation is infinite if and only if the corresponding forest is. Because of the decreasingness result in Prop. 8, infinite paths in the forest are impossible. By construction of the calculus, a node in the forest with infinitely many children can only arise from binary inferences, more precisely, from superposition inferences with the same main premise. Since the number of possible substitutions is essentially finite, by the bounded number of variables, we obtain:

Lemma 9. A $\mathcal{C}$-derivation is infinite if and only if it contains a path with
- an infinite sequence of equations $l \simeq r_i$, up to variable renaming,
- a persistent clause $C[l']$, and
- infinitely many superposition steps $\mathcal{S} \dfrac{l \simeq r_i \quad C[l']}{C[r_i]\sigma\tau}$ with σ, τ fixed,

provided it starts from finite $M \cup \mathcal{T}$.

Note that negative superposition steps cannot be excluded from the characterization: If there is an inference into a positive main clause $s[l'] \simeq t$, then one may also construct one into $s[l'] \not\simeq x \vee x \simeq t$ where x is fresh, and simplify the resolvents $(s[r_i] \not\simeq k \vee k \simeq t)\sigma\tau$ to $(s[r_i] \simeq t)\sigma\tau$.

A first example of an infinite derivation, without any rewriting, was given in the beginning of Sect. 4, on page 78. As in all examples to come, a lexicographic path ordering is employed. The inducing precedence here is $a \succ f \succ b$:

$$\mathcal{I} \frac{a \simeq f(b) \quad f(a) \simeq a}{f(f(b)) \simeq a} \qquad \mathcal{I} \frac{a \simeq f(f(b)) \quad f(a) \simeq a}{f(f(f(b))) \simeq a} \qquad \cdots$$

As already said, all inference steps could be carried out as simplifications, namely by unit rewriting. We would like to know whethe unit rewriting prevents nontermination. Inferencing is not rewriting in the next example, where the precedence is $f \succ h \succ g \succ a$:

$$\mathcal{I} \frac{f(a) \simeq g(x) \quad h(f(y), z) \simeq f(y)}{h(g(1), z) \simeq f(a)} \qquad \mathcal{I} \frac{f(a) \simeq h(g(1), x) \quad h(f(y), z) \simeq f(y)}{h(h(g(1), 1), z) \simeq f(a)} \qquad \cdots$$

However, the second side premise can be rewritten by the first, which is not possible in the following example. We use a signature with the binary function symbols $\cdot$, $+$, and $\_^-$ for exponentiation, and with the unary symbol f. In order to avoid overly many parentheses, the symbol $\_^-$ shall bind tightest, followed by $\cdot$ and $+$, which shall associate to the left. The precedence is $f \succ \cdot \succ \_^- \succ + \succ n \succ \ldots \succ 1$.

$$\mathcal{I} \frac{x^x \cdot z \simeq 1 \cdot x^x + z \cdot x^x \quad x^y \cdot 1 + f(x^x \cdot z') \simeq x^x \cdot z'}{1 \cdot x^x + 1 \cdot x^x + f(x^x \cdot z') \simeq x^x \cdot z'}$$

$$\{x_1 + x_2 + f(x_3 \cdot x_4) \simeq x_1 + x_2 + x_4 \cdot x_3\} \downarrow_Y$$

$$1 \cdot x^x + 1 \cdot x^x + z' \cdot x^x \simeq x^x \cdot z'$$

$$\mathcal{I} \frac{x^x \cdot z \simeq 1 \cdot x^x + 1 \cdot x^x + z \cdot x^x \quad x^y \cdot 1 + f(x^x \cdot z') \simeq x^x \cdot z'}{1 \cdot x^x + 1 \cdot x^x + 1 \cdot x^x + f(x^x \cdot z') \simeq x^x \cdot z'}$$

$$\{x_1 + x_2 + f(x_3 \cdot x_4) \simeq x_1 + x_2 + x_4 \cdot x_3\} \downarrow_Y$$

$$1 \cdot x^x + 1 \cdot x^x + 1 \cdot x^x + z' \cdot x^x \simeq x^x \cdot z'$$

$$\vdots$$

Still unit rewriting enforces termination: From the premise of the first inference into that of the second, there is a superposition inference producing $1 \cdot x^x + 1 \cdot x^x \simeq 1 \cdot x^x + 1 \cdot x^x + 1 \cdot x^x$, which simplifies all higher-index right-hand sides r_i.

This example leads to a general property: Under the conditions given in Lem. 9, there exist superposition inferences from $l \simeq r_1$ and $l \simeq r_2$, and one easily calculates that with respect to these and to $C[r_1]\sigma\tau$, the inference producing $C[r_2]\sigma\tau$ is redundant, though unit rewriting need not always simplify the conclusion. We call a $\mathcal{C}$-derivation *loop-free* if it contains no such inference steps, and satisfies the conditions given in the beginning of this subsection: no repetition of inferences or splits from persisting parent clauses, and no increase in the number of variables when simplifying a clause. Whether an individual inference satisfies these conditions or not can be read off the derivation history, or memoized suitably.

Theorem 10. Loop-free $\mathcal{C}$-derivations decide $\mathcal{T}$-satisfiability of finite clause sets.

This termination result is built on the insight in Prop. 8 that only the decreasing inferences be drawn. In the calculus $\mathcal{C}$, this is achieved with the numbering substitution τ, which is not present in the calculus $\mathcal{S}_B$. Alternatively, one could attach constraints to the clauses and thereby restrict the inference conclusions to the decreasing digit instances.

4.4 Termination by Rewriting

The calculus $\mathcal{C}$ is constructed such that if a clause set is saturated, then the associated model can be read off the set of remaining unit equations, which is ground confluent and reduces every ground term to a digit. Therefore, we set out here a decision procedure with unit rewriting as the major simplification device.

Given a clause set N with unit equations $E \subseteq N$, we say that N *reduces to digits* if $f(\vec{\imath}) \to_E^* j$ for every digit vector $\vec{\imath}$. Inductively every ground term can then be rewritten to a digit as well. Furthermore, a clause is called $[1; n]$-*shallow* if non-digit function symbols occur only at the top-level of positive literals.

One may want to test explicitly whether a given N_k reduces to digits already (and if so, perhaps test immediately whether E_k describes a $\mathcal{T}$-model of M). Notably the property is not always inherited from N_k to N_{k+1}. Consider for example the following simplification steps in the sense of the calculus $\mathcal{C}$:

$$\mathcal{R} \; \frac{f(3) \simeq f(1)}{1 \simeq f(1)} \; 1 \not\simeq 1 \vee f(3) \simeq 1 \qquad \mathcal{R} \; \frac{f(3) \simeq 1 \quad f(1) \simeq 3}{f(2) \simeq 1 \quad f(1) \simeq 2}$$

The term $f(3)$ is E_k-reducible, but not necessarily E_{k+1}-reducible. As the second example shows, this may even occur if unit equations are simplified with respect to E_k only. In case this is not desired, one has to restrict the simplification of unit equations. For example, ordered unit rewriting, instance rewriting, subsumption and tautology elimination are compatible.

Alas, even when a given N_k reduces to digits, such that every term $f(\vec{\imath})$, and every ground term $f(\vec{t})$, is reducible, then unit rewriting on the non-ground level can be inapplicable although it would be possible on every Ω-instance: If $n = 2$ and $N = \{f(1) \simeq 2,\, f(2) \simeq 1,\, f(f(x)) \simeq x\}$, then the third equation cannot be rewritten, but its Ω-instances could be turned into the tautologies $1 \simeq 1$ or $2 \simeq 2$, respectively. Hence we need to combine instantiation and rewriting in that situation. If C is a clause and Γ a set of numbering substitutions with $\operatorname{dom}\tau \subseteq \operatorname{var}(C)$ for every $\tau \in \Gamma$, then we say that Γ *covers* C if every ρ that ground numbers C can be obtained as specialization of some $\tau \in \Gamma$.

Instance rewriting

$$\mathcal{R}\,\frac{C}{\{D_\rho : \rho \in \Gamma\}}\,E \qquad \text{if} \quad \begin{aligned} &\cdot\ E \text{ is a set of unit equations}\\ &\cdot\ \Gamma \text{ covers } C\\ &\cdot\ \text{for every } \rho \in \Gamma \colon C\rho \to_E^{\succeq} \circ \to_E^* D\rho\\ &\quad \text{and } C\rho \to_E^* D\rho \text{ is } \Omega\text{-admissible} \end{aligned}$$

Instance rewriting allows clauses to be replaced eventually by their $[1; n]$-shallow equivalents:

Proposition 11. Consider a complete path $N_1, N_2, \ldots$ in a fair derivation from $M \cup \mathcal{T}'$, where M is finite.
  (i) For some index κ, all $N_{\kappa+i}$ contain $\bot$; or they all reduce to digits.
  (ii) If $C \in N_{\kappa+i}$ is not $[1; n]$-shallow, then C can effectively be simplified into a finite set of $[1; n]$-shallow clauses.

We will require instance rewriting only on newly generated clauses once the clause set reduces to digits. Note also that simplifying a $[1; n]$-shallow clause with respect to other such clauses can arbitrarily increase the number of variables and need not preserve $[1; n]$-shallowness, as for example witnessed by

$$\mathcal{R}\,\frac{f(x) \simeq 2}{g(1) \not\simeq g(1) \vee y_1 \not\simeq y_1 \vee \ldots \vee y_m \not\simeq y_m \vee f(x) \simeq 1}\,2 \simeq 1$$

if $f(1) \succ g(1)$. Clearly this counteracts our efforts towards termination; so a strategy is needed that guides the execution of calculus steps. We say that a $\mathcal{C}$-derivation is a $\mathcal{C}_\kappa$-*derivation* from a clause set M if (i) it is fair, (ii) the root node is $M \cup \mathcal{T}'$, and in every path eventually (iii) simplifications do not increase the number of variables, (iv) $[1; n]$-shallowness is preserved under simplifications, (v) inferences and splits are not repeated, (vi) every fresh inference conclusion which is not $[1; n]$-shallow, is immediately simplified into a set of $[1; n]$-shallow clauses, (vii) no duplicate literals occur in $[1; n]$-shallow clauses, and (viii) $[1; n]$-shallow clauses equal up to variable renaming are identified. Indeed such derivations exist for every finite M: The crucial item (vi) can be satisfied because of Prop. 11.

Theorem 12. $\mathcal{C}_\kappa$-derivations decide $\mathcal{T}$-satisfiability of finite clause sets.

4.5 Extensions

Let us have a short look at a many-sorted setting where $\mathcal{T}$ consists of size restrictions for every sort, each built over an individual set of digits. One has to employ the usual typing constraints for equations, terms and substitutions. Then the calculus $\mathcal{C}$, and the results obtained for it so far, straightforwardly extends to this situation.

Up to now, our calculus did not deal with predicates. Of course one could extend $\mathcal{C}$ with an ordered resolution rule, and consider predicate atoms in the superposition and split rules. Alternatively, we can introduce a two-element sort Bool, say over the digits I and II, and provide a clause I $\not\simeq$ II. As usually we can now encode predicate atoms $P(\vec{t})$ of any other sort as equations $P(\vec{t}) \simeq$ I. Notably $\mathcal{T}'$ need not contain an axiom $P(\vec{x}) \simeq$ I $\vee$ $P(\vec{x}) \simeq$ II: Given an algebra $\mathcal{A}$ such that at some point $P^{\mathcal{A}}$ does not map into $\{$I$^{\mathcal{A}}$, II$^{\mathcal{A}}\}$, let the algebra $\mathcal{B}$ coincide with $\mathcal{A}$ except that $P^{\mathcal{B}}$ maps all such points onto II$^{\mathcal{B}}$. Then $\mathcal{A}$ and $\mathcal{B}$ satisfy the same encoded atoms $P(\vec{t}) \simeq$ I.

As an application, consider the validity problem for a formula $\phi \equiv \forall x_1 \ldots \forall x_n$ $\exists y_1 \ldots \exists y_m \phi'$ where ϕ' is quantifier-free and contains no function symbols. This problem was proven decidable by Bernays and Schönfinkel [9]. Now, ϕ is valid iff $\psi \equiv \forall y_1 \ldots \forall y_m \neg \phi' \{x_1 \mapsto 1, \ldots, x_n \mapsto n\}$ is unsatisfiable iff ψ is $\mathcal{T}$-unsatisfiable. Since no function symbols are present, the set $\mathcal{T}'$ is empty. Notably, no instance rewriting steps are needed in such derivations because all clauses are shallow.

Corollary 13. Both $\mathcal{C}_\kappa$-derivations and loop-free $\mathcal{C}$-derivations decide the Bernays-Schönfinkel class.

Finally, it is often desired to assume the bounded domain digits $1, \ldots, n$ to be different. This can be expressed by n^2 disequations $i \not\simeq j$, where $i \neq j$ and $1 \leq i, j \leq n$. For larger n this is not a desirable solution. Then an additional inference rules that removes equations between digits [33] is a better solution. This approach was already successfully tested for Bernays-Schönfinkel problems over a large number of constants [35].

5   Combinations with Unbounded First-Order Theories

So far, we have only considered the case where the entire Herbrand domain of a formula is finite. The interesting question is whether the techniques developed in the previous sections can be generalized to a setting where the overall Herbrand domain may be infinite, but bounded subsets of the domain are specified. The answer we give in the section is affirmative; the combination can exploit the advanced technology: In every inference, variables over any bounded subset only need to be instantiated to variables and to the finitely many domain representatives. Furthermore no inferences with the axiom expressing boundedness are needed.

The overall approach is to code bounded subsets via monadic predicates, which we also call *soft sorts* [19,36]. In contrast to the use of sorts in algebraic specifications, sorts are represented in the clause set by their monadic relativization predicates enjoying the standard first-order semantics. Sort theories show up in the form of Horn clauses in these monadic predicates and can be dynamically used for simplification. Therefore, soft sorts may be empty, there are no restrictions on the language, sorts are not a priori disjoint, elements of sorts are not necessarily different and sorts may of course also be defined via general clauses. For example, the clause $\neg R(x, f(x)) \vee S(x)$ defines x to be contained in the sort S if the relation $R(x, f(x))$ holds, and the clause $\neg S(x) \vee \neg T(x)$ states that the sorts S and T are disjoint.

Provided a clause C contains a negative literal $\neg S(x)$, we say that x *is of sort* S *in* C. To give an example, any model $\mathcal{A}$ of the clauses $S(1)$, $S(2)$, $\neg S(x) \vee x \simeq 1 \vee x \simeq 2$ must satisfy $1 \leq |S^{\mathcal{A}}| \leq 2$. If we add the clause $1 \not\simeq 2$, then any model $\mathcal{A}$ fulfills $|S^{\mathcal{A}}| = 2$, whereas the alternative extension with $1 \simeq 2$ leads to $|S^{\mathcal{A}}| = 1$. Concerning functions, the clause $\neg S(x) \vee S(f(x))$ declares f to map elements from S into S.

In this section, we study the bounded-domain theory $\mathcal{T}$ for one sort S of cardinality up to n defined by

$$\mathcal{T} = \{S(1), \quad S(2), \quad \ldots, \quad S(n), \quad \neg S(x) \vee x \simeq 1 \vee \ldots \vee x \simeq n\}$$

which is a clausal presentation of the formula $\forall x. \, S(x) \leftrightarrow x \simeq 1 \vee \ldots \vee x \simeq n$. The results can be extended to several bounded-domain sorts in the obvious way.

Similarly to the restricted case of Sect. 3, we would like to instantiate variables of sort S with digits only. All such instances of the clause $\neg S(x) \vee x \simeq 1 \vee \ldots \vee x \simeq n \in \mathcal{T}$ are tautologies. To compensate for this, we introduce an operator $\_^{\circ}$ to be applied to input clauses that replaces every positive literal $S(t)$ by the disjunction $t \simeq 1 \vee \ldots \vee t \simeq n$. Furthermore let in this section

$$\mathcal{T}' = \{S(1), \ldots, S(n)\}.$$

Finally $\Omega_S(C)$ shall denote the set of all clauses obtained from C via instantiation of all variables of sort S in C with digits. In this sense Ω_S is the restriction of Ω to variables of sort S. The following lemma is the analogue of Prop. 2:

Lemma 14. A clause set N is $\mathcal{T}$-satisfiable iff $\Omega_S(N^{\circ})$ is $\mathcal{T}'$-satisfiable.

Proof. "$\Rightarrow$" Let $\mathcal{A}$ be a model for $N \cup \mathcal{T}$, i.e., $\mathcal{A} \models N \cup \mathcal{T}$ and so $\mathcal{A} \models \mathcal{T}'$. Since in particular $\mathcal{A} \models \mathcal{T}$ we know $S^{\mathcal{A}} = \{1^{\mathcal{A}}, \ldots, n^{\mathcal{A}}\}$ and hence $\mathcal{A} \models C$ iff $\mathcal{A} \models \Omega_S(C)$ for any clause C. We show $\mathcal{A} \models C$ implies $\mathcal{A} \models C^{\circ}$ for all $C \in N$. We distinguish the following cases: (i) C does not contain a positive literal $S(t)$. Then $C \equiv C^{\circ}$ and we are done. (ii) Let $C \equiv S(t_1) \vee \ldots \vee S(t_m) \vee D$ and D does not contain a positive literal $S(t)$, $m > 0$. Let σ be any valuation[1] for all variables in all t_i. Then $\mathcal{A}, \sigma \models S(t_i)$ iff $(t_i\sigma)^{\mathcal{A}} \in S^{\mathcal{A}}$ iff $(t_i\sigma)^{\mathcal{A}} = k^{\mathcal{A}}$ for some digit $1 \leq k \leq n$ iff $\mathcal{A}, \sigma \models t_i \simeq k$. Hence if $\mathcal{A} \models C$ so $\mathcal{A} \models C^{\circ}$.

"$\Leftarrow$" Let $\mathcal{A} \models \Omega_S(N^{\circ}) \cup \mathcal{T}'$ and let $\mathcal{A}'$ be identical to $\mathcal{A}$, except that $S^{\mathcal{A}'} = \{1^{\mathcal{A}}, \ldots, n^{\mathcal{A}}\}$. Obviously, $\mathcal{A}' \models \mathcal{T}'$ and $\mathcal{A}' \models \Omega_S(N^{\circ})$ because $\{1^{\mathcal{A}}, \ldots, n^{\mathcal{A}}\} \subseteq S^{\mathcal{A}}$. We need to show $\mathcal{A}' \models N$ and $\mathcal{A}' \models \mathcal{T}$ where the latter holds by construction of $\mathcal{A}'$. By construction $\mathcal{A}', \sigma \models S(t_i)$ iff $\mathcal{A}', \sigma \models t_i \simeq k$ for some digit $1 \leq k \leq n$ and any valuation σ in the variables of t_i. Now assume there is a clause $C \in N$ with $\mathcal{A}', \sigma \not\models C\sigma$ for some valuation σ. Thus, if there is some $\neg S(x)$ in C, then

---

[1] We confuse here substitutions and valuations in the usual way.

$x\sigma \in S^{\mathcal{A}}$ implying $C^{\circ}\sigma \in \Omega_S(N^{\circ})$. Now, since $\mathcal{A}', \sigma \models S(t_i)$ iff $\mathcal{A}', \sigma \models t_i \simeq k$ we have $\mathcal{A}', \sigma \not\models C^{\circ}\sigma$, a contradiction.

Note that the four clauses $N = \{S(1), S(2), \neg S(x) \vee x \simeq 1 \vee x \simeq 2, f^3(x) \not\simeq f(x)\}$ are satisfiable as neither the input, nor the output of f is specified to be of sort S. Adding the declaration $N' = N \cup \{\neg S(x) \vee S(f(x))\}$ lets f map from S into S and hence causes unsatisfiability. For the latter clause, the transformation of Lem. 14 applies. We get $(\neg S(x) \vee S(f(x)))^{\circ} = \neg S(x) \vee f(x) \simeq 1 \vee f(x) \simeq 2$ and we obtain the set

$$\Omega_S((N')^{\circ}) = \{S(1), S(2), f^3(x) \not\simeq f(x),$$
$$\neg S(1) \vee f(1) \simeq 1 \vee f(1) \simeq 2,$$
$$\neg S(2) \vee f(2) \simeq 1 \vee f(2) \simeq 2\}$$

which is unsatisfiable.

Now by the lifting theorem for standard superposition, we know because of Lem. 14 that $N \cup \mathcal{T}$ has a superposition refutation iff $N^{\circ} \cup \mathcal{T}'$ has one. The open question is how we can exploit the fact that we considered solely numbering substitutions for variables of sort S. Note that although S has a bounded domain, the overall domain of N may be infinite. Hence we cannot take the approach of Sect. 3 where we used the numbering substitution available for all variables to require that inferences are only performed on strictly greatest terms and literals. Furthermore, the abstract superposition redundancy notion is no longer effective and satisfiability is of course not decidable anymore. Therefore, the idea is to restrict the range of substitutions for variables of sort S to $\mathcal{V} \cup [1; n]$, and to require that (strict) maximality is preserved under any numbering substitution for the bounded sort S. The superposition calculus including this refinement consists of the standard rules positive and negative superposition, equality resolution and factoring, instantiated by the additional restrictions.

Positive superposition

$$\mathcal{I} \frac{C \vee l \simeq r \quad s[l'] \simeq t \vee D}{(C \vee s[r] \simeq t \vee D)\sigma} \quad \text{if}$$

- $l' \notin \mathcal{V}$ and $\sigma = \mathrm{mgu}(l, l')$
- $\mathrm{ran}\,\sigma|_S \subseteq \mathcal{V} \cup [1; n]$
- there exists a minimally numbering τ of sort S such that l, $l \simeq r$, s, $s \simeq t$ are strictly maximal under $\sigma\tau$ and $(C \vee l \simeq r)\sigma\tau \not\succeq (s \simeq t \vee D)\sigma\tau$

Negative superposition

$$\mathcal{I} \frac{C \vee l \simeq r \quad s[l'] \not\simeq t \vee D}{(C \vee s[r] \not\simeq t \vee D)\sigma} \quad \text{if}$$

- $l' \notin \mathcal{V}$ and $\sigma = \mathrm{mgu}(l, l')$
- $\mathrm{ran}\,\sigma|_S \subseteq \mathcal{V} \cup [1; n]$
- there exists a minimally numbering τ of sort S such that l, $l \simeq r$, s are strictly maximal under $\sigma\tau$, $s \not\simeq t$ is maximal under $\sigma\tau$ or selected

Equality resolution

$$\mathcal{I} \, \frac{C \vee t \not\simeq t'}{C\sigma}$$

if

- $\sigma = \mathrm{mgu}(t, t')$
- $\mathrm{ran}\,\sigma|_S \subseteq \mathscr{V} \cup [1; n]$
- there exists a minimally numbering τ of sort S such that $t \not\simeq t'$ is maximal under $\sigma\tau$ or selected

Equality factoring

$$\mathcal{I} \, \frac{C \vee s \simeq t \vee s' \simeq u}{(C \vee t \not\simeq u \vee s \simeq u)\sigma}$$

if

- $\sigma = \mathrm{mgu}(s, s')$
- $\mathrm{ran}\,\sigma|_S \subseteq \mathscr{V} \cup [1; n]$
- there exists a minimally numbering τ of sort S such that s, s' are strictly maximal under $\sigma\tau$, $s \simeq t$ is maximal under $\sigma\tau$

For the general combination of a bounded sort with arbitrary formulae over potentially infinite domains, we cannot aim at a decision procedure, but only at refutational completeness. Hence, the above rule delays instantiations of bounded-sort variables as long as possible, but applies the underlying restrictions.

We stipulate that the digits $1, \ldots, n$ are minimal in the ordering $\succ$. Furthermore, in every clause $\neg S(t) \vee C$ with a negative sort literal, the argument of S shall always be a digit or a variable. If t is neither of these, then one can apply variable abstraction and obtain $\neg S(x) \vee x \not\simeq t \vee C$. Notably the new variable x needs to be instantiated with digits only, and hence cannot become maximal. Our considerations give rise to the following completeness result:

Theorem 15. A clause set N is $\mathcal{T}$-unsatisfiable iff there is a derivation of the empty clause from $N^\circ \cup \mathcal{T}'$ by the superposition calculus defined above.

Proof. By Lem. 14 $N \cup \mathcal{T}$ is unsatisfiable iff $\Omega_S(N') \cup \mathcal{T}'$ is unsatisfiable. The set $\Omega_S(N') \cup \mathcal{T}'$ is unsatisfiable iff there is a derivation of the empty clause by the standard superposition calculus. As the digits are minimal in the ordering, they might only be replaced by each other. For any clause $\neg S(x) \vee C \in N'$, all instances of $\neg S(x)$ in the proof are generated by substitutions from x into $[1; n]$. Hence, all steps can be lifted to steps of the above refined superposition calculus on N'.

Here is an example for the refined maximality condition. Let $\succ$ denote the lexicographic path ordering induced by the precedence $f \succ g \succ n \succ \ldots \succ 1$. Then in the clause $\neg S(x) \vee g(x, y) \simeq y \vee f(y) \simeq y$, the literal $g(x, y) \simeq y$ is not maximal, because $f(y) \succ g(i, y)$ holds for every $i \in [1; n]$.

6 Three Application Scenarios

6.1 Combination with Theories

There is currently a great interest in combining general purpose reasoning procedures for propositional or first-order logic with theories, such as arithmetic or theories modeling data structures. A combination of this kind has great potential in program analysis and verification. It is also mandatory in the sense that, e.g., the theory of arithmetic is not representable in first-order logic whereas the control flow of a program can hardly be represented by an arithmetic theory.

In addition to the so called SMT (SAT Modulo Theories) approach [30] that combine propositional logic with theories, there are meanwhile also a number of combination approaches between first-order logic and theories available [5,23,1,3,12,18] and (partly) implemented and successfully applied. Following the hierarchic approach [5] if the base theory, e.g. arithmetic [1], is completely separated from the first-order theory, and all first-order sorts are bounded, then our calculus from Sect. 5 yields a decision procedure following the ideas of Sect. 4. However, if theory terms and first-order terms are mixed, then the best we can get is completeness, in general. Already a combination of the Bernays-Schönfinkel fragment with linear arithmetic enables the encoding of the halting problem for two-counter machines [28,17].

Nevertheless it is an interesting question and a starting point for future work whether the inference restrictions introduced here, can in fact be combined with superposition/resolution-based combination ideas starting from first-order logic [3,12].

6.2   Vectors over Finite Domains

We have formalized parts of a LAN infrastructure as a bounded-domain problem, see http://spass-prover.org/prototypes. SPASS saturates this problem in less than one second. Actually, we have currently not integrated the calculus refinements for bounded domains in SPASS. However, by the structure of the clauses in this example, only digit unifiers are considered for inferences. Notably, SPASS even succeeds on this problem extended with the router and firewall configurations of both Max Planck institutes at Saarbrücken, which takes about 30 minutes. You may submit this as a challenge to your favorite instantiation-based provers. The specification includes a theory of IP addresses; these are essentially bitvectors. A vector-level conjunction $AND(\_,\_)$ is defined via recurrence to a bit-level conjunction $\_ \cdot \_$ as follows:

$$AND(IP(x_{31},\dots,x_0), IP(y_{31},\dots,y_0)) \simeq IP(x_{31} \cdot y_{31},\dots,x_0 \cdot y_0)$$

The bit theory has a bounded domain with digits 0 and 1. The theory extension $\mathcal{T}'$ and $\_^\circ$-transformation of the sort declaration for $\cdot$ are (see page 86):

$$Bit(0) \qquad Bit(1) \qquad \forall x,y.\, Bit(x) \wedge Bit(y) \rightarrow x \cdot y \simeq 0 \vee x \cdot y \simeq 1$$

The sort $Bit(\_)$ is needed to prevent confusion with other sorts of the theory, in particular IP addresses. This is a perfect example of our approach to combinations in Sect. 5. Bit-level conjunction is defined by

$$\forall x,y.\, Bit(x) \wedge Bit(y) \rightarrow \begin{array}{l} (x \cdot y \simeq 0 \leftrightarrow x \simeq 0 \vee y \simeq 0) \\ \wedge\, (x \cdot y \simeq 1 \leftrightarrow x \simeq 1 \wedge y \simeq 1) \end{array}$$

Confusion of bits is prevented by the clause $0 \not\simeq 1$, and confusion of IP addresses by the following clauses, where i ranges over all indices:

$$IP(x_{31}, \ldots, x_{i+1}, 0, x_{i-1}, \ldots, x_0) \not\simeq IP(x_{31}, \ldots, x_{i+1}, 1, x_{i-1}, \ldots, x_0)$$

The clausification of the overall theory can be saturated finitely, for example if $\succ$ is an LPO induced by the precedence $AND \succ \cdot \succ IP$. Note that already the minimal model size for this part of the LAN theory is $2^{32} + 2$, and due to classless routing, confusion of IP addresses is not an adequate approach to reduce the size of the model. Now, if we extend this theory to bitvectors of length 64, then the additional effort in saturation is bound by a factor of two, whereas an instantiation-based method has to consider 2^{64} domain elements for the IP addresses. This effect that we already pointed out in [21] was later studied in [32] in a systematic way.

6.3   Proof Obligations from ISABELLE

Recently [10], we have been working on a version of SPASS [37] that in particular supports proof obligations out of SLEDGEHAMMER [31] invocations from ISABELLE [38]. One challenge of the translation of higher-order formulas is that the booleans become explicit, i.e., in almost all obligations there is a clause

$$\neg \mathrm{bool}(x) \vee x \simeq \mathrm{true} \vee x \simeq \mathrm{false}$$

We applied to this clause and the soft sort bool the transformations introduced in Sect. 5 without making use of the extra ordering restrictions. When comparing the runs of the new SPASS [10] with and without the transformation, we gained an average speedup factor of 3 on all examples.

7   Conclusion and Future Work

We have presented a light-weight adaptation of superposition calculi to the first-order theory of bounded domains. The achievement is a superposition calculus for bounded domains that restricts the range of inference unifiers to digits or variables, facilitates the precise calculation of ordering restrictions, introduces an effective general semantic redundancy criterion, incorporates a particular splitting rule for non-Horn clauses, can constitute a decision procedure for any bounded-domain problem, is mostly compatible with all the standard superposition redundancy criteria, and can in particular be embedded via a general dynamic sort discipline based on monadic predicates in any general first-order setting.

We have already done some promising experiments on the basis of ground-level formulations for bounded domains [20], and a partial, light weight integration into SPASS [10] that does not explore the additional ordering restrictions. To this end, ordering computation, inference computation and simplifications need be refined accordingly.

Future work has already started in getting the superposition partial model operator R effective without the need for explicit instantiation. If this works out, then systems can be developed that actually combine the strengths of explicit model search in the style of MACE and automated theorem proving in the style of OTTER. A combination Bill McCune might have already thought of.[2]

Acknowledgements. We are indebted to our reviewers for their valuable and constructive comments that were essential for the eventual quality of the paper.

References

1. Althaus, E., Kruglov, E., Weidenbach, C.: Superposition Modulo Linear Arithmetic SUP(LA). In: Ghilardi, S., Sebastiani, R. (eds.) FroCoS 2009. LNCS, vol. 5749, pp. 84–99. Springer, Heidelberg (2009)
2. Armando, A., Ranise, S., Rusinowitch, M.: Uniform Derivation of Decision Procedures by Superposition. In: Fribourg, L. (ed.) CSL 2001. LNCS, vol. 2142, pp. 513–527. Springer, Heidelberg (2001)
3. Armando, A., Bonacina, M.P., Ranise, S., Schulz, S.: New results on rewrite-based satisfiability procedures. ACM Transactions on Computational Logic 10(1), 4:1–4:51 (2009)
4. Bachmair, L., Ganzinger, H.: Rewrite-based equational theorem proving with selection and simplification. Journal of Logic and Computation 4(3), 217–247 (1994)
5. Bachmair, L., Ganzinger, H., Waldmann, U.: Refutational theorem proving for hierarchic first-order theories. Appl. Algebra Eng. Commun. Comput. 5, 193–212 (1994)
6. Baumgartner, P., Fuchs, A., de Nivelle, H., Tinelli, C.: Computing finite models by reduction to function-free clause logic. In: Ahrendt, W., Baumgartner, P., de Nivelle, H. (eds.) Proceedings of the Third Workshop on Disproving, pp. 82–99 (2006)
7. Baumgartner, P., Furbach, U., Pelzer, B.: The hyper tableaux calculus with equality and an application to finite model computation. Journal of Logic and Computation 20(1), 77–109 (2010)
8. Baumgartner, P., Schmidt, R.A.: Blocking and Other Enhancements for Bottom-Up Model Generation Methods. In: Furbach, U., Shankar, N. (eds.) IJCAR 2006. LNCS (LNAI), vol. 4130, pp. 125–139. Springer, Heidelberg (2006)
9. Bernays, P., Schönfinkel, M.: Zum Entscheidungsproblem der mathematischen Logik. Mathematische Annalen 99, 342–372 (1928)
10. Blanchette, J.C., Popescu, A., Wand, D., Weidenbach, C.: More SPASS with Isabelle—Superposition with Hard Sorts and Configurable Simplification. In: Beringer, L., Felty, A. (eds.) ITP 2012. LNCS, vol. 7406, pp. 345–360. Springer, Heidelberg (2012), http://www4.in.tum.de/~blanchet/more-spass.pdf
11. Bonacina, M.P., Ghilardi, S., Nicolini, E., Ranise, S., Zucchelli, D.: Decidability and Undecidability Results for Nelson-Oppen and Rewrite-Based Decision Procedures. In: Furbach, U., Shankar, N. (eds.) IJCAR 2006. LNCS (LNAI), vol. 4130, pp. 513–527. Springer, Heidelberg (2006)
12. Bonacina, M.P., Lynch, C., Mendonça de Moura, L.: On deciding satisfiability by theorem proving with speculative inferences. Journal of Automated Reasoning 47(2), 161–189 (2011)

---

[2] Actually, in personal communication, one of the authors discussed this idea with Bill.

13. Claessen, K., Sörensson, N.: New techniques that improve MACE-style finite model finding. In: Baumgartner, P., Fermueller, C. (eds.) Proceedings of the Workshop on Model Computation (2003)

14. de Nivelle, H., Meng, J.: Geometric Resolution: A Proof Procedure Based on Finite Model Search. In: Furbach, U., Shankar, N. (eds.) IJCAR 2006. LNCS (LNAI), vol. 4130, pp. 303–317. Springer, Heidelberg (2006)

15. Dershowitz, N.: A Maximal-Literal Unit Strategy for Horn Clauses. In: Okada, M., Kaplan, S. (eds.) CTRS 1990. LNCS, vol. 516, pp. 14–25. Springer, Heidelberg (1991)

16. Fietzke, A., Weidenbach, C.: Labelled splitting. Annals of Mathematics and Artificial Intellelligence 55(1-2), 3–34 (2009)

17. Fietzke, A., Weidenbach, C.: Superposition as a decision procedure for timed automata. In: Ratschan, S. (ed.) MACIS 2011: Fourth International Conference on Mathematical Aspects of Computer and Information Sciences, pp. 52–62 (2011); Journal version to appear in the Journal of Mathematics in Computer Science

18. Fontaine, P., Merz, S., Weidenbach, C.: Combination of Disjoint Theories: Beyond Decidability. In: Gramlich, B., Miller, D., Sattler, U. (eds.) IJCAR 2012. LNCS, vol. 7364, pp. 256–270. Springer, Heidelberg (2012)

19. Ganzinger, H., Meyer, C., Weidenbach, C.: Soft Typing for Ordered Resolution. In: McCune, W. (ed.) CADE 1997. LNCS, vol. 1249, pp. 321–335. Springer, Heidelberg (1997)

20. Hillenbrand, T., Topic, D., Weidenbach, C.: Sudokus as logical puzzles. In: Ahrendt, W., Baumgartner, P., de Nivelle, H. (eds.) Proceedings of the Third Workshop on Disproving, pp. 2–12 (2006)

21. Hillenbrand, T., Weidenbach, C.: Superposition for finite domains. Research Report MPI-I-2007-RG1-002, Max-Planck-Institut für Informatik, Saarbrücken (2007), http://www.mpi-inf.mpg.de/~hillen/documents/HW07.ps

22. Kamin, S., Levy, J.-J.: Attempts for generalizing the recursive path orderings. University of Illinois, Department of Computer Science. Unpublished note (1980), Available electronically from http://perso.ens-lyon.fr/pierre.lescanne/not_accessible.html

23. Kirchner, H., Ranise, S., Ringeissen, C., Tran, D.-K.: On Superposition-Based Satisfiability Procedures and Their Combination. In: Van Hung, D., Wirsing, M. (eds.) ICTAC 2005. LNCS, vol. 3722, pp. 594–608. Springer, Heidelberg (2005)

24. Manthey, R., Bry, F.: Satchmo: A Theorem Prover Implemented in Prolog. In: Lusk, E., Overbeek, R. (eds.) CADE 1988. LNCS, vol. 310, pp. 415–434. Springer, Heidelberg (1988)

25. McCune, W.: Mace4 reference manual and guide. Technical Report ANL/MCS-TM-264, Argonne National Laboratory (2003)

26. McCune, W.: Prover9 and mace4 (2005-2010), http://www.cs.unm.edu/~ccune/prover9/

27. McCune, W.: Otter 3.3 reference manual. CoRR, cs.SC/0310056 (2003)

28. Minsky, M.L.: Computation: Finite and Infinite Machines. Automatic Computation. Prentice-Hall (1967)

29. Nieuwenhuis, R., Rubio, A.: Paramodulation-based theorem proving. In: Robinson, A., Voronkov, A. (eds.) Handbook of Automated Reasoning, vol. I, ch. 7, pp. 371–443. Elsevier (2001)

30. Nieuwenhuis, R., Oliveras, A., Tinelli, C.: Solving SAT and SAT modulo theories: From an abstract Davis–Putnam–Logemann–Loveland procedure to DPLL(T). Journal of the ACM 53, 937–977 (2006)

31. Paulson, L.C., Blanchette, J.C.: Three years of experience with Sledgehammer, a practical link between automatic and interactive theorem provers. In: Sutcliffe, G., Ternovska, E., Schulz, S. (eds.) Proceedings of the 8th International Workshop on the Implementation of Logics (2010)

32. Navarro, J.A., Voronkov, A.: Proof Systems for Effectively Propositional Logic. In: Armando, A., Baumgartner, P., Dowek, G. (eds.) IJCAR 2008. LNCS (LNAI), vol. 5195, pp. 426–440. Springer, Heidelberg (2008)

33. Schulz, S., Bonacina, M.P.: On Handling Distinct Objects in the Superposition Calculus. In: Konev, B., Schulz, S. (eds.) Proc. of the 5th International Workshop on the Implementation of Logics, Montevideo, Uruguay, pp. 66–77 (2005)

34. Slaney, J.: FINDER: Finite Domain Enumerator. In: Bundy, A. (ed.) CADE 1994. LNCS, vol. 814, pp. 798–801. Springer, Heidelberg (1994)

35. Suda, M., Weidenbach, C., Wischnewski, P.: On the Saturation of YAGO. In: Giesl, J., Hähnle, R. (eds.) IJCAR 2010. LNCS, vol. 6173, pp. 441–456. Springer, Heidelberg (2010)

36. Weidenbach, C.: Combining superposition, sorts and splitting. In: Robinson, A., Voronkov, A. (eds.) Handbook of Automated Reasoning, vol. II, ch. 27, pp. 1965–2012. Elsevier (2001)

37. Weidenbach, C., Dimova, D., Fietzke, A., Kumar, R., Suda, M., Wischnewski, P.: SPASS Version 3.5. In: Schmidt, R.A. (ed.) CADE 2009. LNCS, vol. 5663, pp. 140–145. Springer, Heidelberg (2009)

38. Wenzel, M., Paulson, L.C., Nipkow, T.: The Isabelle Framework. In: Mohamed, O.A., Muñoz, C., Tahar, S. (eds.) TPHOLs 2008. LNCS, vol. 5170, pp. 33–38. Springer, Heidelberg (2008)

39. Zhang, J., Zhang, H.: SEM: a system for enumerating models. In: Proceedings of the 14th International Joint Conference on Artificial Intelligence, vol. 1, pp. 298–303. Morgan Kaufmann (1995)

Appendix: Proofs

Proving Proposition 2

Proposition 2. A clause set M is $\mathcal{T}$-satisfiable iff $\Omega(M \cup \mathcal{T}')$ is satisfiable.

Proof. On the one hand, since $M, \mathcal{T} \models \Omega(M \cup \mathcal{T}')$, every $\mathcal{T}$-model of M is a model of $\Omega(M \cup \mathcal{T}')$ as well. On the other hand, consider any model $\mathcal{A}$ of $\Omega(M \cup \mathcal{T}')$. Its restriction to $\{1^{\mathcal{A}}, \ldots, n^{\mathcal{A}}\}$ is a Σ-algebra because of the range restriction on the functions, and it is a $\mathcal{T}$-model by construction. Finally every clause C is $\mathcal{T}$-equivalent to $\bigwedge \Omega(C)$.

Proving Lemma 6

Proposition 6.1. Let N denote a node in a derivation, with successors $N_1, \ldots, N_k$. If $\Omega(N)$ is satisfiable, so is some $\Omega(N_i)$.

Proof. According to the type of calculus step, we distinguish three cases.
- An inference: Here k equals 1, and N_1 is $N \cup \{C\}$ where C is N-valid. Hence N and N_1 are even equivalent.

- A simplification adhering to the form $\mathcal{R}\frac{C}{D}N'$: Again k is 1, but N has a presentation $N = \{C\} \cup N' \cup N''$ such that $N_1 = \{\vec{D}\} \cup N' \cup N''$. The side conditions imply $\Omega(N') \models (\bigwedge \Omega(C)) \leftrightarrow (\bigwedge \Omega(\vec{D}))$, such that the clause sets $\Omega(N)$ and $\Omega(N_1)$ are equivalent.
- A split: In our concrete split rule k equals 2. Let $C' \equiv (C \vee s \simeq t)\tau$ and $D' \equiv (l \simeq r \vee D)\tau$ denote the first and the second conclusion, respectively. Then $C' \vee D'$ is N-valid, and the disjuncts share no variables. If $\mathcal{A}$ is an N-model, then $\mathcal{A}$ satisfies at least one of C' and D', and therefore at least one of $N_1 = N \cup \{C'\}$ and $N_2 = N \cup \{D'\}$.

Proposition 6.2. For every clause set M, the following are equivalent:
  (i) M is $\mathcal{T}$-satisfiable.
 (ii) Every derivation from $M \cup \mathcal{T}'$ contains a complete path $N_1, N_2, \ldots$ such that every $\Omega(N_i)$ is satisfiable.

Proof. If M is $\mathcal{T}$-satisfiable, then by Prop. 2 the set $\Omega(N_1) = \Omega(M \cup \mathcal{T}')$ is satisfiable, from which we can recursively construct a complete path as required by Prop. 6.1. The converse implication follows from $N_1 = M \cup \mathcal{T}$ by Prop. 2.

If a clause C occurs at some point in a path, then the limit N_∞ entails each of its Ω-instances from smaller or equal Ω-instances. Furthermore satisfiability of N_∞ with respect to Ω-instances is the conjunction of this property over all path elements.

Proposition 6.3. Consider a complete path $N_1, N_2, \ldots$ in some derivation.
  (i) If $C \in N_i$ is ground numbered by ρ, then $\Omega(N_\infty)^{\preceq C\rho} \models C\rho$ holds, as well as $\Omega(N_j)^{\preceq C\rho} \models C\rho$ for every $j \geq i$.
 (ii) Every $\Omega(N_i)$ is satisfiable iff $\Omega(N_\infty)$ is.
(iii) N_∞ is saturated in case the derivation is fair.

Proof.
  (i) The proof is by induction on $C\rho$ with respect to $\succ$. Let j denote ∞ or a natural number greater than or equal to i. If $C \in N_j$ we are done. Otherwise there is an index k between i and j such that C is contained in N_i through N_k, but not in N_{k+1}. By definition of simplification we have $\Omega(\vec{D}, M)^{\prec C\rho} \models C\rho$ for appropriate $\vec{D}, M \subseteq N_{k+1}$. Either $\vec{D}, M$ is empty and $C\rho$ is a tautology, or there is a greatest clause D' in $\Omega(\vec{D}, M)^{\prec C\rho}$. Inductively all elements of $\Omega(\vec{D}, M)^{\prec C\rho}$ are valid in $\Omega(N_j)^{\preceq D'}$, and so is $C\rho$.
 (ii) Assume that every $\Omega(N_i)$ is satisfiable. By compactness $\Omega(N_\infty)$ is satisfiable iff each of its finite subsets is. Given one such subset M, for every Ω-instance $C\rho$ within there is an index j such that C is contained in N_j and all successors thereof. Since M is finite, these indices have a finite maximum k. Now $\Omega(N_k)$ comprises M and is satisfiable by assumption. As to the converse implication, consider an Ω-instance $C\rho$ of a clause $C \in N_i$. Then $\Omega(N_\infty)$ entails $C\rho$ by Prop. 6.3 (i). In other words, any model of $\Omega(N_\infty)$ is a model of $\Omega(N_i)$.

(iii) Firstly we consider an inference with premises $\vec{C}$ from N_∞ and conclusion D with ground numbering substitution ρ. Because of fairness $\Omega(N_i)^{\prec\max\{\vec{C}\rho\}} \models D\rho$ holds for some i, which can be rephrased as $C'_1\rho_1, \ldots, C'_k\rho_k \models D\rho$ for clause instances $C'_j\rho_j$ from $\Omega(N_i)$ below $\max\{\vec{C}\rho\}$. By Prop. 6.3 (i) these clause instances are valid in $\Omega(N_\infty)$ below $\max\{\vec{C}\rho\}$, and so is $D\rho$.

Secondly we study a split from a persistent clause $C \equiv C_1 \vee C_2$ with designated partitioning as indicated and minimally numbering substitution τ. Because of fairness, one split conjunct, say $C_1\tau$, is contained in some N_i or redundant with respect to it. So either $C_1\tau$ is persistent, or $C_1\tau$ is redundant with respect to some N_j where $j \geq i$. In the former case the proof is finished. In the latter we have $\Omega(N_j)^{\prec C_1\rho} \models C_1\rho$ for every ground numbering $\rho = \tau\tau'$, which extends to $\Omega(N_\infty)^{\prec C_1\rho} \models C_1\rho$ with an argument like in the preceding paragraph.

For any clause set M, by $\widehat{M}$ is denoted the set of its Ω-instances which are Horn clauses.

Proposition 6.4. $\Omega(M)$ and $\widehat{M}$ are equivalent for $\mathcal{C}$-saturated clause sets M.

Proof. We show by induction on clause instances that every non-Horn clause $C\rho \in \Omega(M)$ is entailed by $\widehat{M}$. Now, C has a presentation $C \equiv C_1 \vee C_2$ such that the partitioning into C_1 and C_2 is designated. Then ρ numbers the clause C such that the subclauses C_1 and C_2 are variable disjoint. More general such substitutions τ have to satisfy $\tau \subseteq \rho$. There exists a $\subseteq$-minimal such τ because all descending $\subseteq$-chains are finite. Then $C \vdash C_1\tau \mid C_2\tau$ is a valid $\mathcal{C}$-split. Because M is saturated, one split conjunct, say $C_1\tau$, is contained in M or redundant with respect to M. In both cases we have $\Omega(M) \models C_1\rho$, and we obtain inductively $\widehat{M} \models C_1\rho$. Finally $C_1\rho$ entails $C\rho$.

The calculus $\mathcal{C}$ is sound and refutationally complete:

Lemma 6. For every clause set M, the following are equivalent:

(i) M is $\mathcal{T}$-satisfiable.

(ii) Every fair derivation from $M \cup \mathcal{T}'$ contains a complete path $N_1, N_2, \ldots$ such that the empty clause is not in N_∞.

Proof. We successively transform the first characterization into the second. By Prop. 6.2 the clause set M is $\mathcal{T}$-satisfiable iff there exists a complete path $N_1, N_2,$ $\ldots$ such that every $\Omega(N_i)$ is satisfiable, or such that $\Omega(N_\infty)$ is, by Prop. 6.3 (ii). Because of Prop. 6.3 (iii) every N_∞ is saturated with respect to $\mathcal{C}$. Hence by Prop. 6.4 the sets $\Omega(N_\infty)$ and $\widehat{N_\infty}$ are equivalent, and the latter is saturated with respect to $\mathcal{G}$ by Prop. 6.4. Since $\mathcal{G}$ is sound and complete, the satisfiability of $\widehat{N_\infty}$ is equivalent to $\perp \notin \widehat{N_\infty}$, which is the same as $\perp \notin N_\infty$.

Proving Lemma 4

We now set out to prove that a ground instance $C\sigma$ of a clause C follows from $\Omega(C, \mathcal{T}')$, and give a criterion when this entailment is from smaller instances.

Proposition 4.5. For every clause C and term t, the following entailment holds:
$C\{x \mapsto 1\}, \ldots, C\{x \mapsto n\}, \mathrm{Dig}(t) \models C\{x \mapsto t\}$

Proof. Consider a model $\mathcal{A}$ of the premises. Then there exists a digit i fulfilling $\mathcal{A} \models t \simeq i$. This identity inductively lifts to term contexts, and as equivalence to clause contexts. In particular $\mathcal{A} \models C\{x \mapsto i\}$ implies $\mathcal{A} \models C\{x \mapsto t\}$.

Proposition 4.6. Let C denote a clause with ground substitution $\sigma = \{x_1 \mapsto t_1, \ldots, x_m \mapsto t_m\}$. Then $\Omega(C), \mathrm{Dig}(t_1), \ldots, \mathrm{Dig}(t_m) \models C\sigma$ holds.

Proof. The proof is by induction on m. If σ is the identity we are done. Otherwise we decompose σ according to $\sigma = \{x_1 \mapsto t_1, \ldots, x_m \mapsto t_m\} \cup \{x_{m+1} \mapsto t_{m+1}\} = \sigma_1 \cup \sigma_2$. Since the substitutions are ground we have $\sigma_1 \cup \sigma_2 = \sigma_1 \circ \sigma_2$. Inductively we obtain $\Omega(C\sigma_1), \mathrm{Dig}(t_1), \ldots, \mathrm{Dig}(t_m) \models C\sigma_1$. Proposition 4.5 gives $C\sigma_1, \mathrm{Dig}(t_{m+1}) \models C\sigma_1\sigma_2$.

Proposition 4.7. Ground terms t obey $\Omega(\mathcal{T}') \models \mathrm{Dig}(t)$.

Proof. We induct on the structure of t. In case $t \equiv i$ the clause $\mathrm{Dig}(t)$ is a tautology. In case $t \equiv f(\vec{t})$ the proposition $\Omega(\mathcal{T}') \models \mathrm{Dig}(t_j)$ is inductively true for every j. Furthermore $\mathcal{T}'$ contains $\mathrm{Dig}(f(\vec{x}))$. Let $\sigma = \{x_1 \mapsto t_1, \ldots, x_m \mapsto t_m\}$, such that $f(\vec{t}) \equiv f(\vec{x})\sigma$. With Prop. 4.6 we obtain $\Omega(\mathrm{Dig}(f(\vec{x}))), \mathrm{Dig}(t_1), \ldots, \mathrm{Dig}(t_m) \models \mathrm{Dig}(f(\vec{x}))\sigma$.

Proposition 4.8. $\Omega(C, \mathcal{T}') \models C\sigma$ is true for every clause C with ground substitution σ.

Proof. Assume $\sigma = \{x_1 \mapsto t_1, \ldots, x_m \mapsto t_m\}$. Then Prop. 4.7 implies that $\Omega(\mathcal{T}') \models \mathrm{Dig}(t_i)$ holds for every i, such that from Prop. 4.6 finally we obtain $\Omega(C), \mathrm{Dig}(t_1), \ldots, \mathrm{Dig}(t_m) \models C\sigma$.

We have seen in Prop. 4.7 that every ground term t is subject to $\Omega(\mathcal{T}') \models \mathrm{Dig}(t)$. In the following we will exploit that usually not all of $\Omega(\mathcal{T}')$ is needed for this entailment. There exist subsets $T \subseteq \Omega(\mathcal{T}')$ such that $T \models \mathrm{Dig}(t)$ holds. By compactness there are finite such T even in case the signature is infinite. Let $\Delta(t)$ denote the smallest of these finite T, with respect to the ordering on clause sets. Let furthermore $\delta(t)$ denote the greatest clause in $\Delta(t) \cup \{\bot\}$, and for ground substitutions σ let $\delta(\sigma)$ stand for the greatest clause in $\delta(\mathrm{ran}\,\sigma) \cup \{\bot\}$. Actually one can construct $\Delta(t)$ recursively, but this is not necessary for our purposes.

Proposition 4.9. Entailment from $\Omega(\mathcal{T}')$ can be restricted by the bounds $\delta(t)$ and $\delta(\sigma)$:
  (i) Every ground term t satisfies $\Omega(\mathcal{T}')^{\preceq\delta(t)} \models \mathrm{Dig}(t)$.
  (ii) If σ is a ground substitution for C, then $\Omega(C), \Omega(\mathcal{T}')^{\preceq\delta(\sigma)} \models C\sigma$ holds.

Proof.
  (i) By definition we have $\Delta(t) \subseteq \Omega(\mathcal{T}')^{\preceq\delta(t)}$ and $\Delta(t) \models \mathrm{Dig}(t)$.

(ii) Let $\sigma = \{x_1 \mapsto t_1, \ldots, x_m \mapsto t_m\}$. Then we obtain $\Omega(\mathcal{T}')^{\preceq \delta(t_i)} \models \mathrm{Dig}(t_i)$ from Prop. 4.9 (i) for every i, and $\Omega(\mathcal{T}')^{\preceq \delta(\sigma)} \models \mathrm{Dig}(t_i)$ by definition of $\delta(\sigma)$. Finally we apply Prop. 4.6 to C and σ.

Proposition 4.10. For ground terms t we have $\delta(t) \equiv \bot$ iff t is a digit.

Proof. In case t is a digit, then $\mathrm{Dig}(t)$ is a tautology and $\Delta(t)$ is empty. Otherwise $\mathrm{Dig}(t)$ is not a tautology.

Proposition 4.11. If t is a ground term and δ a ground substitution, then we can give estimates for $\delta(t)$ and $\delta(\sigma)$ as follows:
  (i) $\delta(t) \equiv \mathrm{Dig}(u)$ implies $t \succeq u$.
  (ii) $\delta(\sigma) \equiv \mathrm{Dig}(u)$ entails $\max(\mathrm{ran}\,\sigma) \succeq u$.

Proof.
  (i) The proof is by induction on the term structure. If t is a digit, then we have $\delta(t) \equiv \bot$ by Prop. 4.10, and there is nothing to show. The case $t \equiv f(\vec{t})$ remains. Let $i_1, \ldots, i_k$ denote exactly the indices for which t_j is not a digit, and let $t' \equiv f(\vec{t})[x_1]_{i_1} \ldots [x_k]_{i_k}$. So t' is obtained from t replacing every non-digit t_j with a fresh variable. Conversely, using $\sigma = \{x_1 \mapsto t_{i_1}, \ldots, x_k \mapsto t_{i_k}\}$ one can instantiate t' back into t again.
  In case $k = 0$ the argument vector $\vec{t}$ contains only digits. Choosing $T = \{\mathrm{Dig}(t)\}$ implies $T \subseteq \Omega(\mathcal{T}')$ and $T \models \mathrm{Dig}(t)$. Therefore we have $T \succeq \Delta(t)$ and $\max T \succeq \max \Delta(t) \equiv \delta(t)$, hence $\mathrm{Dig}(t) \succeq \mathrm{Dig}(u)$ and finally $t \succeq u$.
  In case $k > 0$ every $\delta(t_{i_j})$ is distinct from $\bot$ by Prop. 4.10, and there exists a ground term v such that $\mathrm{Dig}(v) \equiv \max_j \delta(t_{i_j})$. By induction hypothesis and the subterm property of t we obtain $t \succ v$. Here we choose $T = \Omega(\mathrm{Dig}(t')) \cup \Omega(\mathcal{T}')^{\preceq \mathrm{Dig}(v)}$, which satisfies $T \subseteq \Omega(\mathcal{T}')$. By construction $T \models \mathrm{Dig}(t_{i_j})$ holds for every j. Proposition 4.6 yields $\Omega(\mathrm{Dig}(t')), \mathrm{Dig}(t_{i_1}), \ldots, \mathrm{Dig}(t_{i_k}) \models \mathrm{Dig}(t'\sigma)$. Hence we may conclude that $T \succeq \Delta(t)$ and $\max T \succeq \mathrm{Dig}(u)$. Next we compare T with $\{\mathrm{Dig}(t)\}$. We have $\Omega(\mathrm{Dig}(t')) \prec \{\mathrm{Dig}(t)\}$ by minimality of the digits, and furthermore $\Omega(\mathcal{T}')^{\preceq \mathrm{Dig}(v)} \prec \{\mathrm{Dig}(t)\}$ because of $v \prec t$. Hence we obtain that $\mathrm{Dig}(t) \succ \max T \succeq \mathrm{Dig}(u)$ holds, such that $t \succ u$ is true.
  (ii) Let $\sigma = \{x_1 \mapsto t_1, \ldots, x_m \mapsto t_m\}$. Because of $\delta(\sigma) \not\equiv \bot$ we have $\delta(\sigma) \equiv t_i$ for some i. Using Prop. 4.11 (i) we may conclude that $\max_j t_j \succeq t_i \succeq u$ holds. $\qquad\square$

Proposition 4.12. Let C denote a clause with ground substitution σ such that σ is not numbering, and that C, σ is unproblematic. Then $\Omega(C, \mathcal{T}')^{\prec C\sigma} \models C\sigma$ holds.

Proof. We decompose $\sigma = \sigma_1 \cup \sigma_2$ such that the range of σ_1 contains only digits and the range of σ_2 only non-digits. Since the substitutions are ground we have $\sigma = \sigma_1 \circ \sigma_2$. Proposition 4.9 (ii) implies $\Omega(C\sigma_1), \Omega(\mathcal{T}')^{\preceq \delta(\sigma_2)} \models C\sigma_1\sigma_2$. The substitution σ_2 is not empty because σ is not numbering. Hence we have by minimality of the digits $\Omega(C\sigma_1) \prec \{C\sigma_1\sigma_2\}$. We still have to show $\delta(\sigma_2) \prec C\sigma$.

Let t denote the greatest term in $\operatorname{ran}\sigma_2$. By Prop. 4.10 the clause $\delta(t)$ equals $\operatorname{Dig}(f(\bar{\imath}))$ for some term $f(\bar{\imath})$. By Prop. 4.11 (ii) we have $t \succeq f(\bar{\imath})$. If $t \succ f(\bar{\imath})$, then the greatest term of $C\sigma$ is above the greatest of $\delta(\sigma_2)$. Otherwise we obtain $C\sigma \succ \operatorname{Dig}(f(\bar{\imath}))$ from the requirement that C,σ is unproblematic.

Lemma 4. Consider a path in a $\mathcal{C}$-derivation from $M \cup \mathcal{T}'$ to N and a clause C. Then C is redundant with respect to N if one of the following conditions holds, where ρ ranges over all ground numbering substitutions:

(i) $\operatorname{g\mathring{n}d}(N)^{\prec C\rho} \models C\rho$ for all ρ,
(ii) $\operatorname{gnd}(N)^{\prec C\rho} \models C\rho$ for all ρ and C is noncritical.

Proof.

(i) Given an arbitrary ground numbering substitution ρ, there exist clauses $D_1,\ldots,D_m \in N$ and ground substitutions $\sigma_1,\ldots,\sigma_m$ such that every D_i,σ_i is unproblematic and $D_i\sigma_i \prec C\rho$, and that $D_1\sigma_1,\ldots,D_m\sigma_m \models C\rho$. In order to prove $\Omega(N)^{\prec C\rho} \models C\rho$ it suffices to show that $\Omega(N)^{\prec C\rho} \models D_i\sigma_i$ holds for every i. If $D_i\sigma_i$ is a digit instance of D_i, then we have $D_i\sigma_i \in \Omega(N)^{\prec C\rho}$. Otherwise Prop. 4.12 ensures $\Omega(D_i,\mathcal{T}')^{\prec D_i\sigma_i} \models D_i\sigma_i$ because D_i,σ_i is unproblematic. With Prop. 6.3 (i) we get $\Omega(N)^{\preceq D_i\sigma_i} \models D_i\sigma_i$, and therefore $\Omega(N)^{\prec C\rho} \models D_i\sigma_i$.

(ii) Similar to the proof of Lem. 4 (i), for every ground numbering substitution ρ there exist clauses $D_1,\ldots,D_m \in N$ and ground substitutions $\sigma_1,\ldots,\sigma_m$ such that always $D_i\sigma_i \prec C\rho$, and that $D_1\sigma_1,\ldots,D_m\sigma_m \models C\rho$. If $C\rho$ is a tautology we are done. Otherwise we decompose every $\sigma_k = \sigma_k' \cup \sigma_k''$ such that the range of σ_k' contains only digits and the range of σ_k'' only non-digits. Proposition 4.9 (ii) guarantees that $\Omega(D_k\sigma_k'), \Omega(\mathcal{T}')^{\preceq\delta(\sigma_k'')} \models D_k\sigma_k$. By minimality of the digits we obtain $\Omega(D_k\sigma_k') \preceq \{D_k\sigma_k\} \prec \{C\rho\}$.

Next we show that $\delta(\sigma_k'') \prec C\rho$. The clause C is not empty since otherwise $\models \perp$; so $C\rho$ has a greatest term s. Let t denote the greatest term of $D_k\sigma_k$, then we have $s \succeq t$. If $\delta(\sigma_k'') \equiv \perp$ then $\perp \prec C\rho$. Otherwise $\delta(\sigma_k'')$ has the shape $\operatorname{Dig}(f(\bar{\imath}))$. Because of Prop. 4.11 (ii) we have $\max(\operatorname{ran}\sigma_k'') \succeq f(\bar{\imath})$, and because of $t \succeq \max(\operatorname{ran}\sigma_k'')$ we have $s \succeq f(\bar{\imath})$ as well. Now $s \succ f(\bar{\imath})$ directly entails $C\rho \succ \delta(\sigma_k'') \equiv \operatorname{Dig}(f(\bar{\imath}))$. Otherwise s equals $f(\bar{\imath})$, and $C\rho \succ \operatorname{Dig}(f(\bar{\imath}))$ holds because C is noncritical by assumption. Summing it up, we obtain $\Omega(D_k\sigma_k',\mathcal{T}')^{\prec C\rho} \models D_k\sigma_k$ and therefore as well $\Omega(D_k,\mathcal{T}')^{\prec C\rho} \models D_k\sigma_k$. Via Prop. 6.3 (i) we conclude $\Omega(N)^{\prec C\rho} \models D_k\sigma_k$.

Proving Proposition 7

Proposition 7. Inference and split conclusions do not have more variables than one of the premises.

(i) If $\sigma = \operatorname{mgu}(u,v)$ with $\operatorname{ran}\sigma \subseteq \mathcal{V} \cup [1;n]$ and $\operatorname{dom}\sigma \cup \operatorname{cdom}\sigma \subseteq \operatorname{var}(u,v)$, then there is a variant σ' that additionally satisfies $\operatorname{var}(v\sigma') \subseteq \operatorname{var}(v)$.
(ii) If $C \vdash D$ is a unary inference or a split, then $\operatorname{var}(D) \subseteq \operatorname{var}(C)$ holds.
(iii) If $l \simeq r$, $C \vdash D$ is a binary inference, then $|\operatorname{var}(D)| \leq |\operatorname{var}(C)|$ is true.

Proof.

(i) Let $\mathcal{P}(m)$ hold iff there exists an mgu σ of u and v with $\operatorname{ran}\sigma \subseteq \mathcal{V} \cup [1;n]$, $\operatorname{dom}\sigma \cup \operatorname{cdom}\sigma \subseteq \operatorname{var}(u,v)$, and $|\operatorname{var}(v\sigma) \setminus \operatorname{var}(v)| = m$. By assumption $\mathcal{P}$ holds for some $m \geq 0$. We will now show that $\mathcal{P}(j+1)$ implies $\mathcal{P}(j)$.

Assume σ is a witness for $\mathcal{P}(j+1)$. Because of $j+1 > 0$ there exists a variable y in $\operatorname{var}(v\sigma) \setminus \operatorname{var}(v)$. By the shape of σ, this variable is the σ-image of another variable $x \in var(v)$. Consider now the substitutions $\tau = \{x \mapsto y, y \mapsto x\}$ and $\sigma' = \sigma \circ \tau$. The latter is a unifier of u and v. Because of $\sigma'\tau = \sigma\tau^2 = \sigma$, it is even a most general one. The image of a variable z under σ' is x if $z\sigma \equiv y$, and $z\sigma$ otherwise; in particular $x\sigma' \equiv x$ and $y\sigma' \equiv x$. That is, going from σ to σ', the variable x moves from the dom-part to the cdom-part, and y in the opposite direction, which are all effects in terms of dom and cdom. The identity $\operatorname{var}(v\sigma') = (\operatorname{var}(v\sigma) \cup \{x\}) \setminus \{y\}$ concludes the proof of $\mathcal{P}(j)$.

(ii) In case of an equality resolution step $C \vee t \simeq t' \vdash C\sigma$ we have $\operatorname{cdom}\sigma \subseteq \operatorname{var}(t,t')$. Given a split $C \vee s \simeq t \vee l \simeq r \vee D \vdash (C \vee s \simeq t)\tau \mid (l \simeq r \vee D)\tau$, the substitution τ is numbering, such that $\operatorname{cdom}\tau \subseteq [1;n]$.

(iii) We will prove that $\operatorname{var}(D) \subseteq \operatorname{var}(C)$ holds in case the most general unifier is chosen according to Prop. 7 (i). All mgu's are equal up to variable renaming; and the number of variables in a clause is invariant under such renamings. This yields the estimate stated above.

We jointly treat superposition left and right inferences via the pattern $l \simeq r,\ C[l'] \vdash C[r]\sigma\tau \equiv D$. Because of $l'\sigma\tau \equiv l\sigma\tau \succ r\sigma\tau$ we know that $\operatorname{var}(l'\sigma\tau) \supseteq \operatorname{var}(r\sigma\tau)$ is true, and hence $\operatorname{var}(D) \subseteq \operatorname{var}(C\sigma\tau) \subseteq \operatorname{var}(C\sigma)$. Applying Prop. 7 (i), without loss of generality σ can be chosen such that $\operatorname{var}(l'\sigma) \subseteq \operatorname{var}(l')$. Let σ' denote the restriction of σ to $\operatorname{var}(l')$. By this definition we have $\operatorname{cdom}\sigma' \subseteq \operatorname{var}(l'\sigma) \subseteq \operatorname{var}(l') \subseteq \operatorname{var}(C)$. Since the premises are variable disjoint, we obtain $\operatorname{var}(C\sigma) = \operatorname{var}(C\sigma') \subseteq \operatorname{var}(C) \cup \operatorname{cdom}\sigma' = \operatorname{var}(C)$, which completes the proof of $\operatorname{var}(D) \subseteq \operatorname{var}(C)$.

Proving Theorem 12

Theorem 12. $\mathcal{C}_\kappa$-derivations decide $\mathcal{T}$-satisfiability of finite clause sets.

Proof. Consider a $\mathcal{C}_\kappa$-derivation from a finite clause set M. Then M by Lem. 6 is $\mathcal{T}$-satisfiable if and only if the derivation contains a complete path without the empty clause in the limit. The derivation tree is finitely branching. It remains to show that every path $N_1, N_2, \ldots$ is finite. Let $\|N_i\| = \max\{|\operatorname{var}(C)| : C \in N_i\}$.

There exists an index κ such that from N_κ on, the conditions (iii) through (vi) of the definition of $\mathcal{C}_\kappa$-derivation are satisfied. We form a subsequence of $N_1, N_2, \ldots$ that starts from $N_1' = N_\kappa$. If in N_i a new clause C is inferred and, according to condition (vi), immediately simplified into $[1;n]$-shallow clauses $\vec{D}$ until N_{i+k}, then for $(N_j')_j$ all sequence elements but N_{i+k} are dropped, and the latter shows up only if $\vec{D}$ is not empty. Assume now $(N_i)_i$ is infinite. Inferences with empty $\vec{D}$ are not repeated because of condition (v), as well as splits; so there must be infinitely many simplifications or inferences with non-empty $\vec{D}$.

Since simplifications are decreasing with respect to Ω-instances, the latter occur infinitely many times; so $(N'_j)_j$ is then infinite as well.

Inductively $\|N'_j\| \leq \|N'_1\|$ holds for all j: If a clause $C \in N'_j$ is simplified to some non-empty $\vec{D}$, then we know that $|var(D_i)| \leq |var(C)|$ by condition (iii). In case of a split or an inference, we additionally apply Prop. 7 (ii) and Prop. 7 (iii).

Assume now $(N'_j)_j$ were infinite; then we can argue like above for $(N_i)_i$ and obtain that infinitely many inferences are drawn. The inference conclusions are simplified according to condition (vi), such that they become $[1; n]$-shallow and have no more than $\|N'_1\|$ variables. Because of conditions (vii) and (viii), only finitely many such clauses exist. Moreover the number of clauses that are produced from simplification and splitting alone is finite. Therefore, eventually an inference has to be repeated, but this contradicts condition (v). Hence $(N'_j)_j$ is finite, and so is $(N_i)_i$.

MACE4 and SEM: A Comparison
of Finite Model Generators*

Hantao Zhang[1] and Jian Zhang[2]

[1] Department of Computer Science
The University of Iowa
Iowa City, IA 52242, U.S.A.
hantao-zhang@uiowa.edu
[2] State Key Laboratory of Computer Science
Institute of Software, Chinese Academy of Sciences
Beijing 100190, China
zj@ios.ac.cn

Abstract. This article has three objectives: (1) Promote Mace4, a program developed by Bill McCune that searches for finite models of first-order formulas and that is the best way to remember Bill. (2) Promote the research on model generation of first-order formulas. Mace4 remains one of the best model generation programs and we need newcomers who can take over Bill's torch, because model generation is very important to automated reasoning and has many applications. (3) Compare Mace4 with SEM in detail so that the users of these tools or new model generator developers will understand the strengths and weaknesses of both systems and take advantage from this study.

Keywords: Finite models, constraint propagation, backtracking search.

1 Introduction

If automated reasoning tools are regarded as race cars, then the people who created these tools are regarded not only as car designers, but also as car engineers. Engineers pay more attention to details and have their own "small talks". Bill McCune was such an engineer and so are the two authors. In fact, each of us read the source code written by the other two. Sato [48], a tool developed by the first author, still uses some code from Otter [34].

In the fall of 1995, the first author of this article spent four months at Argonne National Laboratory and met Bill daily. At that time, Bill just finished the implementation of the Knuth-Bendix completion procedure modulo AC. We talked about the RRL experience [26] and how to make the AC completion faster [13]. The next year, using the AC completion, Bill successfully solved the long standing open Robbins algebra problem using his EQP (Equational Prover) [37].

In October 2001, Bill drove the second author to the home of Larry Wos, and Larry invited us for lunch. In 2006, Bill was invited to give a speech at the 8th

---

* Supported in part by NSFs of China and USA.

M.P. Bonacina and M.E. Stickel (Eds.): McCune Festschrift, LNAI 7788, pp. 101–130, 2013.

International Conference on Artificial Intelligence and Symbolic Computation (AISC) which was held in Beijing. Bill gave an excellent talk and discussed with the second author during the conference.

The last invited talk given by Bill was on March 30, 2007, at the University of Iowa. After the talk, Bill and the first author went to a famous steak restaurant by the Iowa river. We had a great time there and Bill really liked local beer and drank two bottles. We expressed concern about his early retirement from Argonne and he said that he had enough saving for the rest of his life. Indeed, we no longer need to be concerned about his retirement but also no longer have a chance to offer him the beer he liked.

The objective of this article is three-fold. Bill was not only a great designer of automated reasoning tools, but also a great software engineer. He created many popular tools including Otter, a resolution-based theorem prover, and Mace4, a program that searches for finite models of first-order formulas. Bill is gone but his tools remain. The best way to remember him is to continue to use the tools he created. There are no publications introducing Mace4, except two Argonne National Laboratory technical reports. This article will promote the use of Mace4 and that is the first objective of this article.

The second objective is to promote research on model generation of first-order formulas. If we can generate a model for a formula, we have proved that the formula is satisfiable, which is an important property of logical formulas. The model generation problem, regarded as a special case of the Constraint Satisfaction Problem (CSP), has many applications in AI, computer science and mathematics. For instance, it has been shown in [42,53,50] that model generators can solve various problems from algebra, number theory, and design theory. Model generation is very important to the automation of reasoning. For example, the existence of a model implies the consistency of a theory. A suitable model can also serve as a counterexample which shows some conjecture does not follow from some premises. In this sense, model generation is complementary to classical theorem proving. Finite models help people understand a theory, and they can also guide conventional theorem provers in finding proofs. Thus, it was a major objective of Bill to create Mace4 as a companion of the theorem prover Otter, using the same input files for both Otter and Mace4. Mace4 remains one of the best model generation programs. However, unlike research in propositional satisfiability where we see new solvers every year, we saw very few new model generators for first-order logic. We need newcomers who can take over Bill's torch.

The third objective is to provide a comprehensive comparison of Mace4 and our program SEM (a *System for Enumerating Models*) [59]. From the viewpoint of a theoretician, Mace4 and SEM are identical as they are based on the same principle and use roughly the same strategies. From the viewpoint of an engineer, Mace4 and SEM are different in many technical aspects. We will discuss these differences and their impact on the performance of these programs. Although their development was completed approximately ten years ago, Mace4 and SEM are still two of the most efficient finite model generators. The source codes of

Mace4 and SEM have been made available to the public since they were created. Since then, both tools have been used by other researchers and ourselves to solve various problems.

Before Mace4, there was a model generator called Mace2 [33] which is based on propositional satisfiability, like ModGen [27] and Paradox [17]. Mace4 is the first released program that has been constructed with LADR (Library for Automated Deduction Research) [36], a library of C routines for building automated deduction tools and a larger project developed by Bill. We will give an overview of model generation for first-order formulas, including the design, implementation and main capabilities, as well as some of its applications. We also give an outline of some future improvements.

2   Model Generation: Basic Concepts and Notations

The satisfiability problem (SAT) in propositional logic, i.e., deciding whether an arbitrary propositional formula is satisfiable, is a well-known NP-complete problem. In general, it is undecidable to decide if a set of first-order formulas is satisfiable. However, this satisfiability problem is so important that it cannot be ignored. Consider traditional automated theorem proving (ATP) in first order logic, an important class of theorem provers work by refutation. That is, the negation of the conjecture is added to the hypothesis and the axioms. Then the resulting set of formulas is shown to be unsatisfiable. It may occur that the conjecture is not valid, in which case a conventional theorem prover may not terminate or terminate but cannot give a clear answer. If we can find a model of the formulas including the negation of the conjecture, it will serve as a counter-example showing that the conjecture does not hold. With this application in his mind, McCune developed Mace4 as a companion of the theorem prover Otter and both tools can use the same input files.

Due to the undecidability in the general case and the high complexity in the case of domains of finite size, there was not much progress in creating effective model generators for decades. In the late 1980's and early 1990's, several serious attempts were made to tackle the problem in first-order logic. Some methods are based on first-order reasoning [31,16,46,23,15,40,5,6]; some are based on constraint satisfaction, e.g., Finder [41], Falcon [53] and SEM [59]; while others are based on propositional logic, e.g., ModGen [27] and Mace2 [33]. Later, other powerful tools have been developed, such as Gandalf [47], Mace4 [35], Paradox [17]. These tools have been used to solve a number of challenging problems in discrete mathematics [42,33,53]. The two programs studied in this article, Mace4 and SEM, are still two of the most powerful model generators, especially on problems with deeply-nested terms.

The models of a formula are interesting in themselves, because they may reveal interesting properties of the formula. This can be very useful, especially when the formula is complicated and people do not have a good understanding of it yet. See, for instance, [53].

Given the undecidability of the problem, we can only solve some subcases by putting certain restrictions on the original problem. Alternatively, we may check the satisfiability of the given formula in certain finite domains. If the size of the domain is a fixed positive integer n, we are certainly able to decide whether a formula has a model of size n or not. If we can find a model of fixed size for a first-order formula, we can say that the formula is satisfiable.

A finite model of first-order formulas is an interpretation of the constants, function symbols and predicates over a finite domain of elements that satisfies the formulas [41]. If no such model exists, then the formulas are unsatisfiable in the given domain. Although the decidability in this case is no longer an issue, it is still a very hard problem to find non-trivial finite models, as propositional satisfiability (SAT) is a special finite model generation problem where the domain size is two. In theory, every finite model generation problem can be encoded as an instance of SAT. However, despite the fact that we have very powerful SAT solvers today, many model generation problems cannot be solved by SAT solvers, because converting them into an SAT instance takes too much computer memory.

In the following, we briefly review some relevant concepts and notations in first-order predicate logic. We have function symbols and predicate symbols, variables and constants. We can construct *terms* and *formulas*. As in resolution-based theorem proving, it is often more convenient to transform first-order formulas into clausal form. A *clause* is a disjunction of literals, where a *literal* is a predicate (applied to as many terms as required by its arity) or the negation of a predicate.

A *model* of a set of clauses is an interpretation of the function symbols (including constant names) and predicate symbols in some (nonempty) domain, such that every clause is evaluated to true. A finite model can be conveniently represented by a set of tables, each of which corresponds to the definition of a function/predicate. If the arity of a function symbol is two, its table is two dimensional; if the arity is three, its table is three dimensional, and etc. Suppose the arity of f is k, each entry in the k-dimensional table can be represented by a unit positive clause $f(a_1, a_2, ..., a_k) = a_0$, where $a_i \in D_n$, called *VA (value assignment) clause* and the ground term $f(a_1, a_2, ..., a_k)$ is called a *cell term*. On the other hand, we call unit negative clause $f(a_1, a_2, ..., a_k) \neq a_0$ a *VE (value elimination) clause*, where a_i is an element of the domain.

2.1   Two Small Examples

Example 2.1 A *Latin square* of size n is an $n \times n$ square whose elements are D_n, the integers in the interval $[0..(n-1)]$. Each row and each column of the square is a permutation of the n integers. Latin squares satisfy the following formulas:

$$\forall x, y, z \ (y \neq z \ \rightarrow \ f(x, y) \neq f(x, z));$$
$$\forall x, y, z \ (x \neq y \ \rightarrow \ f(x, z) \neq f(y, z)).$$

where $\rightarrow$ stands for implication, $f(x, y)$ denotes the cell on the x'th row and the y'th column of the Latin square. In abstract algebra, a *quasigroup* is (D_n, f) satisfying the above formulas, and it is well-known that the "multiplication table" of a quasigroup is a Latin square.

Without loss of generality, we can denote an n-element domain by the integer set $D_n = \{0, 1, 2, \ldots, n-1\}$. The following is a 3-element model of the above axioms, over the domain D_3.

f	0	1	2
0	0	1	2
1	1	2	0
2	2	0	1

This model may be represented by the following VA clauses:

$$f(0,0) = 0. \quad f(0,1) = 1. \quad f(0,2) = 2.$$
$$f(1,0) = 1. \quad f(1,1) = 2. \quad f(1,2) = 0.$$
$$f(2,0) = 2. \quad f(2,1) = 0. \quad f(2,2) = 1.$$

In many cases, it is more convenient to use a *many-sorted* language. Sorts are similar to types in programming languages. In such a case, variables, terms, functions are all sorted. The following is a simple example.

Example 2.2 Let us look at the formulas

$$\forall x, y : Dog, \; mating(x, y) \rightarrow gender(x) \neq gender(y)$$
$$\forall x, y : Dog, \; mating(x, y) \rightarrow mating(y, x)$$

where *mating* is a binary predicate, meaning that the two dogs are the mates in a mating, *gender* is a function that maps *Dog* to *Sex*. So we have the sort *Dog* and the sort $Sex = \{Male, Female\}$.

One model of the above formula is given below on the domain $Dog = \{Benji, Bailey\}$ and the function/predicate symbols are interpreted as follows:

gender	Bailey	Benji
	Female	Male

mating	Benji	Bailey
Benji	F	T
Bailey	T	F

Here **F** and **T** are `false` and `true`, respectively. Thus the model says that Benji is male, Bailey is female, and one is the mate of the other.

2.2   Input of Mace4 and SEM

The input of Mace4 usually consists of parameter settings and a list of untyped first-order formulas or clauses, which is compatible with the input syntax of Otter. For Example 2.1, the input of Mace4 may look like this:

```
assign(domain_size, 3).
formulas(theory).
y = z | f(x,y) != f(x,z).
x = y | f(x,z) != f(y,z).
end_of_list.
```

SEM's input file for Example 2.1 is very similar to that of Mace4:

```
3.

y = z | f(x,y) != f(x,z).
x = y | f(x,z) != f(y,z).
```

That is, the first line specifies the size of the domain, and the other lines give the clauses. This syntax applies to problems with one sort of data. In general, the input of SEM consists of the following four parts:

Sorts   Functions   Variables   Clauses

where *Sorts* specify the size of each sort and *Functions* specify the signatures of each function. The *Clauses* often contain some *Variables* which are assumed to be universally quantified over the given finite domains. The third part of the input, i.e., *Variables*, can be omitted. In this case, SEM will try to infer the sorts of the variables from the context.

For Example 2.2, the input of SEM looks like this:

```
% Sorts
( Sex : Female, Male )
( Dog : Bailey, Benji )

% Functions
{ gender: Dog -> Sex }
{ mating: Dog Dog -> BOOL }

% Clauses
[ -mating(x,y) | gender(x) != gender(y) ]
[ -mating(x,y) | mating(y,x) ]
```

Here a line beginning with the character '%' is a comment. The symbol "-" denotes NOT; "|" denotes OR; and the symbol "!=" denotes "not-equal".

To facilitate the use of SEM, the second author of this article wrote (in 2000) some Perl scripts [57]. Given a set of (uni-sorted) first-order formulas, the scripts will call Otter to generate a set of clauses, and call SEM to search for models of increasingly larger sizes. This set of scripts can help Otter users to use SEM, even though SEM and Otter are not tightly integrated.

Mace4 accepts only sortless first-order logic formulas (clauses). For Example 2.2, the user has to provide the predicates for Dog and Sex. On the other hand, because of the simplicity of its input syntax and the popularity of the Otter prover, Mace4's input syntax is more appealing to the community of automated reasoning than that of SEM. Moreover, Mace4 provides many more utility commands to assist the users. For example, the user can specify the command assign(iterate_up_to, n) so that Mace4 will iteratively increase the domain size one by one, from the current domain size to n, unless a model is found during the process; for each different domain size, Mace4 will restart the search. Mace4 also allows the user to use Prolog style variables in the clauses, or specify the precedence of operators for pretty printing. The user can also set the parameter values on the command line when invoking Mace4. For example, if the input file group.in contains the following line:

```
assign(iterate_up_to, 10).
```

then the command

```
% mace4 -n8 -m20 < group.in > group.out
```

tells Mace4 to search for up to 20 models of sizes 8, 9, and 10. There is also an option -c, which allows Mace4 to skip certain unrecognized commands. That way, Mace4 can use directly Otter's input files with -n, -m, and -c.

2.3 Preprocessing

The finite model generation problem can be regarded as a constraint satisfaction problem (CSP), where the constraints are a set of clauses with equality and functions. When finding an n-element model, a general practice is to instantiate the input clauses in all possible ways, substituting each of the n elements in the domain for each variable in the clauses. Then we obtain a set of ground clauses, denoted by GC. If a clause has v variables, then the clause has n^v ground instances. That is, the number of ground clauses grows exponentially with the number of variables in a clause. However, working with ground clauses greatly simplifies the needed inference rules. For instance, unification is not necessary and matching becomes identity checking. All the model generation programs, including Finder [41], Falcon [53], Mace4 and SEM, instantiate the input clauses before searching for the model.

Both Mace4 and SEM start by allocating a table for each function and predicate symbol. Constants are allocated a single cell, function symbols of arity 2 get an $n \times n$ table of cells, and so on. For sortless formulas, the range of values for the cells will be members of the domain $\{0, ..., n-1\}$, and values for predicate symbols will be 0 and 1, the boolean values. For the sorted formulas, the range of each cell is the sort of that cell.

3   Basic Search Algorithm in Mace4 and SEM

In principle, a finite model can be found by exhaustive search. We can enumerate all different assignments to the table cells. For each assignment, we check whether all the input clauses hold. If so, the assignment is a model. Obviously, such a procedure is not efficient. A better alternative is to use a backtracking search procedure.

3.1   A Backtracking Search Procedure

Backtracking search procedures are often used to solve finite domain CSPs, including the SAT problem. In this approach, one starts with an empty assignment and repeatedly assigns a value to a new cell as long as the assignment does not violate the constraints; if a conflict occurs, it backtracks and assigns a different value to the variable. In this way, it explores the search space systematically and exhaustively. The performances of backtracking algorithms can be improved in a number of ways such as forward checking and lookahead [28].

The model generation process in Mace4 and SEM can be described by the recursive procedure in Fig. 1. The procedure uses the following parameters:

- $A \subset CE \times Dom$ is an assignment, i.e., a set of pairs of cells and values, where CE is the set of cells and Dom is the domain of values (elements). A is called a *partial model* and can be represented by a set of VA clauses (i.e., a set of equalities of form $(ce = e)$ where ce is a cell term and e is an element). It should be noted that A is functional, i.e., it may not contain two pairs $(ce = e_1)$ and $(ce = e_2)$ where e_1 and e_2 are different elements.
- $\mathcal{D} \subset CE \times \mathcal{P}(Dom)$, a set of unassigned cells and their possible values, where $\mathcal{P}(Dom)$ is the power set of Dom. In fact, each pair (ce, D) in $\mathcal{D}$ can be represented by a clause $ce = e_1 \vee \ldots \vee ce = e_i \vee \ldots \vee ce = e_k$, where $e_i \in D$. Such clauses are called *PV (possible values) clauses* in [60]. For each cell ce, ce appears uniquely either in A or $\mathcal{D}$, but not both.
- Ψ: constraints (i.e. the ground clauses).

Initially A is empty, Ψ is GC, and $\mathcal{D}$ contains $(ce, Dom(ce))$ for every cell ce. The following procedure will print all models of $(A, \mathcal{D}, \Psi)$ upon termination.

```
proc backtrack_search(A, D, Ψ)
{
    if Ψ = false then return;
    if D = ∅ then /* a model is found */
        { print(A); return; }
    choose and delete (ce_i, D_i) from D;
    for each e ∈ D_i do
        backtrack_search(propagate(A ∪ {(ce_i = e)}, D, Ψ));
}
```

Fig. 1. An abstract backtrack search procedure

The procedure $\texttt{propagate}(A, \mathcal{D}, \Psi)$ propagates assignment A in Ψ: It simplifies Ψ and may force some variables in $\mathcal{D}$ to be assigned. In fact, $\texttt{propagate}$ is essentially a closure operation with respect to a set of sound inference rules. It repeatedly modifies $A, \mathcal{D}$, and Ψ until no further changes can be made. When it exits, it returns the modified triple $(A, \mathcal{D}, \Psi)$.

The procedure $\texttt{propagate}$ repeatedly applies the following basic steps:

1. For each new assignment (ce, e) in A, replace the occurrence of ce in Ψ by e.

2. If there exists an empty clause in Ψ (i.e., each of its literals becomes `false` during the propagation), replace Ψ by `false`, and exit from the procedure. Otherwise, for every unit clause in Ψ (i.e. all but one of its literals become `false`), examine the remaining literal l.

   - l is a cell term ce, $(ce, D) \in \mathcal{D}$: If `true` $\in D$, then delete (ce, D) from $\mathcal{D}$ and add (ce, true) to A; otherwise, replace Ψ by `false`, and exit from the procedure.
   - l is the negation of a cell term, i.e. $\neg ce$, $(ce, D) \in \mathcal{D}$: If `false` $\in D$, delete (ce, D) from $\mathcal{D}$ and add (ce, false) to A; otherwise, exit with $\Psi = \text{false}$.
   - l is of the form $\text{EQ}(ce, e)$ (or $\text{EQ}(e, ce)$), and $(ce, D) \in \mathcal{D}$: If $e \in D$, delete (ce, D) from $\mathcal{D}$ and add (ce, e) to A; otherwise, exit with $\Psi = \text{false}$.
   - l is of the form $\neg\text{EQ}(ce, e)$ (or $\neg\text{EQ}(e, ce)$), and $(ce, D) \in \mathcal{D}$: If $e \in D$, then delete e from D.

3. For each pair $(ce, D) \in \mathcal{D}$, if $D = \emptyset$, then replace Ψ by `false`; if $D = \{e\}$ (i.e. $|D| = 1$), then delete the pair (ce, D) from $\mathcal{D}$, and add the pair $(ce = e)$ to A.

In the next section, we shall give a formal presentation of the above propagation rules as well as more sophisticated rules.

The execution of the search procedure `backtrack_search` can be represented as a search tree. Each node of the tree corresponds to a call of `backtrack_search` and also a partial model, and each edge corresponds to assigning a value to a cell.

3.2   Avoiding Search for Isomorphic Models

The efficiency of the search procedure depends on many factors. One factor is how we choose a cell and assign a value to it; another factor is how we can perform reasoning after we make a choice. Here we address the former issue first. The latter will be discussed in the next section.

During the search process, there are usually more than one unassigned cells, and we need to choose one of them to assign a value. The performance of the procedure `backtrack_search` is mostly affected by the search strategy. A common strategy is to choose the cell which has the fewest possible values (called the most constrained cell). This is usually quite effective for solving various constraint satisfaction problems, and it is also implemented in Mace4 and SEM.

For a typical finite model generation problem, it is very important to avoid the exploration of certain isomorphic subspaces. Two (partial) models are *isomorphic* if there is a permutation of the domain elements such that the models are the same under the permutation. In general, it is very expensive to eliminate all the isomorphic models during the search. In SEM as in Mace4, the least number heuristic (LNH) is used to explore the symmetry between domain values which produce isomorphic models. This technique turns out to be crucial in the success of solving many model generation problems. Other approaches have also been explored, see, e.g., [11,12].

LNH is based on the observation that in typical model generation problems, the domain elements are "equivalent" when the search begins, because each input formula (or axiom) usually specifies that all the elements have to satisfy some property, without naming any element. As we make choices during the search, some elements become "special", i.e., they are no longer equivalent to or symmetric with other elements as they were assigned to some cells. For the uni-sorted case, we can use a single integer, mdn, to denote which domain elements are symmetric. The domain is divided into two disjoint subsets, $[0..mdn]$ and $[(mdn + 1)..(n - 1)]$, where n is the size of the domain. The elements in the second subset have not gotten special status, and thus they are symmetric with each other. When we try to assign values from the second subset to a cell, we need only choose one value and assign it to the cell; the other values from the second subset are equivalent to the chosen value.

The value of mdn is (-1) if no domain element appears in the input formulas. It is increased as the search moves on. To keep as many symmetric elements as possible, we try to avoid giving special status to the domain elements as far as we can. This is usually achieved by trying the smallest elements as values for assignment to the cells. Hence the name *least number heuristic* (LNH).

For instance, suppose we have only one binary function symbol, f, and no domain element appears in the input clauses. Then typically the first level of the search tree has two branches:

1. Let $f(0,0) = 0$. We have $mdn = 0$. Then we check the cells $f(0,1)$, $f(1,0)$, $f(1,1)$ and pick up a cell having the smallest number of possible values.

2. Let $f(0,0) = 1$. We have $mdn = 1$. Then we check the cells $f(0,1)$, $f(1,0)$, $f(1,1)$, $f(0,2)$, $f(1,2)$, $f(2,2)$, $f(2,0)$, $f(2,1)$ and pick up a cell having the smallest number of possible values.

In other words, we search for a table of values for f, starting from the upper-left cell (i.e., the subsquare of size 1×1) and increasing the size of the subsquare gradually. There is no need to consider the assignment $f(0,0) = b$ where $b \neq 0, 1$, because if there exists a model in which $f(0,0) = b$, we may obtain a model in which $f(0,0) = 1$ by switching the names of 1 and b in the model (assuming 1 and b do not appear in the input formula). That is the basic idea of LNH.

Since SEM accepts many-sorted input, an array of integers, mdn_0, mdn_1, ..., is used and one integer of the array for each sort. Since both SEM and Mace4 use the idea of LNH, this will make the performances of SEM and Mace4 similar for many examples. The LNH technique has been used and extended by several researchers, including [3,1]. Instead of using as few individual elements as possible to preserve symmetries, Audemard and Henocque proposed the idea of XLNH [3] which heuristically selects and then fully generates the functions that appear in the problem, using a weighted directed graph of functional dependency. Audemard and Henocque experimented with XLNH in SEM and achieved very good results [3].

3.3   Selecting Cells

The *splitting rule* is implicitly given in the procedure `backtrack_search` of Fig. 1. It tries to extend a partial model by examining a selected unassigned cell. If assigning any one value to the cell leads to a complete model, the procedure terminates. The splitting rule "splits" on a possible values (PV) clause, and transforms the search problem into a number of easier subproblems.

The LNH technique forces a model generator to choose certain cells and values in the `backtrack_search` procedure. We first describe SEM's default rule for choosing cells. For cell selection, the rationale is to keep the value of *mdn* as small as possible. For simplicity, suppose there is only one binary function f and one unary function g in the input file. Then SEM will examine the following cells:

```
f(0,0), g(0),
f(0,1), f(1,0), f(1,1), g(1),
...
```

In each line, if there are several cells whose values are unknown, we select the cell whose domain of values is the smallest.

In Mace4, the cell selection is divided into two stages. The first stage, controlled by the parameter `selection_order`, determines a set of candidate cells. There are three values for `selection_order`: 0, 1, or 2 (default). If the value of `selection_order` is 0, then all unassigned cells are candidates. Let us say the *major* of a cell term ce is $max\{k_i \mid 1 \le i \le n\}$ if ce is $f(k_1, k_2, ..., k_n)$. If the value of `selection_order` is 1, and the major of the first unassigned cell is m, then all unassigned cells whose major is less than or equal to m are candidates; if the value is 2, then all unassigned cells whose major is less than or equal to the major of the current maximum constrained unassigned cell are candidates. The first method defeats the purpose of LNH and the last two methods use the idea of LNH and keep maximum constrained value as low as possible.

The second stage is controlled by the parameter `selection_measure`, which takes a value from 0 to 4 (default). If the value is 0, the first candidate is selected; if the value is 1, select the candidate with the greatest number of occurrences in the current set of ground clauses; if the value is 2, select the candidate that would cause the greatest amount of propagation; if the value is 3, select the candidate that would cause the greatest number of contradictions; if the value is 4, select the candidate with the smallest number of possible values, i.e., select (ce, D) with the minimal $|D|$. When the value of `selection_measure` is 2 or 3, the corresponding method is an expensive lookahead operation as for each cell being considered, all assignments and subsequent propagation are tried (and undone), and the statistics are then collected.

The impact of `selection_order` and `selection_measure` on the performance of Mace4 is huge for many examples. For example, to search for a noncommutative group of size 48, we may use the following input:

```
assign(domain_size, 48).
```

```
formulas(theory).
  e * x = x.
  x' * x = e.
  (x * y) * z = x * (y * z).
  A * B != B * A.
end_of_list.
```

Among 15 pairs of different selections for `selection_order` and `selection_measure`, only four pairs will allow Mace4 to find a model in less than one second. These four pairs are (`selection_order`, `selection_measure`) $= (1, 0), (2, 0), (1, 4), (2, 4)$; each of the other 11 pairs took Mace4 at least 10 minutes before we gave up trying.

As mentioned above, SEM's default selection strategy is to pick (ce, e) where $max(major(ce), e)$ is minimal. This strategy is different from any of Mace4's 15 strategies controlled by the parameters `selection_order` and `selection_measure`. As a result, SEM and Mace4 do not have exactly the same performance for most problems.

4   Constraint Propagation

It is important to obtain useful information (e.g., contradiction or new assignments) when a certain number of cells are assigned values and this process is called *constraint propagation*. We have to address the following issue: How can we implement constraint propagation efficiently?

This issue may be trivial for some constraint satisfaction algorithms because the constraints they accept are often assumed to be unary or binary (i.e., with one or two variables). It is true that n-ary constraints can be converted into an equivalent set of binary constraints; but this conversion usually entails the introduction of new variables and new domains, and hence an increase in problem size. This issue is particularly important to model generation because in this case, the constraints are represented by complex formulas. Thus we have to design special constraint propagation rules which should be very helpful to increase the efficiency of finite model searching. By experimenting with SEM, we identify a set of constraint propagation rules that are both efficient and easy to implement. In this section, we present the rules used in Mace4 and SEM.

As mentioned earlier, our goal is to derive a model (which can be represented by a set of VA clauses) such that all the clauses in Ψ are true. Essentially this can be achieved through a sequence of transformations defined by a set of rules [60], including the splitting rule. In addition, we need the following two types of rules:

- **Simplification Rules:** A current clause is simplified or removed from the clause set Ψ. Tautology deletion and subsumption are such rules.
- **Inference Rules:** A new clause is generated and added into the clause set Ψ without changing or removing old clauses. For instance, resolution and paramodulation are two well-known inference rules.

Roughly speaking, simplification rules are inference rules plus deletion (of original references). In the rest of this section, we describe these rules in detail. We assume that all involved clauses are ground.

4.1   Simplification Rules

Most simplification rules in first order theorem proving, such as tautology deletion, subsumption, and rewriting, can be used to simplify constraints. Let us start with the simplest ones.

Equality Resolution:

$$\textbf{(ER1)} \quad \frac{C \cup \{t \neq t \vee M\}}{C \cup \{M\}} \qquad \textbf{(ER2)} \quad \frac{C \cup \{e = e' \vee M\}}{C \cup \{M\}} \quad (e \neq e')$$

Equality Subsumption:

$$\textbf{(ES1)} \quad \frac{C \cup \{t = t \vee M\}}{C} \qquad \textbf{(ES2)} \quad \frac{C \cup \{e \neq e' \vee M\}}{C} \quad (e \neq e')$$

The above rules are sound, because we have, for all x, $x = x$; and $e \neq e'$ is assumed to be true for any two distinct elements e and e' in the domain.

Next we consider unit resolution, which is a well-known inference rule in theorem proving. In the ground case, it can be used as a simplification rule. When an empty clause is produced by either equality resolution or unit resolution, the entire set of clauses is unsatisfiable.

Unit Resolution:

$$\textbf{(UR1)} \quad \frac{C \cup \{t_1 \neq t_2 \vee M, \ t_1 = t_2\}}{C \cup \{M, \ t_1 = t_2\}}$$

$$\textbf{(UR2)} \quad \frac{C \cup \{t_1 = t_2 \vee M, \ t_1 \neq t_2\}}{C \cup \{M, \ t_1 \neq t_2\}}$$

Both **UR1** and **UR2** for ground clauses are more powerful than the Boolean constraint propagation (BCP for short) for propositional clauses[1], in the sense that more information can be derived during propagation.

Example 4.1 From $(f(1) = 0) \vee (f(1) \neq g(2))$ and $(f(1) = g(2))$, **UR1** can deduce $(f(1) = 0)$. However, $(f(1) = g(2))$ will not be represented by a unit propositional clause in the conventional conversion, which will convert $(f(1) = g(2))$ into a set of binary propositional clauses such as $(f(1) \neq a) \vee (g(2) = a)$ and $(f(1) = a) \vee (g(2) \neq a)$, where a is an element in the domain, and these binary clauses cannot be used in BCP. Thus, BCP cannot deduce new assignments from the propositional representation of the above two clauses. Because it is expensive

---

[1] BCP is the process of repeatedly removing false literals from propositional clauses and making literals in unit clauses to be true.

to perform full subsumption in first-order reasoning, only unit subsumption is considered in SEM.

Unit Subsumption:

$$\textbf{(US1)} \quad \frac{C \cup \{t_1 = t_2 \vee M, \quad t_1 = t_2\}}{C \cup \{t_1 = t_2\}}$$

$$\textbf{(US2)} \quad \frac{C \cup \{t_1 \neq t_2 \vee M, \quad t_1 \neq t_2\}}{C \cup \{t_1 \neq t_2\}}$$

Rewriting is a powerful simplification rule for theorem proving. It can also be used in constraint propagation.

Rewriting:

$$\frac{C \cup \{L[t_1] \vee M, \quad t_1 = t_2\}}{C \cup \{L[t_2] \vee M, \quad t_1 = t_2\}}$$

Using this rule, **UR1** becomes redundant if **ER1** is used. The effect of this rule is to replace t_1 in the clause $L[t_1] \vee M$ by t_2. In our experimentation, we require that t_2 be an element and t_1 be a non-element term, so that the termination of rewriting is guaranteed. In fact, it is often simple and efficient if we require that t_1 be a cell term, i.e., $t_1 = t_2$ is a VA clause.

After a simplification rule is applied, the size of the clause set often becomes smaller. Some of the clauses may be reduced to VA or VE clauses, so that further simplification can be performed. If an empty clause is produced by the simplification process, that result is returned immediately.

4.2   Inference Rules

For the completeness of the algorithm `backtrack_search`, we do not need any new inference rules as the rule of splitting is implicitly given in `backtrack_search`. However, additional inference rules may help us to avoid some fruitless search space and speed up the search. The first inference rule we consider here involves the equality predicate and it is called *dismodulation*.

Dismodulation:

$$\frac{C \cup \{f(t_1, ..., t_m) = t_0 \vee M_1, \quad f(s_1, ..., s_m) \neq s_0 \vee M_2\}}{\begin{array}{c} C \cup \{f(t_1, ..., t_m) = t_0 \vee M_1, \quad f(s_1, ..., s_m) \neq s_0 \vee M_2, \\ \bigvee_{i=0}^{m}(t_i \neq s_i) \vee M_1 \vee M_2\} \end{array}}$$

The soundness of this rule can be easily established. Suppose the derived clause, $\bigvee_{i=0}^{m}(t_i \neq s_i) \vee M_1 \vee M_2$, is false in an interpretation, then both M_1 and M_2 are false, and for each i ($0 \leq i \leq m$), $t_i = s_i$ holds. So one of the parent clauses must be false in the interpretation, since both equalities $f(t_1, ..., t_m) = f(s_1, ..., s_m)$ and $t_0 = s_0$ are true in the interpretation.

The above rule is quite similar to negative paramodulation [38][2] in that new information is derived from inequalities. The dismodulation rule in its general form has a similar problem as the paramodulation rule, i.e., it may generate too many new clauses. From our experiments, we find that it is often beneficial if the above rule produces only VE clauses (of the form $ce \neq e$). In this case, M_1 and M_2 are empty, one of the two terms, $f(t_1, ..., t_m)$ and $f(s_1, ..., s_m)$, is a cell term, and the other term is like a cell term, except one subterm. Formally, the next two rules are an instance of the dismodulation rule.

VE Generation:

$$\textbf{(VEG1)} \quad \frac{C \cup \{f(e_1, .., e_i, .., e_m) = e_0, \quad f(e_1, .., ce, ..., e_m) \neq e_0\}}{C \cup \{f(e_1, .., e_i, .., e_m) = e_0, \quad f(e_1, .., ce, ..., e_m) \neq e_0, \quad ce \neq e_i\}}$$

$$\textbf{(VEG2)} \quad \frac{C \cup \{f(e_1, .., ce, .., e_m) = e_0, \quad f(e_1, .., e_i, ..., e_m) \neq e_0\}}{C \cup \{f(e_1, .., ce, .., e_m) = e_0, \quad f(e_1, .., e_i, ..., e_m) \neq e_0, \quad ce \neq e_i\}}$$

where e_i, $0 \leq i \leq m$, are elements and ce is a cell term.

Note that, to apply the VE generation rules, one parent must be a VA or VE clause, and the complex terms in the two parents differ only in one argument. The resulting new clause is a VE clause. For convenience, we will call the positive unit clause $f(e_1, .., ce, .., e_m) = e_0$ a *near VA clause*, and the negative unit clause $f(e_1, .., e_i, ..., e_m) \neq e_0$ a *near VE clause*.

The above rules of VE generation can be generalized slightly. Let us call a ground term *neat* if either it is a constant, or a cell term or all but one of the top-level function's arguments are elements in the domain and the only non-element argument is also neat. For example, $h(1, g(f(2, 4)), 3)$ is neat, while $h(1, f(2, 4), g(3))$ is not. We may replace the cell term ce in the above rules by a neat term t. The deduced inequality $t \neq e_i$ can participate in further inferences, until a VE clause is produced. The new rule can be decomposed into a sequence of VE generation steps, just as hyperresolution can be decomposed into a sequence of resolution steps. For example, suppose we have the following three clauses:

$$h(1, g(f(2, 4)), 3) \neq 5$$
$$h(1, 2, 3) = 5$$
$$g(1) = 2$$

Then we can conclude that $f(2, 4) \neq 1$ by applying **(VEG1)** twice: from the first two clauses, derive $g(f(2, 4)) \neq 2$; and then from this new inequality and the third clause, derive the VE clause $f(2, 4) \neq 1$.

The VE generation rules are quite useful for some problems having complex terms. One example is the QG5 problem [21,42], which has the following clause: $f(f(f(y, x), y), y) = x$.

The soundness and completeness of the simplification and inference rules can be easily established. Specifically, let $C \Rightarrow C'$ denote that C' is derived from C by either of equality resolution, unit resolution, rewriting by VA clause, VE generation or splitting, and let $\Rightarrow^*$ be the reflexive and transitive closure of $\Rightarrow$, then we have the following theorem.

---

[2] Given two ground clauses $A \vee (a \neq b)$ and $P(a) \vee B$, where A and B are lists of literals, $P(a)$ is a literal containing the term a, negative paramodulation will infer the clause $A \vee \neg P(b) \vee B$.

Theorem 4.1 The set C of ground clauses has a model over D_n iff $C \Rightarrow^* C'$, where C' is satisfiable and contains a VA clause $ce = e$ for every cell ce.

The proof of this theorem is straightfoward following the completeness of the DPLL algorithm for propositional logic as every inference in the DPLL algorithm can be obtained by our inference rules. Note that the satisfiability of C' is very easy to check when it contains a VA clause $ce = e$ for every cell ce.

In Mace4, there are four flags that a user can enable or disable (the default is true for all four flags) to generate VE clauses. When a new VA or VE clause is generated, these flags are checked; if true, the corresponding rules are applied. That is, new VA or VE clauses serve as triggers of these VE generation rules. These four flags are:

- **neg_assign**: If true and a new VA clause is generated, VEG1 applies.
- **neg_elim**: If true and a new VE clause is generated, VEG2 applies.
- **neg_elim_near**: If true and a new near VE clause is generated, VEG1 applies.
- **neg_assign_near**: If true and a new near VA clause is generated, VEG2 applies.

Good data structures are crucial to the performance of many programs. To facilitate the reasoning and search as described in the previous section, relatively sophisticated data structures are used in both SEM and Mace4. These data structures support all the constraint propagation rules as well as backtracking.

When a cell term or a neat term is assigned a value, we need to find which terms in the ground clauses Ψ can be simplified. In SEM, we use some *occurrence lists* to keep track of those terms. For example, the cell $f(0, 1)$ occurs in the following clauses:

$$g(f(0, 1)) = 2. \quad f(f(0, 1), 2) = f(0, f(1, 2)).$$

Such an occurrence list changes dynamically during the search. For example, suppose $f(0, 1)$ is assigned the value 3. Then the cell $f(3, 2)$ will occur in the second clause, although initially it does not. During backtracking, different kinds of information have to be restored, such as the values of cells and the occurrence lists.

In Mace4, in order to quickly locate neat assignments and neat eliminations as well as ordinary assignments and eliminations, Bill used a *complete discrimination tree* [32], that is a discrimination tree in which all possible branches are constructed at the start of the search, and only the leaves are updated for insertions and deletions. For Mace4, this complete discrimination tree is not very big, because the depth of neat terms is limited to 2 (that is, only near VA clauses or near VE clauses are considered). Insertions into the discrimination tree are recorded on the stack so that they are undone on backtracking.

5   Empirical Evaluation

As mentioned in the introduction, satisfiability is an important property of logical formulas. A formula is satisfiable or consistent if we can construct a model. Besides showing the consistency of formulas, a model generation tool can be used in various ways.

5.1   Algebra and Logic

In the introduction, we mentioned that a model can serve as a counterexample showing the non-theoremhood of a formula. Thus a model searcher may be used to identify a false conjecture. In particular, such a tool may be used to check a set of axioms, showing that one axiom is independent of the other axioms. For example, in the mid-1980s, Allen and Hayes proposed axiomatizing temporal intervals by using a set of five axioms. Galton [22] later developed a three-element structure that satisfies the first three axioms but falsifies the fifth. With SEM, we can find a three-element counterexample [55]. The following is another example of such an application.

Example 5.1 A *Skew lattice* is a non-commutative generalization of a lattice which has two binary operators: $\vee$ and $\wedge$. In [30], Jonathan Leech asked if the following two middle distributive identities are independent:

$$x \wedge (y \vee z) \wedge x = (x \wedge y \wedge x) \vee (x \wedge z \wedge x) \tag{1}$$
$$x \vee (y \wedge z) \vee x = (x \vee y \vee x) \wedge (x \vee z \vee x) \tag{2}$$

Spinks showed that for the Skew lattice identities, neither (1) nor (2) implies the other [44]. This was proved by finding some 9-element models using SEM.

To show that (1) does not imply (2), let **f** denote $\wedge$ and **g** denote $\vee$, we give the following input to SEM:

```
9.
f(x,f(y,z)) = f(f(x,y),z).
g(x,g(y,z)) = g(g(x,y),z).
f(x,x) = x.
g(x,x) = x.
f(x,g(x,y)) = x.
g(x,f(x,y)) = x.
f(g(y,x),x) = x.
g(f(y,x),x) = x.
f(f(x,g(y,z)),x) = g(f(f(x,y),x),f(f(x,z),x)).
g(g(a,f(b,c)),a) != f(g(g(a,b),a),g(g(a,c),a)).
```

A model can be found instantly on a personal computer.

The generation of counter-examples is not always so easy.

Example 5.2 Tarski's *High School Problem*: Can the following set of identities form a basis for all the identities which hold for the natural numbers?

$$
\begin{aligned}
x + y &= y + x \\
x + (y + z) &= (x + y) + z \\
x * 1 &= x \\
x * y &= y * x \\
x * (y * z) &= (x * y) * z \\
x * (y + z) &= (x * y) + (x * z) \\
1^x &= 1 \\
x^1 &= x \\
x^{y+z} &= x^y * x^z \\
(x * y)^z &= x^z * y^z \\
(x^y)^z &= x^{y*z}
\end{aligned}
$$

Wilkie gave an identity which holds for the natural numbers, but cannot be derived from the above identities through equational reasoning:

$$(P^x + Q^x)^y * (R^y + S^y)^x = (P^y + Q^y)^x * (R^x + S^x)^y$$

where $P = 1 + x$, $Q = 1 + x + x * x$, $R = 1 + x * x * x$, and $S = 1 + x * x + x * x * x * x$. An interesting question arises: What is the smallest counter-example for Wilkie's identity? Using SEM, we obtained some empirical results, which show that the smallest counter-example has at least 11 elements [58]. But the problem is far from solved. It is still very challenging to find a 12-element counter-example, although it does exist [14].

A model searcher can be useful even when there is no conjecture. In fact, it can be used to formulate conjectures [56]. If a formula is satisfiable, we can learn something about the formula by examining its models. For instance, Kunen [29] obtained some structure theorems for conjugacy closed loops (CC-loops). During that study, he used SEM as well as the theorem prover Otter [34]. The following is quoted from [29]:

> SEM is used to construct finite examples. ... Once one has such an example, it is usually possible to describe it in a more conceptual way ... We originally tried to use SEM to construct a non-group CC-loop of order 15, but this failed, proving that there was no such loop. We then found the proof in this article ... Otter and SEM very useful for quick experimentation and for checking out (often false) conjectures.

From the above, we see that a finite model searcher is complementary to a refutation-based theorem prover. These two classes of tools can be combined to perform more powerful reasoning than a single tool can. In particular, they may form a kind of "decision procedures" for certain logics like propositional modal logics [54].

5.2   Quasigroup Problems

In Example 2.1, we showed how to specify a Latin square. From the algebraic point of view, a Latin square indexed by S defines a cancellative groupoid $(S, *)$, where $*$ is a binary operation on S and the Latin square is the "multiplication table" of $*$. Such a groupoid is called *quasigroup*. For many problems in combinatorics, people are interested in quasigroups with additional constraints. For instance, the following constraints are presented in [21,42,49,50].

Code Name	Identity
QG3	$(x * y) * (y * x) = x$
QG4	$(y * x) * (x * y) = x$
QG5	$((y * x) * y) * y = x$

In [21,42], some previously open problems regarding the existence of quasigroups were solved for the first time, such as QG3.12 and QG4.12, a quasigroup of size 12 satisfying the constraints QG3 or QG4. Mace2 and SEM have been used to solve some previously open quasigroup problems. In particular, Mace2 found for the first time some incomplete Latin squares.

Example 5.3 In an incomplete Latin square, a subsquare is missing. The existence of these incomplete Latin squares is very useful for the construction of larger Latin squares. For instance, an incomplete Latin square with a hole of size two can be specified by the following SEM input file:

```
( elem [10] )
{ h : elem elem -> BOOL }
{ f : elem elem -> elem }
```

```
< x, y, z : elem >
[ h(0,0) ]
[ h(0,1) ]
[ h(1,0) ]
[ h(1,1) ]
[ -h(x,y) | EQ(x,0) | EQ(x,1) ]
[ -h(y,x) | EQ(x,0) | EQ(x,1) ]
[ h(x,z) | h(y,z) | -EQ(f(x,z),f(y,z)) | EQ(x,y) ]
[ h(x,y) | h(x,z) | -EQ(f(x,y),f(x,z)) | EQ(y,z) ]
[ EQ(f(x,x), x) ]
[ h(x,y) | EQ(f(f(y,x),f(x,y)), x) ]
```

where the meaning of $h(x,y)$ is that $h(x,y)$ is true if and only if the cell (x,y) is in the hole.

The problem for the above input is called IQG4.10.2, meaning a Latin square of size 10 with a hole of size two satisfying QG4. Mace2 was the first to find a model of IQG4.14.2. The only remaining open case for all IQG4.$n.k$ is when $(n, k) = (19, 2)$ [62].

5.3   Experimental Results

Now we compare the performance of Mace4 and SEM (all with the default parameters – unless specified otherwise). The results are given in Table 1. The problem name ncg.12 stands for the non-commutative group problem of size 12. That is, the problem asks to find a non-commutative group of size 12.

Table 1. Performance Comparison

Problem	Sat	SEM	Mace4
skewl.9	Yes	0	0.00
ncg.12	Yes	0	0.02
ncg.13	No	1	1.11
ncg.48	Yes	?	0.49
QG3.9	No	?	1.88
QG3.10	No	?	297
QG4.9	Yes	0	0.03
QG4.10	No	128	259.20
IQG4.10.2	Yes	574	9.65
IQG4.11.2	Yes	> 3600	35.01
QG5.10	No	0	0.00
QG5.11	Yes	35	0.72
QG5.12	No	> 600	207.52

It can be seen that Mace4 is usually faster than SEM. But the performance of Mace4 also depends on the settings. For example, if we do not use the negative propagation rule (by specifying clear(negprop)), Mace4 will be much slower on the QG4 problem. For QG4.10, Mace4 does not terminate within 20 minutes.

There are two alternative ways of specifying a Latin square $(Q, *)$. One way is to use the single operator $*$ satisfying

$$((x * z) \neq (y * z)) \vee (x = y)$$
$$((x * y) \neq (x * z)) \vee (y = z)$$

The other way is to use three operators, $*$, $\backslash$, and $/$, satisfying

$$x * (x \backslash y) = y$$
$$x \backslash (x * y) = y$$
$$(x/y) * y = x$$
$$(x * y)/y = x$$

This is because for any Latin square $(Q, *)$, for any $a, b \in Q$, there exists a unique solution for the equation $a * x = b$. The value of $(a \backslash b)$ denotes this unique solution. Similarly, (a/b) denotes the unique solution for the equation $y * a = b$. Once the value of $*$ is decided, the values of $\backslash$ and $/$ can be decided as well.

The experiment shows that the later specification works better for model generators like Mace4 and SEM. For SAT solvers, the first specification is usually used to generate propositional clauses, as the second specification will use two times more propositional variables.

```
% quasigroup axioms (equational)
op(400, infix, [*,\,/]).

clauses(theory).
  x * (x \ y) = y.
  x \ (x * y) = y.
  (x / y) * y = x.
  (x * y) / y = x.
  (x * x) = x.
end_of_list.
```

For instance, to search for IQG4.10.2, Mace4 uses only one tenth of the time of the original formulation. For unsatisfiable problems, the computing times are roughly the same for both formulations. This example suggests that we may try to use different formulations to increase the chance of finding a model.

6   A Challenging Latin Square Problem

In this section, we report a success of solving a long-standing open Latin square problem using the finite model generators SEM and Mace4. Using these tools, we found for the first time a Latin square of size eleven with a missing hole of size three, which is orthogonal to its $(3, 2, 1)$-conjugate and both its main and back diagonals are distinct symbols. With the generation of this Latin square, we completely settled the existence problem for such Latin squares for all sizes.

6.1   Basic Concepts

A *transversal* in a Latin square is a set of positions, one per row and one per column, among which the symbols occur precisely once each. A *diagonal Latin square* is a Latin square whose main and back diagonals are transversals. Two Latin squares $(Q, *)$ and $(Q, \otimes)$ are *orthogonal* if for any $x, y \in Q$, there exists unique $a, b \in Q$ such that $a * b = x$ and $a \otimes b = y$. That is, each symbol x in the first square meets each symbol y in the second square exactly once when the two Latin squares are superposed.

If $(Q, \otimes)$ is a Latin square, we may define on the set Q six binary operations $\otimes_{(1,2,3)}, \otimes_{(1,3,2)}, \otimes_{(2,1,3)}, \otimes_{(2,3,1)}, \otimes_{(3,1,2)}$, and $\otimes_{(3,2,1)}$ as follows: $x \otimes y = z$ if and only if

$$x \otimes_{(1,2,3)} y = z, \quad x \otimes_{(1,3,2)} z = y, \quad y \otimes_{(2,1,3)} x = z,$$
$$y \otimes_{(2,3,1)} z = x, \quad z \otimes_{(3,1,2)} x = y, \quad z \otimes_{(3,2,1)} y = x.$$

These six (not necessarily distinct) quasigroups $(Q, \otimes_{(i,j,k)})$ are called the *conjugates* of $(Q, \otimes)$ ([45,10]).

Example 6.1 Here are the six conjugates of a small Latin square:

(a)	(b)	(c)	(d)	(e)	(f)
1 4 2 3	1 2 4 3	1 3 2 4	1 2 4 3	1 3 4 2	1 3 2 4
2 3 1 4	4 3 1 2	2 4 1 3	3 4 2 1	3 1 2 4	3 1 4 2
4 1 3 2	2 1 3 4	4 2 3 1	2 1 3 4	2 4 3 1	4 2 3 1
3 2 4 1	3 4 2 1	3 1 4 2	4 3 1 2	4 2 1 3	2 4 1 3

(a) a Latin square; (b) its $(2, 1, 3)$-conjugate; (c) its $(3, 2, 1)$-conjugate; (d) its $(2, 3, 1)$-conjugate; (e) its $(1, 3, 2)$-conjugate; (f) its $(3, 1, 2)$-conjugate. Here the domain of elements is $\{1, 2, 3, 4\}$.

For more information on Latin squares and quasigroups, the interested reader may refer to the book of Dénes and Keedwell[18].

For convenience, we denote by (i, j, k)-CODLS(v) a Latin square A of order v, such that A is orthogonal to B, which is the (i, j, k)-conjugate of A and both A and B are diagonal, where $\{i, j, k\} = \{1, 2, 3\}$.

Let $H \subset Q$, $n = |H|$, $v = |Q|$, and $(Q, \otimes)$ be a Latin square with the symbols indexed by H missing from $(Q, \otimes)$ and the missing symbols consist of a missing subsquare right in the center of $(Q, \otimes)$. If we fill the missing square with an (i, j, k)-CODLS(n) over the subset H and the result is an (i, j, k)-CODLS(v) over Q, we say $(Q, \otimes)$ is an *incomplete (i, j, k)-conjugate orthogonal diagonal Latin square of order v with a missing subsquare of size n*, denoted by (i, j, k)-ICODLS(v, n). The subset H as well as the missing subsquare is called the *hole* of $(Q, \otimes)$.

Example 6.2 Here (a) is a $(3, 2, 1)$-ICODLS$(5, 1)$, where $Q = \{0, 1, 2, 3, 4\}$, $H = \{2\}$; if we fill 2 into the hole $\{2\}$, we obtain a $(3, 2, 1)$-CODLS(5). (b) is the $(3, 2, 1)$-conjugate of (a), which is also diagonal. (c) is a $(3, 2, 1)$-ICODLS$(8, 2)$; (d) is the $(3, 2, 1)$-conjugate of (c).

	(c)	(d)
(a)	(b)		

$\otimes$	0 1 2 3 4	$\otimes_{321}$	0 1 2 3 4	0 7 5 6 2 3 4 1	0 3 4 5 1 2 7 6
0	4 2 0 3 1	0	1 3 0 2 4	7 1 6 5 0 4 2 3	4 1 3 7 2 6 5 0
1	0 3 1 4 2	1	2 4 1 3 0	5 6 2 7 1 0 3 4	3 4 2 6 0 7 1 5
2	1 4 _ 0 3	2	3 0 _ 4 1	2 0 1 _ _ 6 7 5	6 5 7 _ _ 0 2 1
3	2 0 3 1 4	3	4 1 3 0 2	1 2 0 _ _ 7 5 6	5 7 6 _ _ 1 0 2
4	3 1 4 2 0	4	0 2 4 1 3	4 3 7 0 6 5 1 2	2 6 0 1 7 5 4 3
				3 5 4 2 7 1 6 0	7 2 1 0 5 3 6 4
				6 4 3 1 5 2 0 7	1 0 5 2 6 4 3 7

It is easy to see that the existence of an (i, j, k)-CODLS(v, n) requires that $v - n$ be even because the missing subsquare must be in the center. The following result is known.

Theorem 6.3 (a) *A $(3, 2, 1)$-CODLS(v) exists for all integers $v \geq 1$, except $v \in \{2, 3, 6\}$ and except possibly $v = 10$.*

(b) For any integer $1 \leq n \leq 6$, a $(3, 2, 1)$-ICODLS(v, n) exists if and only if $v \geq 3n + 2$ and $v - n$ is even, with the possible exception of $(v, n) = (11, 3)$.

(c) For any integer $n \geq 1$, a $(3, 2, 1)$-ICODLS(v, n) exists if $v \geq 13n/4 + 93$ and $v - n$ is even.

We are able to remove the only possible exception of Theorem 6.3(b) by constructing a $(3, 2, 1)$-ICODLS$(11, 3)$ using finite model generators.

6.2 Specification of $(3, 2, 1)$-ICODLS$(11, 3)$

To specify that a Latin square is orthogonal to its $(3, 2, 1)$-conjugate, we may use the constraint $QG1$ as given in [49]:

$$QG1 : (x * y = z * w \wedge u * y = x \wedge u * w = z) \Rightarrow (x = z \wedge y = w)$$

For model generators like SEM and Mace4, the preprocessing process before the search is to obtain ground instances of the first-order formulas. Since the number of ground instances grows exponentially in the terms of the number of variables, it is better to use the equivalent constraint $QG1a$ where we have three variables instead of five:

$$QG1a : (u * y) * y \neq (u * w) * w \vee y = w$$

For SAT solvers, an alternative way is to introduce a new predication Φ and use the following two clauses containing four variables each [49]:

$$QG1b : x * y = s \wedge t * y = x \Rightarrow \Phi(x, s, t),$$
$$\Phi(x, s, t) \wedge \Phi(z, s, t) \Rightarrow x = z.$$

For the $(3, 2, 1)$-ICODLS$(11, 3)$ problem, this new specification will need 4394 propositional variables instead of 2197, but the number of propositional clauses is reduced from 659,994 to 56,004. It is shown in [49] that $QG1b$ is much better than $QG1$ for SAT solvers.

To specify both a Latin square and its $(3, 2, 1)$-conjugate are diagonal, that is, all the symbols on the main and back diagonals are distinct, we may use the following clauses:

$$(x * x) \neq (y * y) \vee x = y$$
$$(x * (n - 1 - x)) \neq (y * (n - 1 - y)) \vee x = y$$
$$z * x \neq x \vee z * y \neq y \vee x = y$$
$$z * (n - 1 - x) \neq x \vee z * (n - 1 - y) \neq y \vee x = y$$

where n is the size of the Latin square and the domain of x and y is assumed to be $\{0, 1, ..., n - 1\}$.

Finally, to specify the hole of a Latin square, we define a binary predicate $h(x, y)$ such that (x, y) is in the hole if and only if $h(x, y)$ is true. For the $(3, 2, 1)$-ICODLS$(11, 3)$ problem, the hole is the set $\{4, 5, 6\}$ and $h(x, y)$ can be specified by the following clauses:

$$\neg h(x,y) \vee h(y,x)$$
$$h(4,4)$$
$$h(4,5)$$
$$h(4,6)$$
$$h(5,5)$$
$$h(5,6)$$
$$h(6,6)$$
$$\neg h(x,y) \vee x = 4 \vee x = 5 \vee x = 6$$
$$\neg h(y,x) \vee x = 4 \vee x = 5 \vee x = 6$$

For every constraint P involving $x * y$, we replace it by $P \vee h(x,y)$. For instance, the diagonal constraint becomes

$$(x * x) \neq (y * y) \vee x = y \vee h(x,x) \vee h(y,y)$$
$$(x * (n - 1 - x) \neq (y * (n - 1 - y) \vee x = y \vee h(x,(n - 1 - x)) \vee h(y,(n - 1 - y))$$
$$z * x \neq x \vee z * y \neq y \vee x = y \vee h(z,x) \vee h(z,y)$$
$$z * (n - 1 - x) \neq x \vee z * (n - 1 - y) \neq y \vee x = y \vee h(x,(n - 1 - x)) \vee h(y,(n - 1 - y))$$

All the formulas in this section can be easily accepted by today's model generators like SEM and Mace4.

6.3 Construction of $(3,2,1)$-ICODLS$(11,3)$

We used the **arithmetic** option of Mace4 so that the built-in functions for arithmetic operations are computed in the ground instances of the formulas. We add a new built-in function named $\text{SUB}(x,y)$ for $x + (-y)$. Because $*$ and $/$ are also built-in functions, we use @ to replace $*$, f for $\backslash$ and g for $/$. The default value for **selection_order** is 0 when **arithmetic** is set; we overwrite it with the command **assign(selection_order, 2)** for better performance. The following is the input file for Mace4.

```
set(arithmetic).
assign(domain_size, 11).
assign(selection_order, 2).

op(400, infix, [@,g,f]).

formulas(theory).

% define hole
h(4,4).
h(4,5).
h(4,6).
h(5,5).
h(5,6).
h(6,6).
-h(y,x) | h(x,y).
-h(x,y) | x = 4 | x = 5 | x = 6.
-h(y,x) | x = 4 | x = 5 | x = 6.
x @ y != z | -h(x,z).
x @ y != z | -h(y,z).
```

```
% define main and back diagonals
x = y | (x @ x) != (y @ y) | h(x,x) | h(y,y).
x = y | (x @ SUB(10, x)) != (y @ SUB(10, y)) | h(x,SUB(10,x)) |
                              h(y, SUB(10, y)).
x = y | z @ x != x | z @ y != y | h(z,x) | h(z,y).
x = y | z @ SUB(10, x) != x | z @ SUB(10, y) != y | h(x,SUB(10,x)) |
                                    h(y, SUB(10, y)).

% quasigroup axioms (equational)
 x @ (x g y) = y | h(x,y).
 x g (x @ y) = y | h(x,y).
(x f y) @ y = x | h(x,y).
(x @ y) f y = x | h(x,y).

% Orthogonal to its (3,2,1)-conjugate
(u @ y) @ y != (u @ w) @ w | y = w | h(u,y) | h(u,w) | h(w,y).
end_of_list.
```

For many model generation problems, once the specification is complete, our job is done and we let the model generator do the work. However, for challenging problems like $(3, 2, 1)$-ICODLS$(11, 3)$, running Mace4 continuously for over two weeks may not find a solution. One of the reasons that SEM and Mace4 are efficient because they can detect certain symmetry of models and reduce the search space. In SEM as well as in Mace4, the least number heuristic (LNH) is used to reduce the search space based on the exploration of symmetries between domain values which produce isomorphic models. This technique is crucial in the success of solving many model generation problems. Since the values $4, 5, 6$ and 10 appear in the specification, it makes the original LNH technique ineffective. However, the extended LNH technique as presented in [2,17] works perfectly by declaring that $0..3$, $4..6$, and $7..10$ are the sets of equivalent values such that each set has its own least used number (for the set of $7..10$, we keep track of the greatest used number).

With the improved built-in function and least number heuristic, Mace4 could find a model of $(3, 2, 1)$-ICODLS$(11, 3)$ after roughly 480 hours of running. On the other hand, we have run the SAT solvers Minisat [20] and Glucose [4] continuously for two weeks without success. The propositional clauses used all the known advanced techniques [17] (4,394 variables and 56,004 clauses).

Here is a part of the output of Mace4.

```
============================== DOMAIN SIZE 11 ========================

============================== MODEL ================================

interpretation( 11, [number=1, seconds=1154566], [

        function(@(_,_), [
               3, 9, 2, 5, 7, 8, 0,10, 6, 4, 1,
               0, 2, 9, 6, 8, 7, 1, 3, 4,10, 5,
               8, 4, 7, 3,10, 1, 2, 5, 9, 6, 0,
               4, 6, 5,10, 9, 0, 3, 2, 8, 1, 7,
               9, 0,10, 8, 0, 0, 0, 1, 7, 3, 2,
               1,10, 0, 2, 0, 0, 0, 9, 3, 7, 8,
```

```
        2, 7, 1, 9, 0, 0, 0, 0,10, 8, 3,
        5, 3, 4, 0, 1,10, 7, 8, 2, 9, 6,
        6, 5, 3, 7, 0, 9, 8, 4, 1, 2,10,
       10, 8, 6, 1, 2, 3, 9, 7, 5, 0, 4,
        7, 1, 8, 4, 3, 2,10, 6, 0, 5, 9 ]),

    function(f(_,_), [
        1, 4, 5, 7, 8, 3, 0, 6,10, 9, 2,
        5,10, 6, 9, 7, 2, 1, 4, 8, 3, 0,
        6, 1, 0, 5, 9,10, 2, 3, 7, 8, 4,
        0, 7, 8, 2,10, 9, 3, 1, 5, 4, 6,
        3, 2, 7,10, 0, 0, 0, 8, 1, 0, 9,
        7, 8, 3, 0, 0, 0, 0, 2, 9,10, 1,
        8, 3, 9, 1, 0, 0, 0,10, 0, 2, 7,
       10, 6, 2, 8, 0, 1, 7, 9, 4, 5, 3,
        2, 9,10, 4, 1, 0, 8, 7, 3, 6, 5,
        4, 0, 1, 6, 3, 8, 9, 5, 2, 7,10,
        9, 5, 4, 3, 2, 7,10, 0, 6, 1, 8 ]),

    function(g(_,_), [
        6,10, 2, 0, 9, 3, 8, 4, 5, 1, 7,
        0, 6, 1, 7, 8,10, 3, 5, 4, 2, 9,
       10, 5, 6, 3, 1, 7, 9, 2, 0, 8, 4,
        5, 9, 7, 6, 0, 2, 1,10, 8, 4, 3,
        1, 7,10, 9, 0, 0, 0, 8, 3, 0, 2,
        2, 0, 3, 8, 0, 0, 0, 9,10, 7, 1,
        7, 2, 0,10, 0, 0, 0, 1, 9, 3, 8,
        3, 4, 8, 1, 2, 0,10, 6, 7, 9, 5,
        4, 8, 9, 2, 7, 1, 0, 3, 6, 5,10,
        9, 3, 4, 5,10, 8, 2, 7, 1, 6, 0,
        8, 1, 5, 4, 3, 9, 7, 0, 2,10, 6 ]),

    relation(h(_,_), [
        0, 0, 0, 0, 0, 0, 0, 0, 0, 0, 0,
        0, 0, 0, 0, 0, 0, 0, 0, 0, 0, 0,
        0, 0, 0, 0, 0, 0, 0, 0, 0, 0, 0,
        0, 0, 0, 0, 0, 0, 0, 0, 0, 0, 0,
        0, 0, 0, 0, 1, 1, 1, 0, 0, 0, 0,
        0, 0, 0, 0, 1, 1, 1, 0, 0, 0, 0,
        0, 0, 0, 0, 1, 1, 1, 0, 0, 0, 0,
        0, 0, 0, 0, 0, 0, 0, 0, 0, 0, 0,
        0, 0, 0, 0, 0, 0, 0, 0, 0, 0, 0,
        0, 0, 0, 0, 0, 0, 0, 0, 0, 0, 0,
        0, 0, 0, 0, 0, 0, 0, 0, 0, 0, 0 ])
]).
```

=============================== end of model ===========================
The above result was checked by Sato [48,51] to ensure its correctness.

Since we have found a $(3, 2, 1)$-ICODLS$(11, 3)$, we can now state an improvement of Theorem 6.3(b) as follows:

Theorem 6.4 *For any positive integer* $1 \leq n \leq 6$, *a* $(3, 2, 1)$-*ICODLS*(v, n) *exists if and only if* $v \geq 3n + 2$ *and* $v - n$ *is even.*

There are sitll many open problems of Latin squares [50]. For instance, the existence of $(3, 2, 1)$-CODLS(10) in Theorem 6.3(a) is still unknown and we believe that it does not exist as we have searched for it for a long time. Other types of (i, j, k)-ICODLS(v, n) have been studied by several researchers. Du [19] and Bennett, Du and Zhang [9] investigated $(2, 1, 3)$-ICODLS(v, n). It is proved that a $(2, 1, 3)$-CODLS(v) exists if and only if $v \notin \{2, 3, 6\}$. For any positive integers v and n, a $(2, 1, 3)$-ICODLS(v, n) exists if and only if $v \geq 3n+2$ and $v-n$ is even, except possibly for $(v, n) \in \{(20, 6), (26, 8), (32, 10)\}$. We believe that modern model generators can play an important role in solving these open problems. On the other hand, attacking these problems will motivate us to develop new techniques for model generation.

7   Conclusion and Future Works

Finite model generation is an important part of automated reasoning. Yet, the progress on new techniques and tools for finite model generation has been slow, as compared with those for solving the propositional satisfiability problem. This article gives an overview of the model generators Mace4 and SEM. In particular, we have compared the two generators on their common parts and differences, and provided some experimental results.

On one hand, we think that Mace4 and SEM can be regarded as successful, because they are quite efficient on some problems and are used by many researchers. On the other hand, their performance on many problems is not so desirable comparing to the SAT based model generators, as the latter can handle well propositional formulas while the former handles well larger equations.

In this section, we discuss several ways in which a finite model generator can be improved, and mention some related works.

- Congruence closure [39]. When designing Mace4 [35], Bill McCune had considered using the ground *congruence closure* algorithm to increase the propagation of constraints. We conducted an experimental study [61] and the results show that using the congruence closure algorithm can reduce the search space for some benchmark problems, as the congruence closure algorithm can deduce some useful information for finite model searching.
- Isomorphism elimination. Although LNH is already very effective in pruning the search space, there can still be some improvements. For example, it is possible to increase the effect of LNH by selecting cells and values appropriately, as proposed by McCune. An extended version of LNH, called XLNH, is suggested in [3]. It is very effective for solving problems having unary bijective functions. Another technique called DASH is described in [25].
- Cell/value selection heuristics. Gandalf [47] uses more sophisticated selection strategies. At each node of the search tree, it examines all the unassigned cells. For each cell, the tool tries every possible value and checks all the consequences. Then the tool either makes an assignment directly, or makes a choice based on a weighted sum of the numbers of consequences.
- Propagation rules. It is still necessary to design and implement more powerful propagation rules. For example, we need better rules for propagating the effects of negative unit clauses. Some work has been done in this area [7,2].

- Learning. Learning from previous failures during the search has been proven to be very useful in increasing the efficiency of SAT solvers. This technique has been investigated for finite model generators in [41] and [24].
- Data structures. It may be worthwhile to design other data structures for representing clauses and cells.

The above list is certainly not exhaustive. With the emergence of new applications, other issues may arise. We hope that more people will get interested in finite model searching, and more powerful tools will be developed.

Acknowledgments. The authors would like to thank the anonymous reviewers for their detailed comments.

References

1. Audemard, G., Benhamou, B.: Reasoning by Symmetry and Function Ordering in Finite Model Generation. In: Voronkov, A. (ed.) CADE 2002. LNCS (LNAI), vol. 2392, p. 226. Springer, Heidelberg (2002)
2. Audemard, G., Benhamou, B., Henocque, L.: Predicting and Detecting Symmetries in FOL Finite Model Search. Journal of Automated Reasoning 36(3), 177–212 (2006)
3. Audemard, G., Henocque, L.: The eXtended Least Number Heuristic. In: Goré, R.P., Leitsch, A., Nipkow, T. (eds.) IJCAR 2001. LNCS (LNAI), vol. 2083, p. 427. Springer, Heidelberg (2001)
4. Audemard, G., Simon, L.: Predicting learnt clauses quality in modern SAT solver. In: Twenty-First International Joint Conference on Artificial Intelligence, IJCAI 2009 (2009)
5. Baumgartner, P., Fuchs, A., De Nivelle, H., Tinelli, C.: Computing finite models by reduction to function-free clause logic. J. of Applied Logic 7(1), 58–74 (2009)
6. Baumgartner, P., Tinelli, C.: The Model Evolution Calculus. In: Baader, F. (ed.) CADE-19. LNCS (LNAI), vol. 2741, pp. 350–364. Springer, Heidelberg (2003)
7. Benhamou, B., Henocque, L.: A new method for finite model search in equational theories: FMSET system. Fundamenta Informaticae 39(1,2), 21–38 (1999)
8. Bennett, F.E., Du, B., Zhang, H.: Existence of conjugate orthogonal diagonal Latin squares. J. Combin. Designs 5, 449–461 (1997)
9. Bennett, F.E., Du, B., Zhang, H.: Existence of self-orthogonal diagonal Latin squares with a missing subsquare. Discrete Math. 261, 69–86 (2003)
10. Bennett, F.E., Zhu, L.: Conjugate-orthogonal Latin squares and related structures. In: Dinitz, J., Stinson, D. (eds.) Contemporary Design Theory: A Collection of Surveys, pp. 41–96. Wiley, New York (1992)
11. Boy de la Tour, T.: Up-to-Isomorphism Enumeration of Finite Models - The Monadic Case. In: Bonacina, M.P., Furbach, U. (eds.) International Workshop First-Order Theorem Proving (FTP 1997). RISC-Linz Report Series No. 97-50, pp. 29–33. Schloss Hagenberg by Linz, Austria (1997)
12. Boy de la Tour, T.: Some Techniques of Isomorph-Free Search. In: Campbell, J., Roanes-Lozano, E. (eds.) AISC 2000. LNCS (LNAI), vol. 1930, pp. 240–252. Springer, Heidelberg (2001)
13. Bürckert, H.-J., Herold, A., Kapur, D., Siekmann, J.H., Stickel, M., Tepp, M., Zhang, H.: Opening the AC-unification race. J. of Automated Reasoning (4), 465–474 (1988)

14. Burris, S., Yeats, K.: The saga of the high school identities. Algebra Universalis 52, 325–342 (2004)
15. Caferra, R., Leitsch, A., Peltier, N.: Automated Model Building. Applied Logic Series, vol. 31. Kluwer Academic Publisher (2004)
16. Caferra, R., Zabel, N.: Extending Resolution for Model Construction. In: van Eijck, J. (ed.) JELIA 1990. LNCS, vol. 478, pp. 153–169. Springer, Heidelberg (1991)
17. Claessen, K., Sörensson, N.: New techniques that improve Mace-style finite model finding. In: Model Computation – Principles, Algorithms, Applications, CADE-19 Workshop W4, Miami, Florida, USA (2003)
18. Dénes, J., Keedwell, A.D.: Latin squares and their applications. Academic Press, New York (1974)
19. Du, B.: Self-orthogonal diagonal Latin square with missing subsquare. JCMCC 37, 193–203 (2001)
20. Een, N., Svrensson, N.: Minisat: A SAT solver with conflict-clause minimization. In: SAT 2005 (2005) (poster paper)
21. Fujita, M., Slaney, J., Bennett, F.: Automatic generation of some results in finite algebra. In: Proc. Int'l Joint Conf. on Artificial Intelligence (IJCAI 1993), pp. 52–57 (1993)
22. Galton, A.: Note on a lemma of Ladkin. Journal of Logic and Computation 6(1), 1–4 (1996)
23. Hasegawa, R., Koshimura, M., Fujita, H.: MGTP: A Parallel Theorem Prover Based on Lazy Model Generation. In: Kapur, D. (ed.) CADE-11. LNCS, vol. 607, pp. 776–780. Springer, Heidelberg (1992)
24. Huang, Z., Zhang, H., Zhang, J.: Improving first-order model searching by propositional reasoning and lemma learning. In: The Seventh International Conference on Theory and Applications of Satisfiability Testing (SAT 2004), Vancouver, BC, Canada (May 2004)
25. Jia, X., Zhang, J.: A Powerful Technique to Eliminate Isomorphism in Finite Model Search. In: Furbach, U., Shankar, N. (eds.) IJCAR 2006. LNCS (LNAI), vol. 4130, pp. 318–331. Springer, Heidelberg (2006)
26. Kapur, D., Zhang, H.: An Overview of RRL: Rewrite Rule Laboratory. In: Dershowitz, N. (ed.) RTA 1989. LNCS, vol. 355, pp. 513–529. Springer, Heidelberg (1989)
27. Kim, S., Zhang, H.: ModGen: Theorem proving by model generation. In: Proc. of National Conference of American Association on Artificial Intelligence (AAAI 1994), Seattle, WA, pp. 162–167. MIT Press (1994)
28. Kumar, V.: Algorithms for constraint satisfaction problems: A survey. AI Magazine 13(1), 32–44 (1992)
29. Kunen, K.: The structure of conjugacy closed loops. Transactions of the American Mathematical Society 352(6), 2889–2911 (2000)
30. Leech, J.: Skew lattices in rings. Algebra Universalis 26, 48–72 (1989)
31. Manthey, R., Bry, F.: SATCHMO: A Theorem Prover Implemented in Prolog. In: Lusk, E., Overbeek, R. (eds.) CADE 1988. LNCS, vol. 310, pp. 415–434. Springer, Heidelberg (1988)
32. McCune, W.: Experiments with discrimination tree indexing and path indexing for term retrieval. J. of Automated Reasoning 9(2), 147–167 (1992)
33. McCune, W.: MACE 2.0 Reference Manual and Guide. Technical Memorandum No. 249, ANL/MCS-TM-249, Argonne National Lab, Argonne, IL, USA (1994), http://www-unix.mcs.anl.gov/AR/mace2/

34. McCune, W.: Otter 3.3 Reference Manual, Technical Memorandum No. 263, Argonne National Laboratory, Argonne, IL, USA (August 2003), http://www-unix.mcs.anl.gov/AR/otter/otter33.pdf
35. McCune, W.: Mace4 reference manual and guide, Technical Memorandum No. 264, Argonne National Laboratory, Argonne, IL, USA (August 2003), http://www.cs.unm.edu/~mccune/prover9/
36. McCune, W.: Library for Automated Deduction Research (2009), http://www.cs.unm.edu/~mccune/prover9/
37. McCune, W.: Solution of the Robbins problem. J. of Automated Reasoning 19(3), 263–276 (1997)
38. McCune, W., Henschen, L.J.: Experiments with semantic paramodulation. J. of Automated Reasoning 1(3), 231–261 (1985)
39. Nelson, G., Oppen, D.C.: Fast decision procedures based on congruence closure. J. ACM 27(2), 356–364 (1980)
40. Pichler, R.: Algorithms on Atomic Representations of Herbrand Models. In: Dix, J., Fariñas del Cerro, L., Furbach, U. (eds.) JELIA 1998. LNCS (LNAI), vol. 1489, pp. 199–215. Springer, Heidelberg (1998)
41. Slaney, J.: Finder: Finite Domain Enumerator. In: Bundy, A. (ed.) CADE-12. LNCS, vol. 814, pp. 798–801. Springer, Heidelberg (1994)
42. Slaney, J., Fujita, M., Stickel, M.: Automated reasoning and exhaustive search: Quasigroup existence problems. Computers & Math. with Appl. 29(2), 115–132 (1995)
43. Slaney, J., Lusk, E.L., McCune, W.: SCOTT: Semantically Constrained Otter (System Description). In: Bundy, A. (ed.) CADE-12. LNCS, vol. 814, pp. 764–768. Springer, Heidelberg (1994)
44. Spinks, M.: On middle distributivity for Skew lattices. Semigroup Forum 61(3), 341–345 (2000)
45. Stein, S.K.: On the foundations of quasigroups. Trans. Amer. Math. Soc. 85, 228–256 (1957)
46. Tammet, T.: Using Resolution for Deciding Solvable Classes and Building Finite Models. In: Barzdins, J., Bjorner, D. (eds.) Baltic Computer Science. LNCS, vol. 502, pp. 33–64. Springer, Heidelberg (1991)
47. Tammet, T.: Finite model building: improvements and comparisons. In: Model Computation – Principles, Algorithms, Applications, CADE-19 Workshop W4, Miami, Florida, USA (2003)
48. Zhang, H.: Sato: An Efficient Propositional Prover. In: McCune, W. (ed.) CADE-14. LNCS, vol. 1249, pp. 272–275. Springer, Heidelberg (1997)
49. Zhang, H.: Specifying Latin squares in propositional logic. In: Veroff, R. (ed.) Automated Reasoning and Its Applications, Essays in Honor of Larry Wos. MIT Press (1997)
50. Zhang, H.: Combinatorial designs by SAT solvers. In: Biere, A., Heule, M., Van Haaren, H., Walsh, T. (eds.) Handbook of Satisfiability, ch. 17. IOS Press (2009)
51. Zhang, H., Stickel, M.: Implementing the Davis-Putnam method. J. of Automated Reasoning 24, 277–296 (2000)
52. Zhang, J.: Problems on the Generation of Finite Models. In: Bundy, A. (ed.) CADE-12. LNCS, vol. 814, pp. 753–757. Springer, Heidelberg (1994)
53. Zhang, J.: Constructing finite algebras with Falcon. J. of Automated Reasoning 17(1), 1–22 (1996)
54. Zhang, J.: On the relational translation method for propositional modal logics. Technical Report ISCAS-LCS-96-12, Laboratory of Computer Science, Institute of Software, Chinese Academy of Sciences (December 1996)

55. Zhang, J.: Showing the independence of an axiom for temporal intervals by model generation. Association for Automated Reasoning Newsletter, No. 40 (1998), http://www.aarinc.org/Newsletters/040-1998-06.html

56. Zhang, J.: System Description: MCS: Model-Based Conjecture Searching. In: Ganzinger, H. (ed.) CADE-16. LNCS (LNAI), vol. 1632, pp. 393–397. Springer, Heidelberg (1999)

57. Zhang, J.: Test problem and Perl scripts for finite model searching. Association for Automated Reasoning Newsletter, No. 47 (April 2000), http://www.aarinc.org/Newsletters/047-2000-04.html

58. Zhang, J.: Computer Search for Counterexamples to Wilkie's Identity. In: Nieuwenhuis, R. (ed.) CADE 2005. LNCS (LNAI), vol. 3632, pp. 441–451. Springer, Heidelberg (2005)

59. Zhang, J., Zhang, H.: SEM: a system for enumerating models. In: Proc. 14th Int'l Joint Conf. on Artif. Intel. (IJCAI), pp. 298–303 (1995)

60. Zhang, J., Zhang, H.: Constraint Propagation in Model Generation. In: Montanari, U., Rossi, F. (eds.) CP 1995. LNCS, vol. 976, pp. 398–414. Springer, Heidelberg (1995)

61. Zhang, J., Zhang, H.: Extending Finite Model Searching with Congruence Closure Computation. In: Buchberger, B., Campbell, J. (eds.) AISC 2004. LNCS (LNAI), vol. 3249, pp. 94–102. Springer, Heidelberg (2004)

62. Zhang, X.: Incomplete perfect Mendelsohn designs with block size four. Discrete Mathematics 254, 565–597 (2002)

Group Embedding of the Projective Plane PG(2, 3)

Eric Ens and Ranganathan Padmanabhan

University of Manitoba
eric.james.ens@gmail.com,
padman@cc.umanitoba.ca

Abstract. An (n, k) configuration is a set of n points and n lines, with k of the points on each line and k of the lines through each point. Motivated by the group law on cubic curves, here we investigate the problem of representing such abstract configurations by abelian groups in such a way that whenever k points P1, P2,..., Pk are collinear in the given configuration then the sum $P_1 + P_2 +,..., + P_k = 0$ in the group. In this note, we show how Bill McCune's Prover9 can be successfully employed to determine the structure of the potential group in which the PG(2, 3), projective plane of order 3, gets embedded.

Keywords: Projective planes, group embeddings, automated deduction, Prover9.

1   Bill McCune

Thanks to the suggestions of Stanley Burris and David Kelly, the second author (= R.P.) wrote his first e-mail to Bill McCune in 1993 proposing some problems of equational nature to his theorem-prover Otter 2.0. Since then R.P. and Bill have exchanged some 2300 emails, written eleven research articles and published one monograph on automated reasoning in equational logic and geometry. If RP wrote A, meant B but it should have been C, Bill would filter out A, consider only B and would ask RP whether he really meant C and then solve the problem! Once the problem was successfully solved, Bill would redesign the proof the way "RP likes it" (in other words, RP can follow the proof) especially when working with problems on uniquely complemented   lattices. Bill always liked Indian food, especially whenever we went together for lunch or dinner. Was it because of his epicurean taste or was it because of his desire to accommodate RP's food habits is an open question. In 1996, RP took his Apple iBook for the first time to Argonne and showed it to Bill. Apart from some Otter proofs, the iBook contained many graphics (mostly cubic curves, geometric configs and some math in art). Then Bill commented: "Scientists use Unix computers; accountants use the PC's but only artists love the Mac". And to RP's pleasant surprise,   within a couple of years, Bill became an artist himself - yes, he had a Mac laptop with GUI version of Otter 4.0. Actually the artist in him really flourished - for example, there will be no Prover10 because the "9" in Prover9 was not the "version number 9" bur rather the mirror reflection of the first letter P so that the word "Prover9" would look symmetric! (also, the "o" and "e" are made symmetric as well). Bill continues to live on in our computer science world and Otter, Prover9 and MACE

M.P. Bonacina and M.E. Stickel (Eds.): McCune Festschrift, LNAI 7788, pp. 131–138, 2013.
© Springer-Verlag Berlin Heidelberg 2013

will continue to inspire to create new theorems, new proofs and new counter-examples.

2   Finite Projective Planes as Groups

Let us recall some basic definitions from the classical projective geometry (for example, see [1], [2]). A projective plane is a geometric structure that extends the concept of the Euclidean plane. In the ordinary Euclidean plane, two lines meet in a single point, but there are some pairs of lines (namely, parallel lines) that do not intersect. A projective plane can be thought of as an extension of the ordinary plane equipped by adding "points at infinity" where parallel lines do meet. Thus any two lines in a projective plane have exactly one common point. Dually, any two points determine a unique straight line. More formally, a projective plane consists of a set of points, a set of lines, and a relation between points and lines called incidence, having the following properties:

1.   Given any two distinct points, there is exactly one line incident with both of them.
2.   Given any two distinct lines, there is exactly one point incident with both of them.
3.   There are four points such that no three of them are collinear.

The above definition of a projective plane is purely combinatorial, but all known constructions of finite projective planes are based on algebraic techniques. In fact, the most useful construction of finite projective planes comes from finite fields (e.g see Chapter 5 in [2]). For example, PG(2,2), the smallest projective plane has seven points and seven lines as shown in the diagram below. The triples of numbers associated with each point serve a dual purpose:

(i)   they are the homogenous coordinates of the projective plane over GF[2], the 2-element field.
(ii)   The sum of three triples, added in the abelian group Z[2]×Z[2]×Z[2], is zero (mod 2) if and only if the corresponding three points are collinear in the geometry.

In this article, we prove that PG(2,3) (e.g. page 101 in [2]) - the finite projective plane of order 3 - can be embedded in an abelian group in such a way that whenever P, Q, R and S are collinear in the projective plane, then $\phi(P) + \phi(Q) + \phi(R) + \phi(S) = 0$ in the abelian group. The motivation for such an embedding obviously comes from the theory of group laws on cubic curves where three points P, Q and R are collinear if and only if $\phi(P) + \phi(Q) + \phi(R) = 0$ in the underlying group structure. Here, using Singer's cyclic difference representation of finite projective planes [7],   we prove that PG(2,3) can be embedded in an abelian group in such a way that whenever four points P, Q, R and S are collinear in the projective plane, then $\phi(P) + \phi(Q) + \phi(R) + \phi(S) = 0$ in the abelian group. Singer has given a beautiful representation for all finite Desarguesian   projective planes.

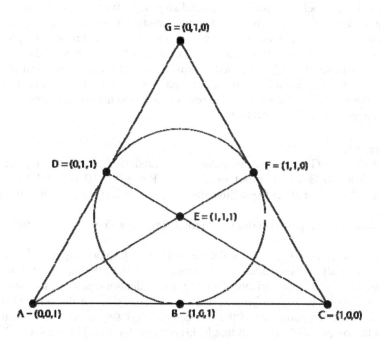

PG(2,2), the Fano Plane

For our present example, the geometry PG(2,3) has $3^2+3+1 = 13$ points and 13 lines. Label the points as $\{0,1,2,3,4,5,6,7,8,9,10,11,12\}$. Then Singer's theorem says that the lines are given cyclically by the block [0,1,3,9]. In other words, the thirteen lines are

$\{[0,1,3,9], [1,2,4,10], [2,3,5,11], [3,4,6,12], [4,5,7,0], [5,6,8,1], [6,7,9,2], [7,8,10,3], [8,9,11,4], [9,10,12,5], [10,11,0,6], [11,12,1,7], [12,0,2,8]\}$

In view of this cyclic nature of the points,   such planes are known as cyclic planes. We use this built-in arithmetic symmetry to find the group representation. First we input these collinearity relations into Prover9 and derive some "tangential relations" among these points. Using these inter-relations among the thirteen points, we determine the actual group in which the geometry gets embedded.

3   The Procedure for Group Embedding of PG(2,3).

Assuming that the projective plane PG(2,3) has a group embedding, we work backwards to find the group and it is here that Prover9 becomes very useful. Suppose f is the mapping which embeds the geometry into the group. Then the collinearity of, say, the block {1,2,4,10} becomes the group equation   f(1)+f(2)+f(4)+f(10) = e, the group identity. Hence f(10) = -f(1)-f(2)-f(4). In other words, the partial ternary function f(x,y,z) behaves like -x-y-z in abelian groups. With this backdrop, we   make an input file for Prover9 giving the identities satisfied by this partial ternary operation as well as the cyclic relations given by Singer and ask Prover9 to prove some tangential relations like "the connection between f(1,1,1) and f(2,2,2)" etc. To our pleasant surprise, Prover9 succeeded in proving that f(i,i,i) is independent of i viz. f(i, i, i) = f(j, j, j) for all i and j and that this common element is of order 3 in any potential target group. This we record as a Theorem and derive some consequences we need for carrying out the group embedding.

Theorem 1.
If f: PG(2,3) --> G is a group embedding as required, then f(i,i,i) = f(j, j, j) for all i, j.
See step # 56733 of the proof obtained by Prover9 f(0,0,0) = f(1,1,1). Because "adding 1" does not affect the collinearity, we have f(i, i, i) = f(j, j, j) for all i, j.

It is clear that the partial ternary operation f(x,y,z) = -x-y-z satisfies the following equations:
f(x,y,z) = f(y,z,x) = f(x,z,y), medial i.e. f(f(x,y,z), f(u,v,w), f(p,q,r)) is completely symmetric in all the nine variables, and finally satisfies the Steiner law f(x, y, f(x,y,z)) = z and their symmetric variants. To verify these identities, just use the interpretation f(x,y,z) = -x-y-z in abelian groups. and expand both sides of the relevant equations (used in the assumption part of the Prover9 proof). Using these, it is not difficult to give a human proof of the   fact that f(i,i,i) = f(j,j,j) for all i, j. Since -3i = -3j for all i and j, we can   simplify matters and take all the elements to be of order 3. So we need a group with at least 13 elements of order 3.   It is well-known that the corresponding affine plane AG(2,3) is simply a manifestation of the nine points of a non-singular cubic curve in the complex plane and that this is isomorphic to the group Z[3]×Z[3]. Since we have thirteen elements in our geometry, the minimal   group candidate is Z[3]×Z[3]×Z[3] and it does work as shown in the table below.

f: PG(2,3) --> $(Z_3)^3$

0 --> f(0) = (1,0,0)	1 --> f(1) = (0,1,0)	2 --> f(2) = (0,0,1)	3 --> f(3) = (1,2,2)
4 --> f(4) = (2,2,0)	5 --> f(5) = (0,2,2)	6 --> f(6) = (2,1,0)	7 --> f(7) = (0,2,1)
8 --> f(8) = (1,2,1)	9 --> f(9) = (1,0,1)	10 --> f(10) = (1,0,2)	11 --> f(11) = (2,2,1)
12 --> f(12) = (1,1,1)			

Now let us verify all the thirteen collinearities (here the addition is done in Z[3] i.e. mod 3):

[0,1,3,9]	f(0) + f(1) + f(3) + f(9)	= (1,0,0) + (0,1,0) + (1,2,2) + (1,0,1) = (0,0,0) ✓
[1,2,4,10]	f(1) + f(2) + f(4) + f(10)	= (0,1,0) + (0,0,1) + (2,2,0) + (1,0,2) = (0,0,0) ✓
[2,3,5,11]	f(2) + f(3) + f(5) + f(11)	= (0,0,1) + (1,2,2) + (0,2,2) + (2,2,1) = (0,0,0) ✓
[3,4,6,12]	f(3) + f(4) + f(6) + f(12)	= (1,2,2) + (2,2,0) + (2,1,0) + (1,1,1) = (0,0,0) ✓
[4,5,7,0]	f(4) + f(5) + f(7) + f(0)	= (2,2,0) + (0,2,2) + (0,2,1) + (1,0,0) = (0,0,0) ✓
[5,6,8,1]	f(5) + f(6) + f(8) + f(1)	= (0,2,2) + (2,1,0) + (1,2,1) + (0,1,0) = (0,0,0) ✓
[6,7,9,2]	f(6) + f(7) + f(9) + f(2)	= (2,1,0) + (0,2,1) + (1,0,1) + (0,0,1) = (0,0,0) ✓
[7,8,10,3]	f(7) + f(8) + f(10) + f(3)	= (0,2,1) + (1,2,1) + (1,0,2) + (1,2,2) = (0,0,0) ✓
[8,9,11,4]	f(8) + f(9) + f(11) + f(4)	= (1,2,1) + (1,0,1) + (2,2,1) + (2,2,0) = (0,0,0) ✓
[9,10,12,5]	f(9) + f(10) + f(12) + f(5)	= (1,0,1) + (1,0,2) + (1,1,1) + (0,2,2) = (0,0,0) ✓
[10,11,0,6]	f(10) + f(11) + f(0) + f(6)	= (1,0,2) + (2,2,1) + (1,0,0) + (2,1,0) = (0,0,0) ✓
[11,12,1,7]	f(11) + f(12) + f(1) + f(7)	= (2,2,1) + (1,1,1) + (0,1,0) + (0,2,1) = (0,0,0) ✓
[12,0,2,8]	f(12) + f(0) + f(2) + f(8)	= (1,1,1) + (1,0,0) + (0,0,1) + (1,2,1) = (0,0,0) ✓

This completes the proof of the embedding theorem for projective plane PG(2,3). See [5, p. 43] for an Otter proof of a similar group representation for the 10-point Desargues configuration (see also [6] for a pure algebraic proof using ring polynomials over GF[2]).

4   An Example of an Impossibility Proof

Let us conclude this paper with an example of a Prover9 proof demonstrating the existence of a configuration having no group embedding. There are 31 distinct (11,3) configurations, nine have group embeddings and 22 of these have no group embeddings. All negative proofs were obtained by a systematic use of Prover9: each of the defined lines of the configuration, given in the opening clauses and ask for the equality of two points with distinct labels. If we obtain a proof, then it shows that there is no embedding. If, on the contrary, we get only some relations among the various points (called "tangential relations" because of the obvious analogy with cubic curve), then we use these relations and find the actual group representing that particular configuration,

Theorem 2. The configuration(11,3)-4 shown below has no group embedding.

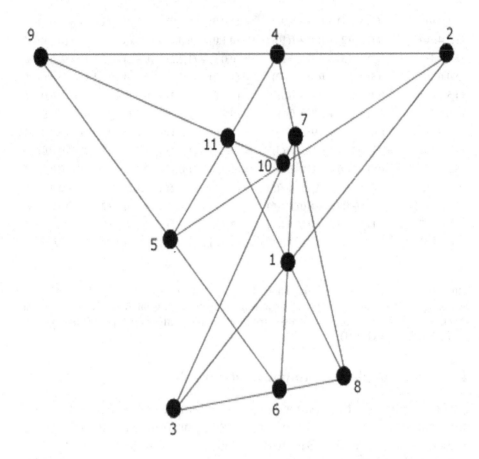

Diagram of Daublebsky's configuration (11, 3)-4 from page 85 in [4]

The 11 given collinearity relations are entered as input clauses in Prover9:

> 1 * 2 = 3. 1 * 6 = 7. 1 * 8 = 11. 2 * 4 = 9. 2 * 5 = 10. 3 * 6 = 8. 3 * 7 = 10.
> 4 * 7 = 8. 4 * 5 = 11. 5 * 6 = 9. 9 * 10 = 11.end_of_list.
> formulas(goals). 5 = 6. end_of_list.

The proof given below shows that the two points "5" and "6" must be mapped onto the same element in any potential group embedding thereby demonstrating the impossibility of a group representation for this particular   planar configuration.

5   Prover9 Steps

Steps for Theorem 1

=============== PROOF =======================

% Proof 1 at 251.04 (+ 3.00) seconds. % Length of proof is 259.
% Level of proof is 18. % Maximum clause weight is 33. % Given clauses 1169.

1 $f(2,2,2) = f(1,1,1)$ # label(non_clause) # label(goal).   [goal].
2 $f(f(x,y,z),u,f(w,v5,v6)) = f(u,f(w,v5,z),f(x,y,v6))$.   [assumption].
3 $f(f(x,y,z),u,f(w,v5,v6)) = f(u,f(x,v5,v6),f(w,y,z))$.   [assumption].
4 $f(x,y,z) = f(x,z,y)$.   [assumption].
5 $f(x,y,f(x,z,f(u,z,w))) = f(u,y,w)$.   [assumption].
6 $f(x,f(x,y,f(z,y,u)),w) = f(z,w,u)$.   [assumption].
7 $f(x,y,f(x,z,f(u,z,w))) = f(w,u,y)$.   [assumption].
8 $f(x,y,z) = f(z,x,y)$.   [copy(7),rewrite([5(3)])].
10 $f(x,y,z) = f(y,x,z)$.   [assumption].
11 $f(x,y,f(x,y,z)) = z$.   [assumption].
12 $f(x,f(x,y,z),z) = y$.   [assumption].
13 $f(x,y,f(x,z,y)) = z$.   [copy(12),rewrite([4(2)])].
14 $f(f(x,y,z),y,z) = x$.   [assumption].
15 $f(x,y,f(z,y,x)) = z$.   [copy(14),rewrite([8(2),4(2)])].
17 $f(0,1,3) = 9$.   [assumption].
18 $f(1,2,4) = 10$.   [assumption].
19 $f(2,3,5) = 11$.   [assumption].
20 $f(3,4,6) = 12$.   [assumption].
21 $f(4,5,7) = 0$.   [assumption].
22 $f(5,6,8) = 1$.   [assumption].
23 $f(6,7,9) = 2$.   [assumption].
24 $f(7,8,10) = 3$.   [assumption].
25 $f(10,7,8) = 3$.   [copy(24),rewrite([4(4),10(4)])].
28 $f(9,10,12) = 5$.   [assumption].
29 $f(10,12,9) = 5$.   [copy(28),rewrite([8(4),4(4),10(4)])].
30 $f(10,11,0) = 6$.   [assumption].
31 $f(0,10,11) = 6$.   [copy(30),rewrite([4(4),10(4)])].
32 $f(11,12,1) = 7$.   [assumption].
33 $f(1,11,12) = 7$.   [copy(32),rewrite([4(4),10(4)])].
34 $f(12,0,2) = 8$.   [assumption].
35 $f(0,12,2) = 8$.   [copy(34),rewrite([10(4)])].
36 $f(1,1,1) \neq f(2,2,2)$.   [deny(1)].

....

56623 **$f(3,3,3) = f(1,1,1)$**.   [para(56622(a,1),11(a,1,3)),flip(a)].
56641 $f(10,10,f(1,1,1)) = 10$.
[para(44595(a,1),13(a,1,3)),rewrite([4(10),3912(10),56623(6)])].
56643 $f(10,10,10) = f(1,1,1)$.   [para(56641(a,1),11(a,1,3))].
56647 $f(2,2,2) = f(12,12,12)$.   [para(56646(a,1),11(a,1,3))].
56648 $f(12,12,12) \neq f(1,1,1)$.   [back_rewrite(37),rewrite([56647(4)])].
56663 $f(10,10,f(5,5,5)) = 10$.   [para(44983(a,1),13(a,1,3)),rewrite([24593(10),4(7)])].
56664 **$f(5,5,5) = f(1,1,1)$**.   [para(56663(a,1),11(a,1,3)),rewrite([56643(4)]),flip(a)].
56732 $f(0,0,f(1,1,1)) = 0$.
[para(45695(a,1),15(a,1,3)),rewrite([10(10),46114(10),5352(10),56664(5),4(7)])].
56733 **$f(1,1,1) = f(0,0,0)$**.   [para(56732(a,1),11(a,1,3)),flip(a)].

56741 f(12,12,12) != f(0,0,0).   [back_rewrite(56648),rewrite([56733(8)])].
56775 **f(12,12,12) = f(0,0,0)**.   [para(56774(a,1),11(a,1,3)),flip(a)].
56776 $F.   [resolve(56775,a,56741,a)].

Steps for Theorem 2

============================== PROOF
==================================

% Proof 1 at 0.37 (+ 0.05) seconds. % Length of proof is 20.
% Level of proof is 5. % Maximum clause weight is 15. % Given clauses 50.

1 5 = 6 # label(non_clause) # label(goal).   [goal].

2 x * y = y * x.   [assumption].

3 x * (y * x) = y.   [assumption].   <---- this simply says that the three points {x, y, x*y} are collinear.

4 (x * y) * (z * u) = (x * z) * (y * u).   [assumption]. <-----medial law for the binary operation "*".

6 1 * 2 = 3.   [assumption].

7 1 * 6 = 7.   [assumption].

9 2 * 4 = 9.   [assumption].

11 3 * 6 = 8.   [assumption].

14 4 * 7 = 8.   [assumption].

15 5 * 6 = 9.   [assumption].

18 5 != 6.   [deny(1)].

19 6 != 5.   [copy(18),flip(a)].

20 x * (x * y) = y.   [para(2(a,1),3(a,1,2))].

42 (1 * x) * (2 * y) = 3 * (x * y).   [para(6(a,1),4(a,1,1)),flip(a)].

77 6 * 9 = 5.   [para(15(a,1),3(a,1,2))].

92 2 * 9 = 4.   [para(9(a,1),20(a,1,2))].

94 3 * 8 = 6.   [para(11(a,1),20(a,1,2))].

992 3 * 5 = 8.   [para(77(a,1),42(a,2,2)),rewrite([7(3),92(4),2(3),14(3)]),flip(a)].

1111 **6 = 5**.   [para(992(a,1),20(a,1,2)),rewrite([94(3)])].

1112 $F.   [resolve(1111,a,19,a)].

============================== end of proof ==========================

References

[1] Albert, A.A., Sandler, R.: An introduction to finite projective planes. Holt, Rinehart and Winston, New York-Toronto, Ont.-London (1968)

[2] Bennett, M.K.: Affine and projective geometry. John Wiley & Sons, Inc., New York (1995)

[3] Ens, E.: Group Embeddings of (n, k)-Configurations, M.Sc Thesis, University of Manitoba (2011)

[4] Grünbaum, B.: Configurations of Points and Lines. Graduate Studies in Mathematics. AMS, Rhode Island (2009)

[5] McCune, W., Padmanabhan, R.: Automated Deduction in Equational Logic and Cubic Curves. LNCS, vol. 1095. Springer, Heidelberg (1996)

[6] Mendelsohn, N.S., Padmanabhan, R., Wolk, B.: Placement of the Desargues configuration on a cubic curve. Geom. Dedicata 40, 165–170 (1991)

[7] Singer, J.: A theorem of finite projective geometry and some applications to number theory. Trans. Amer. Math. Soc. 43, 377–385 (1938)

A Geometric Procedure with Prover9

Ranganathan Padmanabhan[1,*] and Robert Veroff[2]

[1] University of Manitoba, Winnipeg, Manitoba, Canada
padman@cc.umanitoba.ca
[2] University of New Mexico, Albuquerque, New Mexico, USA
veroff@cs.unm.edu

Abstract. Here we give an automated proof of the fact that a cubic curve admits at most one group law. This is achieved by proving the tight connection between the chord-tangent law of composition and any potential group law (as a morphism) on the curve. An automated proof of this is accomplished by implementing the rigidity lemma and the Cayley-Bacharach theorem of algebraic geometry as formal inference rules in Prover9, a first-order theorem prover developed by Dr. William McCune.

1 In Memory of Bill McCune

This book is dedicated to the memory of our friend and colleague, Bill McCune. Both authors of this article collaborated with Bill for many years before his untimely passing in 2011. R. P. has included comments about Bill in the article, *Group Embedding of the Projective Plane PG(2, 3)*, that also appears in the book. We preface this article with the following personal reflections from Bob.

> *After Bill moved to New Mexico in 2006, we would get together regularly to discuss specific math applications, new features for his programs and implementation strategies for some of the more complex features. We preferred to meet at my house rather than at the university, and when we did meet, Bill would stay for dinner with my family and an evening of shooting pool and playing games. He enjoyed cooking, and sometimes he would bring bread or a dessert that he had made. He especially liked cooking with New Mexico green chiles; during chile harvest we'd buy a year's supply of fresh roasted chiles and spend an afternoon peeling and bagging them for freezing.*

> *Bill enjoyed taking hikes on the many mountain trails in the Albuquerque area. He often hiked alone, but I would join him occasionally. We had our share of adventures—bushwhacking to find our way back to hard-to-follow trails and hiking through unexpected long stretches of calf-deep snow—but the hikes always were great fun.*

> *Bill was very reserved in public, but he appreciated the humor of situations such as our hiking adventures or realizing that something we*

---

* Partially supported by a University of Manitoba research leave grant.

M.P. Bonacina and M.E. Stickel (Eds.): McCune Festschrift, LNAI 7788, pp. 139–150, 2013.
© Springer-Verlag Berlin Heidelberg 2013

*were working on was a really bad idea. He'd smile skeptically at some
of my suggestions for new features for his programs, but sometimes he
would implement them anyway, just to humor me.*

2   A First-Order Property for Cubic Curves

Lying at the crossroads of algebra, arithmetic and geometry, the modern theory
of elliptic curves has many fascinating applications. For example, Andrew Wiles's
celebrated 1995 proof of Fermat's Last Theorem was accomplished using the
theory of elliptic curves. In this paper, we show how the modern technology of
automated deduction can be successfully employed in understanding some of the
inference rules enjoyed by elliptic curves.

An elliptic curve is simply the solution set of a non-singular cubic equation
defined in a projective plane over an algebraically closed field. A remarkable fact
from algebraic geometry is that elliptic curves admit a group law (see Figure 1)
and such a group law is unique and commutative. This is a simple consequence
of the so-called "rigidity lemma", a deep theorem in classical algebraic geometry [4, p. 43]. Also, the associativity of the group law on an elliptic curve—the
most non-trivial of the group properties—is a simple consequence of an intersection theorem known as the Cayley-Bacharach theorem. This is a far-reaching
generalization of the Pappus-Pascal theorem of classical projective geometry.

During the late 1980s, the rigidity lemma and the Cayley-Bacharach theorem
were formalized as inference rules in the language of first-order logic with equality
[5]. The resulting formal rules were successfully implemented by the late Dr.
William McCune in his theorem prover Otter [2]. Now there are new powerful
theorem provers, like Prover9 [3], also created by McCune. In this paper, we
show how one can use Prover9 to prove incidence theorems on non-singular cubic
curves within the framework of first-order logic with equality. We use a scheme
of inference rules derived from the rigidity lemma and the Cayley-Bacharach
theorem, incorporate these rules in Prover9 and then obtain first-order proofs
of several well-known incidence theorems including the associativity and the
uniqueness of the group law on cubic curves.

3   Implementing gL in Prover9

In [6], Padmanabhan and McCune describe the implementation of a first-order
property for cubic curves, called gL, as an inference rule in McCune's theorem prover Otter. Although gL is not supported directly as an inference rule in
Prover9, we can implement applications of gL using hyperresolution with appropriate hyperresolution nuclei included as additional input clauses. For example,
consider the gL clause

```
(z * x0) * x1 != (z * y0) * y1 | (w * x0) * x1 = (w * y0) * y1.
```

This clause is an implication which simply says that if $(z * x_0) * x_1 = (z * y_0) * y_1$ for some z (which occurs in the same location on both sides of the equation), then the common element z can be replaced with an arbitrary variable w (see Figure 2, the basic gL). In other words, resolving with the negative literal of this clause has the effect of replacing the subterm that matches the z in the first literal with a newly introduced variable w.

Since gL operates at the term level, there needs to be a separate gL clause for every combination of relevant functions and every possible term position. As a matter of practicality, when working on a specific problem, we limit the nesting level of the gL clauses that are included in the input file. To facilitate these studies, we have written a gL clause generator [9] that takes as arguments a signature and a maximum nesting level and generates all of the relevant gL clauses up to the specified nesting-level.

As an example, here is the output of the gL clause generator for a signature with two functions, $f(x, y)$ and $g(x)$, and a maximum nesting level of 1.

```
% ----------------------------------------------------------
% The following is output from the gL clause generator.
% ----------------------------------------------------------

formulas(assumptions).

% Signature:  [['f', 2], ['g', 1]]
% Clauses to implement gL to nesting level 1

f(z,x0) != f(z,y0) | f(w,x0) = f(w,y0).
f(f(z,x0),x1) != f(f(z,y0),y1) | f(f(w,x0),x1) = f(f(w,y0),y1).
f(x0,f(z,x1)) != f(y0,f(z,y1)) | f(x0,f(w,x1)) = f(y0,f(w,y1)).
g(f(z,x0)) != g(f(z,y0)) | g(f(w,x0)) = g(f(w,y0)).
f(x0,z) != f(y0,z) | f(x0,w) = f(y0,w).
f(f(x0,z),x1) != f(f(y0,z),y1) | f(f(x0,w),x1) = f(f(y0,w),y1).
f(x0,f(x1,z)) != f(y0,f(y1,z)) | f(x0,f(x1,w)) = f(y0,f(y1,w)).
g(f(x0,z)) != g(f(y0,z)) | g(f(x0,w)) = g(f(y0,w)).
g(z) != g(z) | g(w) = g(w).
f(g(z),x0) != f(g(z),y0) | f(g(w),x0) = f(g(w),y0).
f(x0,g(z)).!= f(y0,g(z)) | f(x0,g(w)) = f(y0,g(w)).
g(g(z)) != g(g(z)) | g(g(w)) = g(g(w)).

end_of_list.
```

In each case, the subterm matching the z in the first literal is replaced by the newly introduced variable w. See [8] for additional discussion and examples.

In some cases, we may wish to restrict the use of the gL clauses to be used *only* in applications of the inference rule gL. In particular, we may want to prevent their use in applications of paramodulation. This is easily done in Prover9 by setting the flag para_units_only.

We may also wish to produce proofs that are strictly-forward derivations of the theorems and to produce proofs that are demodulation free, that is, having proof steps that consist only of a single application of an inference rule. These can be accomplished with the Prover9 directives set(restrict_denials) and clear(back_demod) respectively.

It typically is more difficult for Prover9 to find a proof with restrictions such as these. In these cases, we can appeal to the method of proof sketches [11] by first finding proofs with fewer restrictions and then using these proofs as hints to help find the desired proofs. The proof presented for Theorem 3 in this article, for example, is the result of a run using previously found proofs as hints. The original, unrestricted proof is included on the article's associated Web page [8].

There are tradeoffs between the approaches used by Otter and Prover9 for the implementation of gL. The advantages of the built-in approach (Otter) include execution speed of a single application of the rule and convenience to the user of just setting the appropriate flag. A disadvantage is that the rule is applied exhaustively when it is built in, which can have negative implications for search performance. In the Prover9 approach, we can restrict the application of gL by deciding which hyperresolution nuclei to include as input and can try and include just the ones that are needed (more realistically, a set that we believe suffices). The corresponding disadvantage, of course, is that we have to know (or guess) in advance which ones to include.

4   Group Law on Non-singular Cubic Curves

Given two points P and Q on a non-singular cubic curve C, the line joining P and Q meets the curve again at a unique third point, say R. If $Q = P$, then the unique third point is where the tangent meets the curve again. Thus, we have a well-defined binary algebra $(C; *)$. The binary operation $*$ of chord-tangent construction is not a group operation—it is neither associative nor admits an identity element. However, if we define addition $x \oplus y$ as the derived binary term $(x * y) * e$ where "e" is some fixed (rational inflection) point, then the addition becomes a group in such way that three points P, Q and R are collinear if and only if $P \oplus Q \oplus R = e$, the identity of the group law (see Figure 1). In fact, this is the only way to define a group morphism, i.e., if $x \oplus y$ and $x \otimes y$ are two group laws sharing the same identity, then they are indeed equal and coincide with the geometrically defined group operation $(x * y) * e$ (see Theorem 3). In other words, it is the geometry of the cubics that determines the algebra!

Cayley-Bacharach Theorem. [10, p. 240]. If P is a set of mn points that is a complete intersection cycle of two curves $C1$ and $C2$ of degrees m and n respectively, then any plane curve of degree $m + n - 3$ passing through all but one point of P necessarily contains the whole of P.

Theorem 1. *Let C be a non-singular cubic curve in the complex projective plane, and let $*$ be the binary morphism of the chord-tangent procedure. Let a, b, c, d and e be given points, and let x be an arbitrary point on the curve C.*

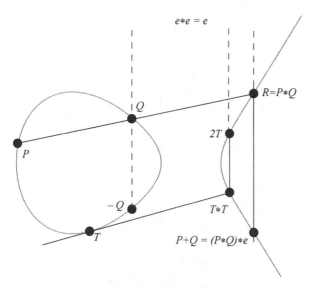

Fig. 1. Geometric Definition of the Group Law

Then the binary algebra $(C; *)$ satisfies the implication $(a * b) * c = (a * d) * e \Rightarrow (x * b) * c = (x * d) * e$.

Proof (taken from Padmanabhan and McCune [7]). Let $C1$ be the quartic formed by the four lines $\{1 \cup 2 \cup 3 \cup 4\}$; let D be the quartic formed by the four lines $\{5 \cup 6 \cup 7 \cup 8\}$; and let C be the given nonsingular cubic curve (see Figure 2). We have

$$C \cap C1 = \{a, d, a * d, c, a * b, (a * b) * c, e, x * d, (x * d) * e, x, b, x * b\}$$
$$C \cap D = \{e, a * d, (a * d) * e, x, d, x * d, a, b, a * b, c, x * b, (x * b) * c\}.$$

Hence, if $(a * b) * c = (a * d) * e$, then both $C1$ and D share 11 common points with the base cubic C. Here, both $C1$ and D are quartics; and in the notation of the Cayley-Bacharach theorem, we have $m = 4$, $n = 3$ and the degree of the curve D is 4, which is, of course, $4 + 3 - 3$. So, by the Cayley-Bacharach theorem, the two quartics $C1$ and D must share the 12th common point as well. In other words, we have $(x * b) * c = (x * d) * e$. This completes the proof of the validity of the gL implication $(a * b) * c = (a * d) * e \Rightarrow (x * b) * c = (x * d) * e$.

In what follows, we call this implication the "basic gL for cubics" or simply "basic gL". This is only a special case of the full version of the rule gL, which in turn is a modern avatar of the powerful rigidity lemma of complete (projective) varieties: $f(x, b) = c \Rightarrow f(x, z) = f(y, z)$ for all n-ary terms f in the projective variety (see [4, p. 43] and [6, p. 37] for more details and references about the related recently discovered conditions like the term condition). Such varieties are known as abelian varieties.

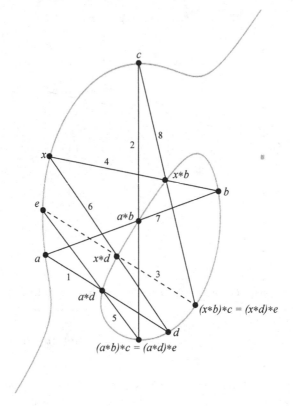

Fig. 2. Basic gL

One of the most nontrivial facts about the addition of points on a cubic is the proof of the associative law. Usual proofs of associativity range from using the "8 ⇒ 9 theorem" of Chasles, the Bezout theorem, the explicit formulas for the coordinates of $P+Q$, the rigidity lemma and function theory via the elliptic functions of Weierstrass. The proof closest to the spirit of universal algebra is given by Knapp [1, p. 67]): the associativity of the group law is equivalent to the medial law $(x * y) * (z * u) = (x * z) * (y * u)$ (see Figure 3). Consistent with the spirit of our approach, we will first derive this as a consequence of the basic gL. The idea is very simple: to prove $f(x, y, ..., z) = g(x, y, ..., z)$, where f and g are unifiable, apply a suitable gL clause to the composite morphism $f * (u * g)$, where u is a variable not occurring in f or g. This is the essential thrust of this communication.

Theorem 2. *The binary algebra* $(C; *)$ *satisfies the following:*

(1) $(x * y) * (z * u) = (x * z) * (y * u)$ *(Figure 3)*
(2) $((x * y) * z) * u = ((u * y) * z) * x$ *([1, p. 68])*
(3) *the binary addition defined by* $x + y = e * (x * y)$ *is associative (Figure 4)*

Prover9 proof of (1).

```
1 (x * y) * (z * u) = (x * z) * (y * u)
    # label(non_clause) # label(goal).   [goal].
2 (x * y) * x = y.  [assumption].
4 (x * y) * z != (x * u) * w | (v5 * y) * z = (v5 * u) * w
    # label("gL").  [assumption].
9 (c1 * c3) * (c2 * c4) != (c1 * c2) * (c3 * c4).   [deny(1)].
11 ((x * (y * z)) * x) * y = z.  [para(2(a,2),2(a,1,1))].
13 (x * ((y * z) * u)) * y = (x * z) * u.  [hyper(4,a,2,a)].
32 ((x * y) * x) * (z * y) = z.  [para(2(a,1),11(a,1,1,1,2))].
792 (x * y) * (z * u) = (x * z) * (y * u).
    [para(32(a,1),13(a,1,1,2))].
793 $F.  [resolve(792,a,9,a)].
```

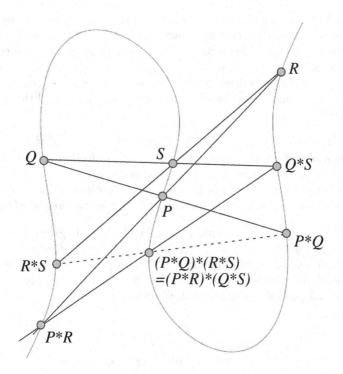

Fig. 3. Medial Law

Proof of (2).

$$((x * y) * z) * u = (z * (y * x)) * u \text{ by the commutativity of } *$$
$$= (z * (y * x)) * (v * (u * v))$$
$$= (z * v) * ((y * x) * (u * v)) \text{ by the medial law (1)}$$

$$= (z * v) * ((y * u) * (x * v)) \text{ by the medial law (1)}$$
$$= (z * (y * u)) * (v * (x * v)) \text{ by the medial law (1)}$$
$$= (z * (y * u)) * x$$
$$= ((u * y) * z) * x \text{ by the commutativity of } *.$$

Proof of (3).

$$x + (y + z) = e * (x * (e * (y * z)))$$
$$= e * (z * (e * y * x)) \text{ by the identity (2)}$$
$$= z + (x + y) = (x + y) + z$$

since the commutativity of $+$ is obvious from that of $*$.

As another application of gL clauses (which are just special cases of the powerful rigidity lemma), we derive the well-known uniqueness of the group law of a non-singular cubic curve. In other words, if e, an inflection point, is chosen as the identity element of a group law, say $x + y$, then we have $x + y = e * (x * y)$. The Prover9 proof included below employs an appropriate gL clause to prove associativity and commutativity of $+$, the inverse property and the uniqueness of the group law.

Theorem 3. *Let $+ : C \times C \to C$ be a binary morphism of the cubic algebra $(C; *)$ such that the binary operation $+$ has e, an inflection point of the curve, as an identity, i.e., $x + e = x$ and $e + y = y$ for all points x, y in C. Then $x + y = e * (x * y)$ is an identity in C. Moreover, the algebra $(C; +)$ is an abelian group: $+$ is commutative and associative, and $e * x$ is the inverse of x.*

Proof. Consider the ternary morphism $\varphi : C \times C \times C \to C$ given by the formula $\varphi(x, y, z) := (x+y)*(x*z)$. Now $\varphi(x, e, z) = (x+e)*(x*z) = x*(x*z) = z$. So we have $\varphi(x, e, z) = \varphi(u, e, z)$ and hence, by an appropriate gL clause—a complete Prover9 proof is included below—we have the stronger equality $\varphi(x, y, z) = \varphi(u, y, z)$ for all points x, y, z and u in C. In particular,

$$\varphi(x, y, z) = \varphi(e, y, z)$$
$$= (e + y) * (e * z)$$
$$= y * (e * z)$$

since, by assumption, $e + y = y$. Thus, we have $(x + y) * (x * z) = y * (e * z)$. Hence,

$$x + y = (y * (e * z)) * (x * z) \text{ post-multiply by } (x * z)$$
$$= (y * x) * ((e * z) * z) \text{ by the medial law}$$
$$= (y * x) * e = (x * y) * e.$$

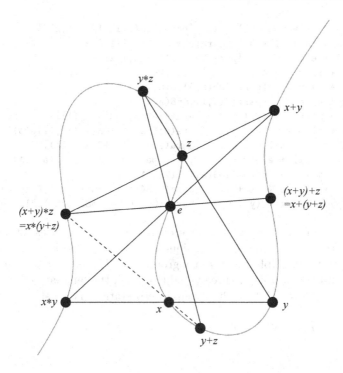

Fig. 4. Associativity

Of course, the term "$(x * y) * e$" is the celebrated group law obtained by taking the vertical reflection of $x * y$ (vertical reflection because the point "e" in the classical Weierstrass curve is $(0, 1, 0)$, the point at infinity (see Figures 1 and 4).

Following are Prover9 derivations of the group law (clause 267), group inverse (clause 521), commutativity of $+$ (clause 1581) and associativity of $+$ (clause 2209).

```
1 x * (y * x) = y.  [assumption].
2 x * y = y * x.  [assumption].
3 e * e = e.  [assumption].
4 x + e = x.  [assumption].
5 e + x = x.  [assumption].
6 (x + y) * z != (x + u) * w | (v5 + y) * z = (v5 + u) * w
    # label("gL").  [assumption].
12 x * (x * y) = y.  [para(2(a,1),1(a,1,2))].
35 (x + y) * (z + u) = (x + u) * (z + y).  [hyper(6,a,2,a)].
99 (x + y) * (z + e) = x * (z + y).
    [para(4(a,1),35(a,1,1)),flip(a)].
103 (e + x) * (y + z) = z * (y + x).
    [para(5(a,1),35(a,1,1)),flip(a)].
125 (x + y) * z = x * (z + y).  [para(4(a,1),99(a,1,2))].
163 x * (y + z) = y * (x + z).  [para(125(a,1),2(a,1))].
```

```
178 (x + y) * (x * (z + y)) = z.   [para(125(a,1),12(a,1,2))].
224 (x + y) * (x * y) = e.   [para(5(a,1),178(a,1,2,2))].
267 (x * y) * e = x + y.   [para(224(a,1),1(a,1,2))].
305 x + (y * x) = y * e.   [para(1(a,1),267(a,1,1)),flip(a)].
468 x + (e * x) = e.   [para(305(a,2),3(a,1))].
521 x + (x * e) = e.   [para(2(a,1),468(a,1,2))].
815 x * (y + z) = z * (y + x).   [para(5(a,1),103(a,1,1))].
927 x * (y + z) = z * (x + y).   [para(815(a,1),163(a,1)),flip(a)].
1109 x * (y * (x + z)) = z + y.   [para(927(a,1),12(a,1,2))].
1134 (x * (y + z)) * e = z + (x + y).   [para(927(a,2),267(a,1,1))].
1581 x + y = y + x.   [para(1109(a,1),12(a,1))].
2032 x + (y + z) = z + (x + y).   [para(1134(a,1),267(a,1)),flip(a)].
2209 (x + y) + z = x + (y + z).   [para(2032(a,2),1581(a,1)),flip(a)].
```

As our final example, we give a Prover9 proof, using gL clauses, of a theorem involving ternary analogues of the uniqueness of the group law and the group inverse. Take, for example, the ternary group term $f(x, y, z) = -x - y - z$, and write a set of two-variable or three-variable laws in the language of one ternary operation f that are valid for this group interpretation. For instance, we have

$$f(x, y, z) = -x - y - z = -x - z - y = f(x, z, y)$$

and

$$f(x, y, f(x, y, z)) = -x - y - (-x - y - z) = -x - y + x + y + z = z.$$

It turns out that there are at least two interpretations for such a ternary group term satisfying these conditions, namely, $f(x, y, z) = -x - y - z$ and $f(x, y, z) = -y - z$. Under both interpretations, the group law $+$ with identity e is expressible by the formula $y + z = f(x, e, f(x, y, z))$, and the group inverse is given by the formula $-x = f(e, x, e)$.

Theorem 4. *If $f(x, y, z)$ is a ternary morphism of an elliptic curve satisfying $f(e, e, e) = e$, $f(x, y, z) = f(x, z, y)$ and $f(x, y, f(x, y, z)) = z$, and if $+$ is a group law with e as its identity, then $x + y = f(e, e, f(e, x, y))$ and $-x = f(e, x, e)$.*

We can prove this theorem using appropriate gL clauses. In this case, however, rather than proving the stated form of the unary inverse $(-x)$ explicitly, we use Prover9 to "discover" it as follows. The theorem that states the existence of the unary inverse is $\forall x \exists y (x + y = e)$. Its negation (for proof by contradiction) is $\neg \forall x \exists y (x + y = e)$, which is logically equivalent to $\exists x \forall y \neg (x + y = e)$. Replacing the existentially quantified variable with Skolem constant A results in the clause A + y != e. We note then that if a clause conflicts with this clause, the term matching the variable y will indeed be the unary inverse of the arbitrary element A, and we can use a Prover9 **answer** literal to identify this term.

In the following Prover9 proof, clause 3789 identifies the unary inverse of the arbitrary element A. (In fact, clause 3788 shows the most general formula for the inverse under the given interpretations.) It also happens that clause 498 in this proof suffices to establish the first part of the theorem, $x + y = f(e, e, f(e, x, y))$, so the proof of the theorem is complete.

```
1 f(e,e,e) = e.  [assumption].
2 f(x,y,z) = f(x,z,y).  [assumption].
3 f(x,y,f(x,y,z)) = z.  [assumption].
4 x + e = x.  [assumption].
5 e + x = x.  [assumption].
29 x + (y + z) != u + (y + w) | x + (v5 + z) = u + (v5 + w)
      # label("gL").  [assumption].
32 f(x,y + z,u) != f(w,v5 + z,v6) | f(x,y + v7,u) = f(w,v5 + v7,v6)
      # label("gL").  [assumption].
36 A + x != e # answer(x).  [assumption].
41 f(e,e,f(x,y,f(x,y,e))) = e.  [para(3(a,2),1(a,1,3))].
86 e + (x + (y + z)) = y + (x + z).  [hyper(29,a,5,a)].
90 f(x,y + z,u + w) = f(x,u + z,y + w).  [hyper(32,a,2,a)].
207 x + (y + z) = y + (x + z).  [para(86(a,1),5(a,1))].
234 x + (y + e) = y + x.  [para(4(a,1),207(a,1,2)),flip(a)].
256 x + y = y + x.  [para(4(a,1),234(a,1,2))].
298 f(x,y + z,e + u) = f(x,z,y + u).
      [para(5(a,1),90(a,1,2)),flip(a)].
380 f(x,y + z,e + u) = f(x,y,z + u).  [para(256(a,1),298(a,1,2))].
401 f(x,y + z,u) = f(x,y,z + u).  [para(5(a,1),380(a,1,3))].
458 f(x,e,y + z) = f(x,y,z).  [para(5(a,1),401(a,1,2)),flip(a)].
498 f(x,e,f(x,y,z)) = y + z.  [para(458(a,1),3(a,1,3))].
3788 x + f(e,x,e) = e.  [para(41(a,1),498(a,1)),flip(a)].
3789 $F # answer(f(e,A,e)).  [resolve(3788,a,36,a)].
```

5 Conclusion

We have just scratched the surface of the area. As explained at the beginning of this article, gL clauses apply to the whole range of clone of terms of a given gL algebra. The depth of the problems assigned to the gL clauses can be increased by using more operations, higher arities and higher nesting levels. In this article, we have just employed types (2), (2,2) and (3,2) and gL clauses to nesting level 1. A possible new direction of exploiting automated deduction with gL clauses is to prove new theorems on quartic curves, Mal'cev type conditions, the uniqueness of Mal'cev terms in loops and groups, closure conditions of Reidemeister and Thomsen for group-like algebras, group embeddings of planar configurations, etc. Some of these will be explored in future papers.

As we mentioned, the results presented in this paper are well-known and are usually proved by using either the Bezout theorem (for associativity of the group law) or the rigidity lemma (for the uniqueness of the group law). This is the first time these results are proved by a first-order theorem prover (and hence employ only first-order logic with equality). Thus, this opens up a new vista of techniques and models for computer scientists designing theorem provers and for mathematicians working in equational logic and geometry.

References

1. Knapp, A.: Elliptic Curves. Princeton University Press (1992)
2. McCune, W.: Otter 3.0 Reference Manual and Guide. Tech. Report ANL-94/6, Argonne National Laboratory, Argonne, IL (1994), http://www.mcs.anl.gov/AR/otter/
3. McCune, W.: Prover9, version 2009-02A, http://www.cs.unm.edu/~mccune/prover9/
4. Mumford, D.: Abelian varieties. Tata Institute of Fundamental Research. Studies in Mathematics, vol. 5. Oxford University Press, London (1970)
5. Padmanabhan, R.: Logic of equality in geometry. Discrete Mathematics 15, 319–331 (1982)
6. Padmanabhan, R., McCune, W.: Automated reasoning about cubic curves. Computers and Mathematics with Applications 29(2), 17–26 (1995)
7. Padmanabhan, R., McCune, W.: Uniqueness of Steiner laws on cubic curves. Beiträge Algebra Geom. 47(2), 543–557 (2006)
8. Padmanabhan, R., Veroff, R.: A geometric procedure with Prover9 (Web support) (2012), http://www.cs.unm.edu/~veroff/gL_Paper/
9. Padmanabhan, R., Veroff, R.: A gL clause generator (2012), http://www.cs.unm.edu/~veroff/gL_Paper/gL_gen.html
10. Silverman, J., Tate, J.: Rational Points on Elliptic Curves. Springer (1992)
11. Veroff, R.: Solving open questions and other challenge problems using proof sketches. J. Automated Reasoning 27, 157–174 (2001)

Loops with Abelian Inner Mapping Groups:
An Application of Automated Deduction[*]

Michael Kinyon[1], Robert Veroff[2], and Petr Vojtěchovský[1,**]

[1] Department of Mathematics, University of Denver,
Denver, CO 80208 USA
{mkinyon,petr}@math.du.edu
www.math.du.edu/~mkinyon
www.math.du.edu/~petr
[2] Department of Computer Science, University of New Mexico,
Albuquerque, NM 87131 USA
veroff@cs.unm.edu
www.cs.unm.edu/~veroff

Abstract. We describe a large-scale project in applied automated deduction concerned with the following problem of considerable interest in loop theory: If Q is a loop with commuting inner mappings, does it follow that Q modulo its center is a group and Q modulo its nucleus is an abelian group? This problem has been answered affirmatively in several varieties of loops. The solution usually involves sophisticated techniques of automated deduction, and the resulting derivations are very long, often with no higher-level human proofs available.

1 Introduction

A *quasigroup* $(Q, \cdot)$ is a set Q with a binary operation $\cdot$ such that for each $x \in Q$, the *left translation* $L_x : Q \to Q; y \mapsto yL_x = xy$ and the *right translation* $R_x : Q \to Q; y \mapsto yR_x = yx$ are bijections. A quasigroup is a *loop* if, in addition, there exists $1 \in Q$ satisfying $1 \cdot x = x \cdot 1 = x$ for all $x \in Q$. Standard references for the theory of quasigroups and loops are [1], [3] and [45].

Example 1. The above definition merely says that the multiplication table of a loop is a *Latin square* (that is, every symbol occurs in every row and in every column precisely once), in which the row labels are duplicated in column 1 and the column labels are duplicated in row 1.

For instance, the following multiplication table defines a loop Q with elements $\{1, \ldots, 6\}$.

---

[*] Dedicated to the memory of William McCune (1953–2011).
[**] Partially supported by Simons Foundation Collaboration Grant 210176.

M.P. Bonacina and M.E. Stickel (Eds.): McCune Festschrift, LNAI 7788, pp. 151–164, 2013.
© Springer-Verlag Berlin Heidelberg 2013

$$
\begin{array}{c|cccccc}
\cdot & 1 & 2 & 3 & 4 & 5 & 6 \\
\hline
1 & 1 & 2 & 3 & 4 & 5 & 6 \\
2 & 2 & 1 & 4 & 3 & 6 & 5 \\
3 & 3 & 4 & 5 & 6 & 1 & 2 \\
4 & 4 & 3 & 6 & 5 & 2 & 1 \\
5 & 5 & 6 & 2 & 1 & 3 & 4 \\
6 & 6 & 5 & 1 & 2 & 4 & 3 \\
\end{array}
$$

Note that Q is neither commutative (as $3 \cdot 5 = 1$ while $5 \cdot 3 = 2$), nor associative (as $3 \cdot (3 \cdot 4) = 2$ while $(3 \cdot 3) \cdot 4 = 1$). The left translation L_2 is the permutation $(1, 2)(3, 4)(5, 6)$, the right translation R_3 is the permutation $(1, 3, 5, 2, 4, 6)$.

A quasigroup can be equivalently defined as a set Q with *three* binary operations $\cdot$ (multiplication), $\backslash$ (left division), and $/$ (right division) satisfying the axioms

$$
x \cdot (x \backslash y) = y = x \backslash (x \cdot y), \qquad (x \cdot y)/y = x = (x/y) \cdot y.
$$

Starting with a quasigroup $(Q, \cdot)$, one merely needs to set $x \backslash y = y L_x^{-1}$ and $x / y = x R_y^{-1}$. Conversely, starting with a three operation quasigroup $(Q, \cdot, \backslash, /)$, the above axioms guarantee that all translations L_x, R_x are bijections of Q. For more details on the equivalence of the two definitions, see [9].

Since quasigroups and loops can be equationally defined, they have been fallow ground for the techniques and tools of automated deduction. They appeared already in the milestone paper of Knuth and Bendix [24], and interest in them has continued in the theorem-proving community [12] [49]. In more recent years, mathematicians specializing in quasigroups and loops have been making significant use of automated deduction tools [15–29] [36–44]. With the exception of [39] and [40], all of the aforementioned references used Bill McCune's PROVER9 [31] or its predecessor OTTER [30].

The purpose of this paper is to report on progress in a large-scale project in the application of automated deduction to loop theory. The primary tool has been PROVER9, and the search for proofs has relied heavily on the method of *proof sketches* [48]. Proof sketches have been especially effective in this project, in part because of the large number of closely related problems being considered.

In §2 we give the mathematical background and the Main Conjecture and then explain why it is a problem suitable for equational reasoning tools. In §3 we discuss our strategy for attacking the problem with PROVER9. As an example, in §4 we discuss one particular class of loops for which we were able to solve the problem. Finally in §5 we mention other classes of loops for which the solution is known.

2   The Main Conjecture and the Project

The left and right translations in a loop Q generate the *multiplication group* $\mathrm{Mlt}(Q) = \langle L_x, R_x \mid x \in Q \rangle$, a subgroup of the group of all bijections on Q. The *inner mapping group* $\text{Inn}(Q) = \{\varphi \in \text{Mlt}(Q) \mid 1\varphi = 1\}$ is a subgroup of $\text{Mlt}(Q)$ consisting of all bijections in $\text{Mlt}(Q)$ that fix the element 1.

A loop Q is said to be an *AIM loop* (for **A**belian **I**nner **M**appings) if $\text{Inn}(Q)$ is an abelian group. AIM loops are the main subject of this investigation.

A nonempty subset S of a loop Q is a *subloop* ($S \le Q$) if it is closed under the three operations $\cdot, \backslash, /$. A subloop S of Q is *normal* ($S \trianglelefteq Q$) if $S\varphi = \{s\varphi \mid s \in S\}$ is equal to S for every $\varphi \in \text{Inn}(Q)$.

To save space, we will often denote $x \cdot y$ by xy, and we will use $\cdot$ to indicate the priority of multiplication. For instance, $x \cdot yz$ stands for $x \cdot (y \cdot z)$.

The *left, right,* and *middle nucleus* of a loop Q are defined, respectively, by

$$N_\lambda(Q) = \{a \in Q \mid ax \cdot y = a \cdot xy, \quad \forall x, y \in Q\},$$
$$N_\rho(Q) = \{a \in Q \mid xy \cdot a = x \cdot ya, \quad \forall x, y \in Q\},$$
$$N_\mu(Q) = \{a \in Q \mid xa \cdot y = x \cdot ay, \quad \forall x, y \in Q\},$$

and the *nucleus* is $N(Q) = N_\lambda(Q) \cap N_\rho(Q) \cap N_\mu(Q)$. It is not hard to show that each of these nuclei is a subloop.

The *commutant* of a loop Q is the set

$$C(Q) = \{a \in Q \mid ax = xa \quad \forall x \in Q\},$$

which is not necessarily a subloop. The *center* of Q is

$$Z(Q) = C(Q) \cap N(Q).$$

The center is always a normal subloop.

Thus the nucleus $N(Q)$ consists of all elements $a \in Q$ that associate with all $x, y \in Q$, the commutant $C(Q)$ consists of all elements $a \in Q$ that commute with all $x \in Q$, and the center $Z(Q)$ consists of all elements $a \in Q$ that commute and associate with all $x, y \in Q$.

For a general loop Q, the inclusions $N_\lambda(Q) \supseteq N(Q), N_\rho(Q) \supseteq N(Q), N_\mu(Q) \supseteq N(Q), N(Q) \supseteq Z(Q)$ and $C(Q) \supseteq Z(Q)$ hold, but not necessarily the equalities.

Given a normal subloop S of a loop Q, denote by Q/S the *factor loop* Q modulo S whose elements are the subsets (left cosets) $xS = \{xs \mid s \in S\}$ for $x \in Q$, and where we multiply according to $xS \cdot yS = (x \cdot y)S$.

Now define $Z_0(Q) = \{1\}$, and for $i \ge 0$ let $Z_{i+1}(Q)$ be the preimage of $Z(Q/Z_i(Q))$ under the canonical projection $\pi_i : Q \to Q/Z_i(Q); x \mapsto xZ_i(Q)$. Note that $\{1\} = Z_0(Q) \le Z_1(Q) \le Z_2(Q) \le \cdots \le Q$. The loop Q is *(centrally) nilpotent of class n*, written $\text{c}\ell(Q) = n$, if $Z_{n-1}(Q) \ne Q$ and $Z_n(Q) = Q$.

Example 2. Let Q be the loop from Example 1. A short computer calculation shows that the multiplication group $\text{Mlt}(Q)$ has size 24, and the inner mapping group $\text{Inn}(Q)$ consists of the permutations (), $(3,4)$, $(5,6)$, $(3,4)(5,6)$. In particular, $\text{Inn}(Q)$ is a group of size 4, hence an abelian group, and Q is therefore an AIM loop. Q has only three subloops, namely the subsets $\{1\}$, $\{1,2\}$ and $\{1,2,3,4,5,6\}$. It so happens that all four nuclei, the commutant and the center of Q are equal to the normal subloop $\{1,2\}$. We have $Z_0(Q) = 1$, $Z_1(Q) = Z(Q) = \{1,2\}$ and $Z_2(Q) = \{1,2,3,4,5,6\}$ (because $Q/Z_1(Q)$ is an abelian group). Thus $\text{c}\ell(Q) = 2$.

Recall (or see [46, Thm. 7.1]) that if Q is a group then

$$\text{Inn}(Q) \text{ is isomorphic to } Q/Z(Q). \tag{1}$$

This result cannot be generalized to loops. For instance, we saw in Example 2 that there is a loop Q of size 6 with $\text{Inn}(Q)$ of size 4 and $Q/Z(Q)$ of size $6/2 = 3$.

Now, if Q is a group, we deduce from (1) that

$$c\ell(\text{Inn}(Q)) \leq n \text{ if and only if } c\ell(Q) \leq n+1. \tag{2}$$

Neither of the two implications in (2) generalizes to loops. Indeed, Vesanen found a loop Q of size 18 with $c\ell(Q) = 3$ such that $\text{Inn}(Q)$ is not even nilpotent, much less of nilpotency class at most 2; see [14]. To falsify the other implication, we will see below that there exist loops with $c\ell(\text{Inn}(Q)) = n$ but with $c\ell(Q) \neq n+1$. (Note, however, that Niemenmaa was able to prove recently that if Q is finite and $\text{Inn}(Q)$ is nilpotent then Q is at least nilpotent [34].)

Using $n = 1$ in (2), we see that for a group Q

$$\begin{array}{l} \text{Inn}(Q) \text{ is abelian (that is, } Q \text{ is an AIM loop)} \\ \text{if and only if } c\ell(Q) \leq 2. \end{array} \tag{3}$$

Does this statement generalize to loops? The answer is of importance in loop theory because it sheds light on loops of nilpotency class 2, in some sense the most immediate generalization of abelian groups to loops.

Bruck showed in [2] that one implication of (3) holds for all loops: if Q is a loop with $c\ell(Q) \leq 2$ then Q is an AIM loop. For a long time, it was conjectured that the other implication of (3) also holds, and much work in loop theory was devoted to this problem [6] [13] [35]. However, in 2004, Csörgő [4] disproved (3) by constructing an AIM loop Q (of size 128) with $c\ell(Q) = 3$.

Consequently, AIM loops Q with $c\ell(Q) \geq 3$ have come to be called *loops of Csörgő type*. Additional constructions of loops of Csörgő type followed in rapid succession [7] [8] [33]. We remark that it is still not known if there exists an AIM loop Q with $c\ell(Q) > 3$.

What can be salvaged from the statement (3) for loops? Based on the structure of all known loops of Csörgő type, the first-named author has been offering the following structural conjecture in various talks and presentations. This is the first published statement of the conjecture.

Main Conjecture. *Let Q be an AIM loop. Then $Q/N(Q)$ is an abelian group and $Q/Z(Q)$ is a group. In particular, Q is nilpotent of class at most 3.*

Three remarks are worth making here. First, the primary assertion of the Main Conjecture is actually somewhat stronger than the "in particular" part, that is, having nilpotency class 3 does not necessarily imply $Q/N(Q)$ is abelian or $Q/Z(Q)$ is a group. Second, note that the Main Conjecture makes no reference to cardinality of the loop Q. It is certainly conceivable that the conjecture holds for all finite loops but that there is some infinite counterexample. Finally, it is tacit in the statement of the conjecture that the nucleus $N(Q)$ is a normal subloop of Q (else the factor loop $Q/N(Q)$ cannot be formed). This is not true for arbitrary loops but is easy to show for AIM loops.

From the discussion so far, it may seem that the Main Conjecture is too high order to be attacked fruitfully by the methods of automated deduction. However, the hypotheses and conclusions of the conjecture can be given purely equationally as we now describe.

Bruck showed [3] that the inner mapping group is generated by three kinds of mappings that measure deviations from associativity and commutativity, namely,

$$\mathrm{Inn}(Q) = \langle R_{x,y}, T_x, L_{x,y}; \ x, y \in Q \rangle,$$

where

$$R_{x,y} = R_x R_y R_{xy}^{-1}, \qquad T_x = R_x L_x^{-1}, \qquad L_{x,y} = L_x L_y L_{yx}^{-1}.$$

Since a group is abelian if and only if any two of its generators commute, we immediately obtain the following characterization of AIM loops:

Lemma 1. *A loop Q is an AIM loop if and only if the following identities hold:*

$$T_x T_y = T_x T_y,$$
$$L_{x,y} L_{z,w} = L_{z,w} L_{x,y},$$
$$R_{x,y} R_{z,w} = R_{z,w} R_{x,y},$$
$$L_{x,y} T_z = T_z L_{x,y},$$
$$R_{x,y} T_z = T_z R_{x,y},$$
$$L_{x,y} R_{z,w} = R_{z,w} L_{x,y}$$

for all $x, y, z, w \in Q$.

To encode the conclusions of the Main Conjecture, it is useful to introduce two more functions, the *associator*

$$[x, y, z] = (x \cdot yz) \backslash (xy \cdot z)$$

and the *commutator*

$$[x, y] = (yx) \backslash (xy),$$

that is, $(x \cdot yz)[x, y, z] = xy \cdot z$ and $yx \cdot [x, y] = xy$. The former is a measure of nonassociativity and the latter is a measure of noncommutativity. Different conventions are possible for each of these functions, *e.g.* one could use $(xy \cdot z)/(x \cdot yz)$ as a definition of associator. Our convention is the traditional one [3]. Note that an element $a \in Q$ is in the left nucleus $N_\lambda(Q)$ if and only if $[a, x, y] = 1$ for all $x, y \in Q$, and that the other nuclei are similarly characterized.

The following is now a straightforward consequence of the definitions.

Lemma 2. *Let Q be a loop. Then*

(i) If $N(Q) \trianglelefteq Q$ then $Q/N(Q)$ is an abelian group if and only if the following identities hold:

$$[[x, y, z], u, v] = [u, [x, y, z], v] = [u, v, [x, y, z]] = 1,$$
$$[[x, y], z, u] = [z, [x, y], u] = [z, u, [x, y]] = 1$$

for all $x, y, z, u, v \in Q$;

(ii) $Q/Z(Q)$ is a group if and only if the following identities hold:

$$[[x, y, z], u, v] = [u, [x, y, z], v] = [u, v, [x, y, z]] = 1,$$
$$[[x, y, z], u] = 1$$

for all $x, y, z, u, v \in Q$.

Proof. Let $S \trianglelefteq Q$. The following conditions are equivalent: $(xS \cdot yS) \cdot zS = xS \cdot (yS \cdot zS)$, $(xy \cdot z)S = (x \cdot yz)S$, $((x \cdot yz)\backslash(xy \cdot z))S = S$, $[x, y, z]S = S$, $[x, y, z] \in S$. Thus Q/S is a group if and only if $[x, y, z] \in S$ for every x, y, $z \in Q$. Similarly, Q/S is commutative if and only if $[x, y] \in S$ for every x, $y \in Q$.

The condition $[[x, y, z], u, v] = [u, [x, y, z], v] = [u, v, [x, y, z]] = 1$ says that $[x, y, z] \in N(Q)$ for all x, y, $z \in Q$, and the condition $[[x, y], z, u] = [z, [x, y], u] = [z, u, [x, y]] = 1$ says that $[x, y] \in N(Q)$ for all x, $y \in Q$. This proves (i).

Concerning (ii), the condition $[[x, y, z], u] = 1$ says that $[x, y, z] \in C(Q)$ for every x, y, $z \in Q$. Since $Z(Q) = N(Q) \cap C(Q)$, we are done. $\qquad\square$

In Figure 1 we give a basic PROVER9 input file for the Main Conjecture. To encode loops, we use the definition with three binary operations $\cdot$, $\backslash$ and $/$. The clauses labeled "obvious compatibility" are not necessary, but are included to help the search. Note how the assumptions on AIM loops correspond to the identities of Lemma 1, and how the goals correspond to the identities of Lemma 2.

A resolution one way or the other of the full Main Conjecture would be a major milestone in loop theory, but, alas, the answer remains elusive. Nevertheless we managed to confirm the Main Conjecture for several classes of loops. We describe one of the successful cases in §4 and list others in §5.

3   The Strategy

Our search for proofs for the various AIM cases involves sequences of PROVER9 experiments that rely heavily on the use of hints [47] and on the method of proof sketches [48]. Under the hints strategy, a generated clause is given special consideration if it *matches* (subsumes) a user-supplied hint clause. In PROVER9, the actions associated with hint matching are controllable by the user, but the most typical action is to give hint matchers the highest priority in the proof search.

A *proof sketch* for a theorem T is a sequence of clauses giving a set of conditions *sufficient* to prove T. In the ideal case, a proof sketch consists of a sequence of lemmas, where each lemma is fairly easy to prove. In any case, the clauses

```
formulas(assumptions).
  % loop axioms
  1 * x = x.              x * 1 = x.
  x \ (x * y) = y.       x * (x \ y) = y.
  (x * y) / y = x.       (x / y) * y = x.

  % associator
  (x * (y * z)) \ ((x * y) * z) = a(x,y,z).

  % commutator
  (x * y) \ (y * x) = K(y,x).

  % inner mappings
  % L(u,x,y) = u L(x) L(y) L(yx)^{-1}
  (y * x) \ (y * (x * u)) = L(u,x,y).

  % R(u,x,y) = u R(x) R(y) R(xy)^{-1}
  ((u * x) * y) / (x * y) = R(u,x,y).

  % T(u,x) = u R(x) L(x)^{-1}
  x \ (u * x) = T(u,x).

  % obvious compatibility
  a(x,y,z) = 1 -> L(z,y,x) = z.    L(x,y,z) = x -> a(z,y,x) = 1.
  T(x,y) = x -> T(y,x) = y.        T(x,y) = x -> K(x,y) = 1.
  K(x,y) = 1 -> T(x,y) = x.

  % abelian inner mapping group (AIM loop)
  T(T(u,x),y) = T(T(u,y),x).
  L(L(u,x,y),z,w) = L(L(u,z,w),x,y).
  R(R(u,x,y),z,w) = R(R(u,z,w),x,y).
  T(L(u,x,y),z) = L(T(u,z),x,y).
  T(R(u,x,y),z) = R(T(u,z),x,y).
  L(R(u,x,y),z,w) = R(L(u,z,w),x,y).
end_of_list.

formulas(goals).
  a(K(x,y),z,u) = 1            # label("aK1").
  a(x,K(y,z),u) = 1            # label("aK2").
  a(x,y,K(z,u)) = 1            # label("aK3").
  K(a(x,y,z),u) = 1            # label("Ka").
  a(a(x,y,z),u,w) = 1          # label("aa1").
  a(x,a(y,z,u),w) = 1          # label("aa2").
  a(x,y,a(z,u,w)) = 1          # label("aa3").
end_of_list.
```

Fig. 1. A PROVER9 input file for the problem "In an AIM loop Q, is $Q/N(Q)$ an abelian group and is $Q/Z(Q)$ a group?"

of a proof sketch identify potentially notable milestones on the way to finding a proof. From a strategic standpoint, it is desirable to recognize when we have achieved such milestones and to adapt the continued search for a proof accordingly. In particular, we want to focus our attention on such milestone results and pursue their consequences sooner rather than later. The hints mechanism provides a natural and effective way to take full advantage of proof sketches in the search for a proof.

The use of hints is additive in the sense that hints from multiple proof sketches or from sketches for different parts of a proof can all be included at the same time. For this reason, hints are particularly valuable for "gluing" subproofs together and completing partial proofs, even when wildly different search strategies were used to find the individual subproofs.

In [48], we consider how the generation and use of proof sketches, together with the sophisticated strategies and procedures supported by an automated reasoning program such as PROVER9, can be used to find proofs to challenging theorems, including open questions. The general approach is to find proofs with additional assumptions and then to systematically eliminate these assumptions from the input set, using all previous proofs as hints. It also can be very effective to include as proof sketches proofs of related theorems in the same area of study. In the AIM study, for example, proofs for the LC case (see Section 4) were found by first proving the LC case with the additional assumption "left inner maps preserve inverses",

```
L(x,y,z) \ 1 = L(x \ 1,y,z).
```

The resulting proofs, together with proofs of other previously proved cases, were included as proof sketches in the search that found the proofs for LC alone.

Our strategy for searching for AIM proofs is based on the following two observations.

- Proofs for the various special cases tend to share several key steps. This makes these problems especially amenable to the use of previously found proofs as hints.
- Proof searches in this problem area tend to be especially sensitive to the underlying lexical ordering of terms. This is due primarily to the resulting effect on the demodulation (rewriting) of deduced clauses.

Addressing the second observation, we can run multiple searches, trying each of several term orderings in turn. But rather than simply running each of these searches as independent attempts, we can leverage PROVER9's hints mechanism to take advantage of any apparent progress made in previous searches. In particular, after running a search with one term ordering, we can gather the derived clauses that match hints and include these as input assumptions for the following runs.

The second-named author has automated this approach with a utility called p9loop. P9loop takes as input an ordinary PROVER9 input file (including hints) and a list of term ordering directives and proceeds as follows.

- Run PROVER9 with a term ordering from the list until either a proof is found or a user-specified processing limit is reached.
- If no proof has been found, restart PROVER9 with the next term ordering from the list, including all of the previous hint matchers as additional input assumptions.

We have had numerous successes using this approach, sometimes finding proofs that rely on several p9loop iterations. We have, for example, found proofs after iterating through over 50 term orderings, with as many as 30 of the iterations contributing to the found proof.

We note that the final iteration of a successful p9loop execution generally results in a proof that includes assumptions derived in previous PROVER9 runs and that we do not immediately have the derivations of these assumptions. Furthermore, this property may be nested in that an assumption from a previous run may in turn rely on assumptions from even earlier runs. In order to get a complete proof of the final theorem, we use PROVER9 to recover the missing derivations by systematically eliminating these assumptions in a way that is analogous to the general proof sketches method.

4   LC Loops

A loop Q is said to be an *LC loop* if it satisfies any of the following equivalent identities:

$$x(x(yz)) = (x(xy))z,$$
$$x(x(yz)) = ((xx)y)z,$$
$$(xx)(yz) = (x(xy))z,$$
$$x(y(yz)) = (x(yy))z$$

for all $x, y, z \in Q$. LC loops were introduced by Fenyves [10] as one of the loops of *Bol-Moufang type*. They were studied in more detail in [41] where the equivalence of the identities above was proven.

We have been able to establish the Main Conjecture in the special case of LC loops, that is, we have the following theorem.

Theorem 1. *Let Q be an AIM LC loop. Then $Q/N(Q)$ is an abelian group and $Q/Z(Q)$ is a group. In particular, Q is nilpotent of class at most 3.*

It sometimes is desirable to obtain a humanized proof of a result first found by means of automated deduction tools. Generally speaking, the process of "humanization" can give some higher-level insights into the structure being studied and can often result in a simpler (from a human perspective) proof than the one originally generated by an automated deduction tool. Many of the references in the bibliography feature humanized proofs.

There are various varieties of loops with abelian inner mappings for which the proof of the conjecture will be worth humanizing. We list a few of these in the next section. However, that desire for a human proof is tempered by the law of diminishing returns. In some cases, a human proof may not be worth the time and effort put into translating the automated proof. For LC loops, we run right into this issue.

We often prefer to produce automated proofs that are strictly-forward derivations of the theorems and that are free of any applications of demodulation. This sometimes is for purposes of presentation, but most often it's because we find that these proofs make better proof sketches for future searches, especially in a larger study such as the AIM project. In PROVER9, these constraints can be satisfied with the directives

    `set(restrict_denials)` and

    `clear(back_demod)`

respectively. If the input list `formulas(demodulators)` is empty, including the directive `clear(back_demod)` ensures that no clause—input or derived—will be used as a demodulator.

Here is the data for the strictly-forward, demodulation-free proofs of the seven LC goals:

Goal	Length	Level
Ka	2192	222
aK1	2191	221
aK2	2199	224
aK3	2394	276
aa1	2842	321
aa2	2842	321
aa3	2841	320

Of course, the seven proofs have many clauses in common and some proofs are virtually identical. For example, it is known that in LC loops, the left and middle nuclei coincide. The goals aa1 and aa2 state that associators are in those nuclei, so it is not surprising that their proof lengths and levels are the same. The proofs (and corresponding input file) can be found on this paper's associated Web page [23].

Since at present, the authors think that the classes of loops discussed in the next section will be more worthy of human translation, any attempt to do so for LC loops has been postponed.

5   Further Remarks

Within certain varieties of loops, it is known that there are no loops of Csörgő type. In other words, given an AIM loop Q in that variety, Q is necessarily nilpotent of class at most 2.

In order to formulate the problem of showing that an AIM loop in a particular variety is centrally nilpotent of class at most 2, it is only necessary to add the defining equations of the variety to the assumptions and to add one more goal:

$$[[x, y], z] = 1$$

for all x, y, z. Indeed, we already know from the identities of Lemma 2 that $Q/Z(Q)$ is a group, that $[x,y] \in N(Q)$, and the last goal says that $[x,y] \in C(Q)$, so $[x,y] \in Z(Q) = C(Q) \cap N(Q)$ and $Q/Z(Q)$ is an abelian group, i.e., $c\ell(Q) \leq 2$.

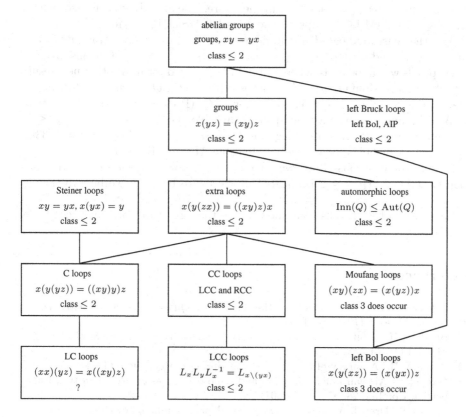

Fig. 2. Well-studied varieties of loops in which the Main Conjecture is true

Figure 2 summarizes what is known about the Main Conjecture and about the above-mentioned problem in several well-studied varieties of loops. The varieties include abelian groups, groups, *Steiner loops* (defined by the identities $xy = yx$, $x(yx) = y$), *extra loops* (defined by $x(y(zx)) = ((xy)z)x$), *automorphic loops* (loops where all inner mappings are automorphisms), *C loops* (defined by $((xy)y)z = x(y(yz))$), *conjugacy closed* (or *CC*) *loops* (loops in which $L_x L_y L_x^{-1}$ is a left translation and $R_x R_y R_x^{-1}$ is a right translation for every x, y), LC loops, *left Bol loops* (defined by $x(y(xz)) = (x(yx))z$), *left conjugacy closed* (or *LCC*) *loops* (loops in which $L_x L_y L_x^{-1}$ is a left translation for every x, y), and *left Bruck loops* (that is, left Bol loops with the *automorphic inverse property* (or *AIP*) $(xy)^{-1} = x^{-1} y^{-1}$). To keep the figure legible, we omitted all dual varieties (RC loops, right Bol loops, RCC loops, right Bruck loops) for which analogous results hold.

The varieties in Figure 2 are listed with respect to inclusion, with smaller varieties higher up. For instance, every Moufang loop is a left Bol loop.

All varieties in Figure 2 satisfy the Main Conjecture. Moreover, some varieties have the stronger property discussed above that an AIM loop Q satisfies $c\ell(Q) \leq 2$. This is indicated by "class ≤ 2" in the figure. The already known cases where $c\ell(Q) \leq 2$ are AIM LCC loops [5] and AIM left Bruck loops [40].

All other cases indicated in the figure—automorphic loops, Moufang loops, left Bol loops and C loops—comprise new results which are part of this project. Their proofs will appear elsewhere, often in humanized form. Another new result is that an AIM Moufang loop Q satisfies $c\ell(Q) \leq 2$ if Q is *uniquely 2-divisible*, that is, the mapping $x \mapsto x^2$ is a bijection of Q. (Previously it was known that an AIM Moufang loop that is uniquely 2-divisible and *finite* satisfies $c\ell(Q) \leq 2$, and that there is an AIM Moufang loop Q of size 2^{14} satisfying $c\ell(Q) = 3$ [33].) The same result holds for uniquely 2-divisible AIM left Bol loops.

Despite some serious effort, we have not been able to decide if an AIM LC loop Q satisfies $c\ell(Q) \leq 2$ (indicated in the figure by "?"); we believe that it does.

Finally, all other claims contained within the figure are consequences of the inclusions.

References

1. Belousov, V.D.: Foundations of the Theory of Quasigroups and Loops. Izdat. Nauka, Moscow (1967) (in Russian)
2. Bruck, R.H.: Contributions to the theory of loops. Trans. Amer. Math. Soc. 60, 245–354 (1946)
3. Bruck, R.H.: A Survey of Binary Systems. Springer, Heidelberg (1971)
4. Csörgő, P.: Abelian inner mappings and nilpotency class greater than two. European J. Combin. 28, 858–867 (2007)
5. Csörgő, P., Drápal, A.: Left conjugacy closed loops of nilpotency class two. Results Math. 47, 242–265 (2005)
6. Csörgő, P., Kepka, T.: On loops whose inner permutations commute. Comment. Math. Univ. Carolin. 45, 213–221 (2004)
7. Drápal, A., Kinyon, M.K.: Buchsteiner loops: associators and constructions. arXiv/0812.0412
8. Drápal, A., Vojtěchovský, P.: Explicit constructions of loops with commuting inner mappings. European J. Combin. 29, 1662–1681 (2008)
9. Evans, T.: Homomorphisms of non-associative systems. J. London Math. Soc. 24, 254–260 (1949)
10. Fenyves, F.: Extra loops II. On loops with identities of Bol-Moufang type. Publ. Math. Debrecen 16, 187–192 (1969)
11. The GAP Group: GAP – Groups, Algorithms, and Programming, Version 4.4.10 (2007), http://www.gap-system.org
12. Hullot, J.-M.: A catalogue of canonical term rewriting systems. Technical Report CSC 113, SRI International (1980)
13. Kepka, T.: On the abelian inner permutation groups of loops. Comm. Algebra 26, 857–861 (1998)

14. Kepka, T., Phillips, J.D.: Connected transversals to subnormal subgroups. Comment. Math. Univ. Carolin. 38, 223–230 (1997)
15. Kepka, T., Kinyon, M.K., Phillips, J.D.: The structure of F-quasigroups. J. Algebra 317, 435–461 (2007)
16. Kinyon, M.K., Kunen, K., Phillips, J.D.: Every diassociative A-loop is Moufang. Proc. Amer. Math. Soc. 130, 619–624 (2002)
17. Kinyon, M.K., Kunen, K., Phillips, J.D.: A generalization of Moufang and Steiner loops. Algebra Universalis 48, 81–101 (2002)
18. Kinyon, M.K., Kunen, K., Phillips, J.D.: Diassociativity in conjugacy closed loops. Comm. Algebra 32, 767–786 (2004)
19. Kinyon, M.K., Kunen, K.: The structure of extra loops. Quasigroups and Related Systems 12, 39–60 (2004)
20. Kinyon, M.K., Kunen, K.: Power-associative, conjugacy closed loops. J. Algebra 304, 679–711 (2006)
21. Kinyon, M.K., Phillips, J.D., Vojtěchovský, P.: C-loops: extensions and constructions. J. Algebra Appl. 6, 1–20 (2007)
22. Kinyon, M.K., Phillips, J.D., Vojtěchovský, P.: When is the commutant of a Bol loop a subloop? Trans. Amer. Math. Soc. 360, 2393–2408 (2008)
23. Kinyon, M.K., Veroff, R., Vojtěchovský, P.: Loops with abelian inner mapping groups: an application of automated deduction, Web support (2012), http://www.cs.unm.edu/~veroff/AIM_LC/
24. Knuth, D.E., Bendix, P.B.: Simple word problems in universal algebras. In: Leech, J. (ed.) Proceedings of the Conference on Computational Problems in Abstract Algebras, pp. 263–298. Pergamon Press, Oxford (1970)
25. Kunen, K.: Moufang quasigroups. J. Algebra 183, 231–234 (1996)
26. Kunen, K.: Quasigroups, loops, and associative laws. J. Algebra 185, 194–204 (1996)
27. Kunen, K.: Alternative loop rings. Comm. Algebra 26, 557–564 (1998)
28. Kunen, K.: G-Loops and permutation groups. J. Algebra 220, 694–708 (1999)
29. Kunen, K.: The structure of conjugacy closed loops. Trans. Amer. Math. Soc. 352, 2889–2911 (2000)
30. McCune, W.: Otter 3.0 Reference Manual and Guide. Tech. Report ANL-94/6, Argonne National Laboratory, Argonne, IL (1994), http://www.mcs.anl.gov/AR/otter/
31. McCune, W.: Prover9, version 2009-02A, http://www.cs.unm.edu/~mccune/prover9/
32. Nagy, G., Vojtěchovský, P.: LOOPS: Computing with quasigroups and loops in GAP – a GAP package, version 2.0.0 (2008), http://www.math.du.edu/loops
33. Nagy, G.P., Vojtěchovský, P.: Moufang loops with commuting inner mappings. J. Pure Appl. Algebra 213, 2177–2186 (2009)
34. Niemenmaa, M.: Finite loops with nilpotent inner mapping groups are centrally nilpotent. Bull. Aust. Math. Soc. 79, 109–114 (2009)
35. Niemenmaa, M., Kepka, T.: On connected transversals to abelian subgroups in finite groups. Bull. London Math. Soc. 24, 343–346 (1992)
36. Phillips, J.D.: A short basis for the variety of WIP PACC-loops. Quasigroups Related Systems 14, 73–80 (2006)
37. Phillips, J.D.: The Moufang laws, global and local. J. Algebra Appl. 8, 477–492 (2009)
38. Phillips, J.D., Shcherbacov, V.A.: Cheban loops. J. Gen. Lie Theory Appl. 4, Art. ID G100501, 5 p. (2010)

39. Phillips, J.D., Stanovský, D.: Automated theorem proving in quasigroup and loop theory. AI Commun. 23, 267–283 (2010)
40. Phillips, J.D., Stanovský, D.: Bruck loops with abelian inner mapping groups. Comm. Alg. (to appear)
41. Phillips, J.D., Vojtěchovský, P.: The varieties of loops of Bol-Moufang type. Algebra Universalis 54, 259–271 (2005)
42. Phillips, J.D., Vojtěchovský, P.: The varieties of quasigroups of Bol-Moufang type: an equational reasoning approach. J. Algebra 293, 17–33 (2005)
43. Phillips, J.D., Vojtěchovský, P.: C-loops: an introduction. Publ. Math. Debrecen 68, 115–137 (2006)
44. Phillips, J.D., Vojtěchovský, P.: A scoop from groups: equational foundations for loops. Comment. Math. Univ. Carolin. 49, 279–290 (2008)
45. Pflugfelder, H.O.: Quasigroups and Loops: Introduction. Sigma Series in Pure Math., vol. 8. Heldermann Verlag, Berlin (1990)
46. Rotman, J.J.: An Introduction to the Theory of Groups, 4th edn. Springer, New York (1995)
47. Veroff, R.: Using hints to increase the effectiveness of an automated reasoning program: case studies. J. Automated Reasoning 16, 223–239 (1996)
48. Veroff, R.: Solving open questions and other challenge problems using proof sketches. J. Automated Reasoning 27, 157–174 (2001)
49. Zhang, H., Bonacina, M.P., Hsiang, J.: PSATO: a distributed propositional prover and its application to quasigroup problems. J. Symbolic Computation 21, 543–560 (1996)

(Dual) Hoops Have Unique Halving

Rob Arthan and Paulo Oliva

Queen Mary University of London
School of Electronic Engineering and Computer Science
Mile End Road, London E1 4NS

Abstract. Continuous logic extends the multi-valued Łukasiewicz logic by adding a halving operator on propositions. This extension is designed to give a more satisfactory model theory for continuous structures. The semantics of these logics can be given using specialisations of algebraic structures known as hoops and coops. As part of an investigation into the metatheory of propositional continuous logic, we were indebted to Prover9 for finding proofs of important algebraic laws.

1 Introduction

(Like its title, this chapter begins with a parenthesis concerning notation. It is common practice to order truth-values by decreasing logical strength, but the opposite, or *dual*, convention is used in the literature that motivates the present work. So in this chapter $A \geq B$ means that A is logically stronger than B. Accordingly, in the algebraic structures we will study, 0 models truth rather than falsehood and conjunction corresponds to an operation written as addition rather than multiplication. The halves alluded to in the title would otherwise be square roots.)

Around 1930, Łukasiewicz and Tarski [16] instigated the study of logics admitting models in which the truth values are real numbers drawn from some subset T of the interval $[0,1]$. In these models, with the notational conventions discussed above, conjunction is capped addition: $x \dotplus y = \inf\{x+y, 1\}$. Boolean logic is the special case when $T = \{0,1\}$. These Łukasiewicz logics have been widely studied, e.g., as instances of fuzzy logics [11].

In recent years, Ben Yaacov has used a Łukasiewicz logic with an infinite number of truth values as a building block in what is called continuous logic [3]. Continuous logic unifies work of Henson and others [14] that aims to overcome shortfalls of classical first-order model theory when applied to continuous structures such as metric spaces and Banach spaces. A detailed discussion of these shortfalls would be out of place here, but a few remarks are in order. In functional analysis there is a well-accepted notion of ultraproduct that takes into account metric structure and is an important tool for constructing Banach spaces. By contrast, the class of Banach spaces is not closed under the standard model-theoretic notion of ultraproduct. Continuous logic aims to capture properties that are preserved under the good notion of ultraproduct for continuous

M.P. Bonacina and M.E. Stickel (Eds.): McCune Festschrift, LNAI 7788, pp. 165–180, 2013.

structures [14]. From another point of view, continuous logic mitigates the fact that ordinary first-order logic for continuous structures tends to be unexpectedly strong, the first-order theory of Banach spaces being strictly stronger than second-order arithmetic [20].

The motivation for ordering truth values by increasing logical strength in continuous logic stems from the fact that in a metric space with metric d, $x = y$ iff $d(x, y) = 0$. In first-order continuous logic, one wishes to treat d as a two-place predicate symbol analogous to equality in classical first-order logic. Representing truth by 0 is then the natural choice.

A difficulty with both the Łukasiewicz logics and continuous logic is that it requires considerable ingenuity to work with the known axiomatisations of their propositional fragments. Work on algebraic semantics for Łukasiewicz logic begun by Chang [8,9] has helped greatly with this, but basic algebraic laws in the algebras involved are often quite difficult to prove. This chapter reports on ongoing work to gain a better understanding of both the proof theory and the semantics of continuous logic that is benefitting from the use of automated theorem proving to find counterexamples and to derive algebraic properties.

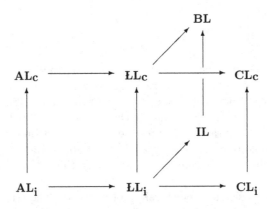

Fig. 1. Eight Logics and the Relationships between Them

Our work began with the observation that both Łukasiewicz logic, $\mathbf{LL_C}$, and Ben Yaacov's continuous logic, $\mathbf{CL_C}$, are extensions of a very simple intuitionistic substructural logic $\mathbf{AL_i}$. In Section 2 of this chapter we show how $\mathbf{CL_C}$ may be built up via a system of extensions of $\mathbf{AL_i}$. We also show how the Brouwer-Heyting intuitionistic propositional logic, $\mathbf{IL}$, and Boolean logic, $\mathbf{BL}$, fit into this picture. The relationships between the eight logics in this system of extensions are depicted in Figure 1. In Section 3, we describe a class of monoids called pocrims that have been quite widely studied in connection with $\mathbf{AL_i}$ and sketch a proof of a theorem asserting that each of the eight logics is sound and complete with respect to an appropriate class of pocrims. The sketch is easy to complete apart from one tricky lemma concernng the continuous logics.

In Section 4, we discuss our use of Bill McCune's Mace4 and Prover9 to assist in these investigations, in particular to prove the lemma needed for the

$$\frac{\Gamma, A \vdash B}{\Gamma \vdash A \multimap B} \, [\multimap\!I] \qquad \frac{\Gamma \vdash A \quad \Delta \vdash A \multimap B}{\Gamma, \Delta \vdash B} \, [\multimap\!E]$$

$$\frac{\Gamma \vdash A \quad \Delta \vdash B}{\Gamma, \Delta \vdash A \otimes B} \, [\otimes I] \qquad \frac{\Gamma \vdash A \otimes B \quad \Delta, A, B \vdash C}{\Gamma, \Delta \vdash C} \, [\otimes E]$$

Fig. 2. The Inference Rules

theorem of Section 2. Our application seems to be a "sweet spot" for this kind of technology: the automatic theorem prover found a proof of a difficult problem that can readily be translated into a human readable form.

In Section 5 we discuss some other results that Prover9 has proved for us. Section 6 gives some concluding remarks.

2   The Logics

We work in a language $\mathcal{L}_{\frac{1}{2}}$ whose atomic formulas are the propositional constants 0 (truth) and 1 (falsehood) and propositional variables drawn from the set Var $= \{P, Q, \ldots\}$. If A and B are formulas of $\mathcal{L}_{\frac{1}{2}}$ then so are $A \multimap B$ (implication), $A \otimes B$ (conjunction) and $A/2$ (halving). We adopt the convention that implication associates to the right and has lower precedence than conjunction, which in turn has lower precedence than halving. So, for example, the brackets in $(A \otimes (B/2)) \multimap (C \multimap (D \otimes F))$ are all redundant, while those in $(((A \to B) \to C) + D)/2$ are all required. We denote by $\mathcal{L}_1$ the language without halving. We write $A^{\perp}$ as an abbreviation for $A \multimap 1$, a form of negation.

The judgements of the eight logics that we will consider are sequents $\Gamma \vdash A$, where A is an $\mathcal{L}_{\frac{1}{2}}$-formula and Γ is a multiset of $\mathcal{L}_{\frac{1}{2}}$-formulas. The inference rules are the introduction and elimination rules for the two binary connectives[1] shown in Figure 2.

The axiom schemata for the logics are selected from those shown in Figure 3. These are the axiom of assumption [ASM], ex-falso-quodlibet [EFQ], double negation elimination [DNE], commutative weak conjunction [CWC], commutative strong disjunction [CSD], the axiom of contraction [CON], and two axioms giving lower and upper bounds for the halving operator: [HLB] and [HUB].

[ASM], [EFQ], [DNE] and [CON] are standard axioms of classical logic. [CON] asserts that A is a strong as $A \otimes A$ and is equivalent to the rule of *contraction* allowing us to infer $\Gamma, A \vdash B$, from $\Gamma, A, A \vdash B$. [CON] allows one to think of the contexts Γ as sets rather than multisets. The significance of [CWC], [CSD],

---

[1] Omitting disjunction from the logic greatly simplifies the algebraic semantics. While it may be unsatisfactory from the point of view of intuitionistic philosophy, disjunction defined using de Morgan's law is adequate for our purposes.

$$\frac{}{\Gamma, A \vdash A} \text{ [ASM]} \qquad\qquad \frac{}{\Gamma, 1 \vdash A} \text{ [EFQ]}$$

$$\frac{}{\Gamma, A^{\perp\perp} \vdash A} \text{ [DNE]} \qquad\qquad \frac{}{\Gamma, A \otimes (A \multimap B) \vdash B \otimes (B \multimap A)} \text{ [CWC]}$$

$$\frac{}{\Gamma, (A \multimap B) \multimap B \vdash (B \multimap A) \multimap A} \text{ [CSD]} \qquad \frac{}{\Gamma, A \vdash A \otimes A} \text{ [CON]}$$

$$\frac{}{\Gamma, A/2, A/2 \vdash A} \text{ [HLB]} \qquad\qquad \frac{}{\Gamma, A/2 \multimap A \vdash A/2} \text{ [HUB]}$$

Fig. 3. The Axiom Schemata

[HLB] and [HUB] will be explained below as we introduce the logics that include them and as we give the semantics for those logics.

The definitions of the eight logics are discussed in the next few paragraphs and are summarised in Table 1. In all but $\mathbf{CL_i}$ and $\mathbf{CL_c}$, halving plays no rôle and the logical language may be taken to be the sublanguage $\mathcal{L}_1$ in which halving does not feature.

Intuitionistic affine logic [4], $\mathbf{AL_i}$, has for its axiom schemata [ASM] and [EFQ]. All our other logics include $\mathbf{AL_i}$. The contexts Γ, Δ are multisets because we wish to keep track of how many times each of the assumptions in Γ is used in order to derive the conclusion A in $\Gamma \vdash A$. This is not relevant if formulas can be duplicated or contracted (i.e. if A is equivalent to $A \otimes A$). We will, however, mainly work with so-called *substructural logics* where such equivalences are not valid in general. $\mathbf{AL_i}$ serves as a prototype for such substructural logics.

Under the Curry-Howard correspondence between proofs and λ-terms, the proof system $\mathbf{AL_i}$ corresponds to a λ-calculus with pairing and paired abstractions, so in this calculus, if t, u and v are terms, then so are (t, u), $(t, (u, v)), \lambda(x, y) \bullet t$, $\lambda((x, y), z) \bullet u$, $\lambda(x, (y, z)) \bullet v$ etc. where x, y and z are variables. Proofs in $\mathbf{AL_i}$ then correspond to *affine* λ-terms: terms in which each variable is used at most once. So for example $\lambda f \bullet \lambda x \bullet \lambda y \bullet f(x, y)$ is an affine λ-term corresponding to a proof of the sequent $\vdash (A \otimes B \multimap C) \multimap A \multimap B \multimap C$.

Classical affine logic [10], $\mathbf{AL_c}$, extends $\mathbf{AL_i}$ with the axiom schema [DNE]. It can also be viewed as the extension of the so-called multiplicative fragment of Girard's linear logic by allowing weakening and the axiom schema [EFQ].

What we will call *intuitonistic Łukasiewicz logic*, $\mathbf{LL_i}$, extends $\mathbf{AL_i}$ with the axiom schema [CWC]. $\mathbf{LL_i}$ is known by a variety of names in the literature. The name we use reflects its position in Figure 1. For any formulas A and B, $A \otimes (A \multimap B)$ implies both A and B and so can be thought of as a weak form of conjunction. In $\mathbf{LL_i}$ we have commutativity of this weak conjunction. [CWC] turns out to be a surprisingly powerful axiom. However, it often requires considerable ingenuity to use it.

Classical Łukasiewicz logic [13], $\mathbf{LL_c}$, extends $\mathbf{AL_i}$ with the axiom schema [CSD]. Just as $A \otimes (A \multimap B)$ can be viewed as a form of conjunction,

Table 1. The Logics and their Models

Logic	Axioms	Models
$\mathbf{AL_i}$	[ASM] + [EFQ]	bounded pocrims
$\mathbf{AL_c}$	$\mathbf{AL_i}$ + [DNE]	bounded involutive pocrims
$\mathbf{LL_i}$	$\mathbf{AL_i}$ + [CWC]	bounded hoops
$\mathbf{LL_c}$	$\mathbf{AL_i}$ + [CSD]	bounded Wajsberg hoops
$\mathbf{IL}$	$\mathbf{AL_i}$ + [CON]	bounded idempotent pocrims
$\mathbf{BL}$	$\mathbf{IL}$ + [DNE]	bounded involutive idempotent pocrims
$\mathbf{CL_i}$	$\mathbf{LL_i}$ + [HLB] + [HUB]	bounded coops
$\mathbf{CL_c}$	$\mathbf{LL_c}$ + [HLB] + [HUB]	bounded involutive coops

$(A \multimap B) \multimap B$ can be viewed as a form of disjunction that may be stronger than the one defined by the usual intuitionistic rules for disjunction. In $\mathbf{LL_c}$ we have commutativity of this strong disjunction. This gives the widely-studied multi-valued logic of Łukasiewicz. Like [CWC], [CSD] is powerful but not always easy to use.

Intuitionistic propositional logic, $\mathbf{IL}$, extends $\mathbf{AL_i}$ with the axiom schema of contraction [CON]. This gives us the conjunction-implication fragment of the well-known Brouwer-Heyting intuitionistic propositional logic.

Classical propositional logic (or boolean logic), $\mathbf{BL}$, extends $\mathbf{IL}$ with the axiom schema [DNE]. This is the familiar two-valued logic of truth tables.

What we have termed *intuitionistic continuous logic*, $\mathbf{CL_i}$, allows the halving operator and extends $\mathbf{LL_i}$ with the axiom schemas [HLB] and [HUB], which effectively give lower and upper bounds on the logical strength of $A/2$. They imply the surprisingly strong condition that $A/2$ is equivalent to $A/2 \multimap A$. This is an intuitionistic version of the continuous logic of Ben Yaacov [3].

Classical continuous logic, $\mathbf{CL_c}$ extends $\mathbf{CL_i}$ with the axiom schema [DNE]. This gives Ben Yaacov's continuous logic. The motivating model takes truth values to be real numbers between 0 and 1 with conjunction defined as capped addition.

Our initial goal was to gain insight into $\mathbf{CL_c}$ by investigating the relations amongst $\mathbf{AL_i}$, $\mathbf{LL_c}$ and $\mathbf{CL_c}$. The other logics came into focus when we tried to decompose the somewhat intractable axiom [CSD] into a combination of [DNE] and an intuitionistic component. It can be shown that the eight logics are related as shown in Figure 1. In the figure, an arrow from T_1 to T_2 means that T_2 extends T_1, i.e., the set of provable sequents of T_2 contains that of T_1. In each square, the north-east logic is the least extension of the south-west logic that contains the other two. For human beings, at least, the proof of this fact is quite tricky for the $\mathbf{AL_i}$-$\mathbf{LL_c}$ square, see [11, chapters 2 and 3].

The routes in Figure 1 from $\mathbf{AL_i}$ to $\mathbf{IL}$ and $\mathbf{BL}$ have been quite extensively studied, as may be seen from [5,18,15] and the works cited therein. We are not aware of any work on $\mathbf{CL_i}$, but it is clearly a natural object of study in connection with Ben Yaacov's continuous logic. It should be noted that $\mathbf{IL}$ and $\mathbf{CL_i}$ are incompatible: as we will see at the end of this section, any formula is provable given the axioms [CON], [HLB] and [HUB].

3   Algebraic Semantics

We give algebraic semantics for the logics of Section 2 using *pocrims*: partially ordered, commutative, residuated, integral monoids.

Definition 1. *A* pocrim[2] *is a structure for the signature* $(0, +, \rightarrow; \geq)$ *of type* $(0, 2, 2; 2)$ *satisfying the following laws:*

$$(x + y) + z = x + (y + z) \qquad [\mathsf{m_1}]$$
$$x + y = y + x \qquad [\mathsf{m_2}]$$
$$x + 0 = x \qquad [\mathsf{m_3}]$$
$$x \geq x \qquad [\mathsf{o_1}]$$
$$\text{if } x \geq y \text{ and } y \geq z, \text{ then } x \geq z \qquad [\mathsf{o_2}]$$
$$\text{if } x \geq y \text{ and } y \geq x, \text{ then } x = y \qquad [\mathsf{o_3}]$$
$$\text{if } x \geq y, \text{ then } x + z \geq y + z \qquad [\mathsf{o_4}]$$
$$x \geq 0 \qquad [\mathsf{b}]$$
$$x + y \geq z \text{ iff } x \geq y \rightarrow z \qquad [\mathsf{r}]$$

Intuitively, $\rightarrow$ is the semantic counterpart of the syntactic implication $\multimap$, whereas $+$ corresponds to the syntactic conjunction $\otimes$. As with the syntactic connectives, we adopt the convention that $\rightarrow$ associates to the right and has lower precedence than $+$. Note that $=$ and $\geq$ are predicate symbols and so necessarily have lower precedence than the function symbols $\rightarrow$ and $+$: the only valid reading of $a \rightarrow b \geq c + d$ is as $(a \rightarrow b) \geq (c + d)$.

Let $\mathbf{M} = (M, 0, +, \rightarrow; \geq)$ be a pocrim. The laws $[\mathsf{m_i}]$, $[\mathsf{o_j}]$ and $[\mathsf{b}]$ say that $(M, 0, +; \geq)$ is a partially ordered commutative monoid with the identity 0 as least element. Law $[\mathsf{r}]$, the *residuation property*, says that for any x and z the set $\{y \mid x + y \geq z\}$ is non-empty and has $x \rightarrow z$ as least element. $\mathbf{M}$ is said to be *bounded* if it has a (necessarily unique) *annihilator*, i.e., an element 1 such that for every x we have:

$$1 = x + 1 \qquad [\mathsf{ann}]$$

Let us assume $\mathbf{M}$ is bounded. Then $1 = x + 1 \geq x \geq 0$ for any x and $(M; \geq)$ is indeed a bounded ordered set. Let $\alpha : \mathsf{Var} \rightarrow M$ be an interpretation of logical variables as elements of M and extend α to a function $v_\alpha : \mathcal{L}_1 \rightarrow M$ by interpreting 0, 1, $\otimes$ and $\multimap$ as 0, 1, $+$ and $\rightarrow$ respectively. If $\Gamma = C_1, \ldots, C_n$, we say that α *satisfies* the sequent $\Gamma \vdash A$, iff $v_\alpha(C_1) + \ldots + v_\alpha(C_n) \geq v_\alpha(A)$. We say that $\Gamma \vdash A$ is *valid in* $\mathbf{M}$ if it is satisfied by every assignment $\alpha : \mathsf{Var} \rightarrow M$.

---

[2] Strictly speaking, this is a *dual* pocrim, since we order it by increasing logical strength and write it additively.

We say **M** is a *model* for a logic **L** if every sequent provable in **L** is valid in **M**. If C is a class of pocrims, we say $\Gamma \vdash A$ is *valid* if it is valid in every $\mathbf{M} \in C$.

We will need some special classes of pocrim. We write $\neg x$ as an abbreviation for $x \to 1$, a semantic analogue of the derived syntactic operator $\perp$. We say a bounded pocrim is *involutive* if it satisfies $\neg\neg x = x$. We say a pocrim is *idempotent* if it is idempotent as a monoid, i.e., it satisfies $x + x = x$.

Definition 2 (Büchi & Owens[7]). *A* hoop[3] *is a pocrim that is naturally ordered, i.e., whenever $x \geq y$, there is z such that $x = y + z$.*

It is a nice exercise in the use of the residuation property to show that a pocrim is a hoop iff it satisfies the identity

$$x + (x \to y) = y + (y \to x) \qquad [\mathsf{cwc}]$$

In any pocrim, $x \leq x + (x \to y) \geq y$, so we can view $x + (x \to y)$ as a weak form of conjunction, but in general this conjunction is not commutative and there need be no least z such that $x \leq z \geq y$. In a hoop, the weak conjunction is commutative and $x + (x \to y)$ can be shown to be the least upper bound of x and y.

Definition 3 (Blok & Ferreirim[5]). *A* Wajsberg hoop *is a hoop satisfying the identity*

$$(x \to y) \to y = (y \to x) \to x \qquad [\mathsf{csd}]$$

We may view $(x \to y) \to y$ as a form of disjunction. In a Wajsberg hoop this disjunction is commutative and can be shown to give a greatest lower bound of x and y. See [5] for more information on hoops and Wajsberg hoops.

Definition 4. *A* continuous hoop, *or* coop, *is a hoop where for every x there is a unique y such that $y = y \to x$. In this case we write $y = x/2$.*

In a coop, for any x, we have $x \geq x/2 \to x = x/2$, whence, by [cwc], $x = x + 0 = x + (x \to x/2) = x/2 + (x/2 \to x) = x/2 + x/2$, justifying our choice of notation. Here, as with the syntactic connectives, we take halving to have higher precedence than conjunction.

If **M** is a coop, we extend the function $v_\alpha : \mathcal{L}_1 \to M$ induced by an interpration $\alpha : \mathsf{Var} \to M$ to a function $v_\alpha : \mathcal{L}_{\frac{1}{2}} \to M$ by interpreting $A/2$ as $v_\alpha(A)/2$. The notions of validity and satisfaction extend to interpretations of $\mathcal{L}_{\frac{1}{2}}$ in a coop in the evident way.

We say that a logic L is sound for a class of pocrims C if every sequent that is provable in L is valid in C. We say that L is complete for C if the converse holds. We then have:

---

[3] Büchi and Owens [7] write of hoops that "their importance ... merits recognition with a more euphonious name than the merely descriptive "commutative complemented monoid"". Presumably they chose "hoop" as a euphonious companion to "group" and "loop".

Theorem 1. *Each of the logics* $\mathbf{AL_i}$, $\mathbf{AL_c}$, $\mathbf{LL_i}$, $\mathbf{LL_c}$, $\mathbf{IL}$, $\mathbf{BL}$, $\mathbf{CL_i}$ *and* $\mathbf{CL_c}$ *is sound and complete for the class of pocrims listed for it in the column headed "Models" in Table 1.*

Proof: The proof follows a standard pattern and, with one exception, filling in the details is straightforward. Soundness is a routine exercise. For the completeness, one defines an equivalence relation $\simeq$ on formulas such that $A \simeq B$ holds iff both $A \vdash B$ and $B \vdash A$ are provable in the logic. One then shows that the set of equivalence classes becomes a pocrim in the indicated class, the *term model*, under operators $+$ and $\rightarrow$ induced on the equivalence classes by $\otimes$ and $\multimap$. As the only sentences valid in the term model are those provable in the logic, completeness follows. The difficult detail is showing that the term models for the continuous logics satisfy our definition of a coop: it is easy to see that for any $x = [A]$, one has that $y = [A/2]$ satisfies $y = y \rightarrow x$, but is this y unique? We shall answer this question in the affirmative in the next section. If the equation $y = y \rightarrow x$ did not uniquely determine y, halving would not be well-defined on the term model and the completeness proof would fail.   ∎

Using Theorem 1, we can give an algebraic proof of the claim made earlier that $\mathbf{IL}$ and $\mathbf{CL_i}$ are incompatible. By dint of the theorem, this is equivalent to the claim that a bounded idempotent coop is the trivial coop $\{0\}$. We may prove this as follows: if a is an element of a coop and $a/2$ is idempotent, so that $a/2 = a/2 + a/2$, then $a/2 \geq a/2 + a/2 = a$, so by the residuation property, $a/2 \rightarrow a = 0$. Now $a/2 = a/2 \rightarrow a$ by the definition of a coop, so we have $a = a/2 + a/2 = (a/2 \rightarrow a) + (a/2 \rightarrow a) = 0 + 0 = 0$.

4   Automated Proofs and Counterexamples

In our early attempts to understand the relationships represented in Figure 1, we spent some time devising finite pocrims with interesting properties. This can be a surprisingly difficult and error-prone task. Verifying associativity, in particular, is irksome. Having painstakingly accumulated a small stock of examples, a conversation with Alison Pease reminded us of the existence of Bill McCune's Mace4 tool [17] that automatically searches for finite counter-examples to conjectures in a finitely axiomatised first-order theory.

It was fascinating to see Mace4 recreate examples similar to those we had already constructed. The following input asks Mace4 to produce a counterexample to the conjecture that all bounded pocrims are hoops:

```
op(500, infix, "==>").
formulas(assumptions).
    (x + y) + z = x + (y + z).        % monoid law 1
    x + y = y + x.                    % monoid law 2
    x + 0 = x.                        % monoid law 3
    x >= x.                           % ordering law 1
    x >= y & y >= z -> x >= z.        % ordering law 2
```

```
    x >= y & y >= x -> x = y.          % ordering law 3
    x >= y -> x + z >= y + z.          % ordering law 4
    x >= 0.                            % boundedness law
    x + 1 = 1.                         % annihilator law
    x + y >= z <-> x >= y ==> z.       % residuation law
end_of_list.
formulas(goals).
    x + (x ==> y) = y + (y ==> x).     % can we derive cwc?
end_of_list.
```

Here we use '==>' and '>=' to represent '$\to$' and '$\geq$' in the pocrim and '&', '->' and '<->' are Mace4 syntax for logical conjunction, implication and bi-implication. Given the above, Mace4 quickly prints out the diagram of a pocrim on the ordered set $0 < p < q < 1$ with $x + y = 1$ whenever $\{x, y\} \subseteq \{p, q, 1\}$, a counter-example which we had already come up with over the course of an afternoon. That led us to test Mace4 on yet other conjectures which we had already refuted with some small counter-examples. Mace4, again and again, came up with similar counter-models to the ones we had contrived.

Some weeks later we wanted to show that the two axiom schemata [HLB] and [HUB] uniquely determine the halving operator over the logic $\mathbf{LL_i}$, which would conclude the proof of Theorem 1. That would give us an intuitionistic counterpart $(\mathbf{CL_i})$ to continuous logic $\mathbf{CL_C}$. In logical terms, we wanted to show that the rule shown in Figure 4 is derivable in $\mathbf{LL_i}$:

$$\frac{A \multimap B \vdash A \qquad A \vdash A \multimap B \qquad C \vdash C \multimap B \qquad C \multimap B \vdash C}{A \vdash C}$$

Fig. 4. A Conjectured Inference Rule

After several failed attempts to find a proof, we had started to wonder whether the rule was not derivable. That is when we thought of using Prover9 to look for a proof. We gave Prover9 the input shown below comprising the laws for a hoop, the assumptions $a \to b = a$ (corresponding to $A \multimap B \vdash A$ and $A \vdash A \multimap B$) and $c \to b = c$ (corresponding to $C \vdash C \multimap B$ and $C \multimap B \vdash C$) and the goal $a = c$. (Because the conjectured inference rule is symmetric in A and C, if the rule is valid, then the antecedents imply that A and C are equivalent).

```
op(500, infix, "==>").
formulas(assumptions).
    (x + y) + z = x + (y + z).         % monoid law 1
    x + y = y + x.                     % monoid law 2
    x + 0 = x.                         % monoid law 3
    x >= x.                            % ordering law 1
```

```
x >= y & y >= z -> x >= z.              % ordering law 2
x >= y & y >= x -> x = y.               % ordering law 3
x >= y -> x + z >= y + z.               % ordering law 4
x >= 0.                                 % boundedness law
x + y >= z <-> x >= y ==> z.            % residuation law
x + (x ==> y) = y + (y ==> x).          % cwc
a ==> b = a.                            % assumption 1
c ==> b = c.                            % assumption 2
end_of_list.
formulas(goals).
    a = c.
end_of_list.
```

To our surprise Prover9 took just a few seconds to produce the proof shown in the appendix. The proof that Prover9 found seems perplexingly intricate at first glance, but after studying it for a little while, we found we could edit it into a form fit for human consumption. From a human perspective, the proof involves the 9 intermediate claims given in the following lemma. Once these are proved, we will see that the desired result is an easy consequence of claim (9).

Lemma 2. *Let* $\mathbf{M} = (M, 0, +, \rightarrow; \geq)$ *be a hoop and let* $a, b, c, x, y \in M$. *Assume that, (i),* $a \rightarrow b = a$ *and, (ii),* $c \rightarrow b = c$. *Then the following hold:*

$$
\begin{aligned}
&(1) \quad b \geq a \text{ and } b \geq c, \\
&(2) \quad a + a = b, \\
&(3) \quad a \rightarrow (a \rightarrow c) = 0, \\
&(4) \quad (x \rightarrow y) + z \geq x \rightarrow (y + (y \rightarrow x) + z), \\
&(5) \quad c \rightarrow (a + a + x) \geq c, \\
&(6) \quad c \rightarrow a \geq a \rightarrow c, \\
&(7) \quad c \rightarrow a = a \rightarrow c, \\
&(8) \quad c + (c \rightarrow a) + ((a \rightarrow c) \rightarrow a) = b, \\
&(9) \quad a + c = b.
\end{aligned}
$$

Proof: In the proof below (in)equalities which are not labelled as following from one of the assumptions (i) and (ii) or an earlier part of the lemma follow immediately from the axioms of a pocrim.

(1) We have $b \geq a \rightarrow b$ and, by (i), $a \rightarrow b = a$). So $b \geq a$ and similarly $b \geq c$ using (ii).

(2) By (1) we have $b \rightarrow a = 0$. Therefore

$$
\begin{aligned}
a + a &= a + (a \rightarrow b) & (i) \\
&= b + (b \rightarrow a) & [\text{cwc}] \\
&= b.
\end{aligned}
$$

(3) By (i) and (1) we have $a = a \rightarrow b \geq a \rightarrow c$ and hence $0 \geq a \rightarrow (a \rightarrow c)$, which implies (3).

(4) By [cwc] $x + (x \rightarrow y) + z = y + (y \rightarrow x) + z$, whence (4) follows.
(5) We have

$$c \rightarrow (b + x) \geq c \rightarrow b$$
$$= c \qquad (ii)$$

and then using (2) we obtain (5).
(6) By (5), as $(c \rightarrow a) + a \geq c \rightarrow (a + a)$, we have $(c \rightarrow a) + a \geq c$ and hence (6).
(7) Our assumptions are symmetric in a and c. Hence, (6) holds with a and c
interchanged, i.e., $a \rightarrow c \geq c \rightarrow a$, which taken with (6) gives (7).
(8) We have

$$c + (c \rightarrow a) + ((a \rightarrow c) \rightarrow a) = a + (a \rightarrow c) + ((a \rightarrow c) \rightarrow a) \qquad \text{[cwc]}$$
$$= a + a + (a \rightarrow (a \rightarrow c)) \qquad \text{[cwc]}$$
$$= b + (a \rightarrow (a \rightarrow c)) \qquad (2)$$
$$= b. \qquad (3)$$

(9) We have

$$b = c + (c \rightarrow a) + ((a \rightarrow c) \rightarrow a) \qquad (8)$$
$$= c + (a \rightarrow c) + ((a \rightarrow c) \rightarrow a) \qquad (7)$$
$$= c + a + (a \rightarrow (a \rightarrow c)) \qquad \text{[cwc]}$$
$$= c + a. \qquad (3)$$

This completes the proof of the lemma.   ∎

It is interesting to note the complexity of the proof in terms of uses of [cwc]
(used 6 times!) and the important sub-lemma (2) (used twice) as depicted in the
outline proof tree shown in Figure 5.

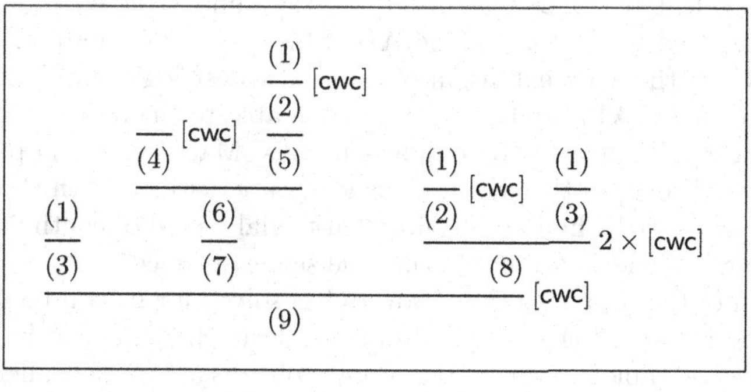

Fig. 5. Outline of the Proof of Lemma 2

Finally, from part (9) of Lemma 2 we have the theorem that the equation
$a \rightarrow b = a$ uniquely determines a in terms of b:

Theorem 3. *In any hoop, if $a \rightarrow b = a$ and $c \rightarrow b = c$ then $a = c$.*

Proof: Since the assumptions are symmetric in a and c it is enough to show $c \geq a$, from which we can immediately conclude $a \geq c$ and hence $a = c$. By Lemma 2 (9) we have $c \geq a \rightarrow b$ and hence $c \geq a$.　■

We already have the part of Theorem 1 that gives soundness and completeness of $\mathbf{LL_i}$ for bounded hoops. Theorem 3 now gives us that the continuous logic axioms [HLB] and [HUB] uniquely determine halving given the other axioms of $\mathbf{LL_i}$ and that is exactly what we need to complete the proof of Theorem 1.

5 Subsequent Work

The importance of Theorem 3 is that it provides a powerful method for proving statements of the form $a = b/2$ in a coop: to prove $a = b/2$, one proves that $a = a \rightarrow b$. Very frequently one has to prove statements of the forms $a \geq b/2$ and $a \leq b/2$. The result on equality suggests that sufficient conditions for these should be $a \geq a \rightarrow b$ and $a \leq a \rightarrow b$ respectively. In logical terms, this means that it is valid to omit either the first or the last of the antecedents in the inference rule of Figure 4. Encouraged by our success with Theorem 3, we presented these two problems to Prover9, which, in just under 4 minutes and just over 20 minutes respectively, found proofs, that turned out to be even simpler than that of Theorem 3. Once one has these basic tools for reasoning about the halving operator, a deeper investigation of the algebra of coops becomes possible. One finds for example, that a coop is simple (in the sense of universal algebra) iff it is isomorphic to a coop of real numbers under capped addition. See [2] for more information and for the lovely proofs found by Prover9 of the rules for $a \geq b/2$ and $a \leq b/2$.

Prover9 has also found some other intricate proofs in this area. For example, it can prove a lemma on pocrims implying that the axiom schemata [CWC] + [DNE] is equivalent to [CSD] over intuitionistic affine logic $\mathbf{AL_i}$. This implies the aforementioned result that in the $\mathbf{AL_i}$-$\mathbf{LL_c}$ square of Figure 1, the north-east logic $\mathbf{LL_c}$ is the least extension of the south-west logic $\mathbf{AL_i}$ that contains the other two logics $\mathbf{AL_c}$ and $\mathbf{LL_i}$. Prover9 is able to prove analogous results for each square in Figure 1. To complement this, Mace4 can also produce the examples needed to show that the various logics are distinct, with the exception of the logics in the right-hand column: a non-trivial model of continuous logic is necessarily infinite and hence not within the scope of Mace4.

A selection of the problems that Prover9 has solved for us will be included in a forthcoming release of the TPTP Problem Library [21]. As can be seen from the CPU times in Table 2, some of the proof problems are quite challenging. The timings were taken on an Apple iMac with a 3.06 GHz Intel Core 2 Duo processor using Prover9's "auto" settings. The only tuning we have done is with the choice of axiomatization. Most of the problems use a straightforward translation into first-order logic of the various equations and Horn clauses given above as the axioms for pocrims, hoops etc. For hoops, a purely equational axiomatization is

Table 2. CPU Times for Theorems Contributed to TPTP

TPTP Name	Problem Statement	Seconds
LCL888+1.p	Halving is unique: rule for $a = b/2$	3.38
LCL889+1.p	Halving is unique: rule for $a \geq b/2$	229.13
LCL890+1.p	Halving is unique: rule for $a \leq b/2$ (i)	1,216.69
LCL891+1.p	Halving is unique: rule for $a \leq b/2$ (ii)	12,724.08
LCL892+1.p	Halving is unique: rule for $a \leq b/2$ (iii)	51,876.82
LCL893+1.p	$x/2 = x$ implies $x = 0$	0.01
LCL894+1.p	Weak conjunction is l.u.b. in a hoop (Horn)	1.90
LCL895+1.p	Weak conjunction is l.u.b. in a hoop (Equational)	14.41
LCL896+1.p	Associativity of weak conjunction implies [cwc]	5.95
LCL897+1.p	Weak conjunction is associative in a hoop	0.10
LCL898+1.p	An involutive hoop has [csd]	66.30
LCL899+1.p	A bounded pocrim with [csd] is involutive	0.01
LCL900+1.p	A bounded pocrim with [csd] is a hoop	7.21
LCL901+1.p	An idempotent pocrim with [csd] is boolean	0.74
LCL902+1.p	A boolean pocrim is involutive	0.02
LCL903+1.p	A boolean pocrim is idempotent	1.42

known and, in one case (LCL897+1.p), we were unable to obtain a proof using the Horn axiomatization but obtained a proof very rapidly with the equational axioms. In other cases (LCL894+1.p, LCL895+1.p), the Horn axiomatization gives quicker results.

The three axiomatizations we tried for the rule for proving $a \leq b/2$ displayed an interesting phenomenon: in the first axiomatization we tried (LCL890+1.p), we included the annihilator axiom $1 + x = x$, but the proof, which has 53 steps and was found in about 20 minutes, makes no use of this. When we tried again without the unnecessary axiom (LCL891+1.p), the search took an order of magnitude longer and found a proof with 154 steps. When we put the axiom back in, but this time at the end of the list of axioms (LCL892+1.p), the search took over 14 hours and gave a proof with 283 steps. Presumably, in our fortunate first attempt the annihilator axiom had a beneficial influence on the subsumption process and eliminated a lot of blind alleys.

When the TPTP formulation of the problems were tried on a selection of automated theorem provers, only Prover9 was able to find a proof for the first two problems in less than 300 seconds. Each problem has been proved by at least one other prover given enough time. From our perspective as users of this technology, this is very remarkable: Prover9 delivered a proof of a key lemma (LCL888+1.p) in just over 3 seconds. Encouraged by that, we were prepared to be patient when we tried the two important refinements of that lemma (LCL889+1.p and LCL890+1.p). These three lemmas have been invaluable in our subsequent theoretical work on the algebra of coops. We suspect our progress would have been very different if the first lemma had severely tested our patience.

6   Final Remarks

We are by no means the first to apply automated theorem proving technology in the area of Łukasiewicz logics. In 1990, a conjecture of Łukasiewicz was proposed by Wos as a challenge problem in automated theorem proving [23] that was successfully attacked by Anantharaman and Bonacina [1,6]. Others to apply automated theorem proving to Łukasiewicz logics include Harris and Fitelson [12] and Slaney [19]. Veroff and Spinks [22] used Otter to find a remarkable direct algebraic proof of a property of idempotent elements in hoops that had previously only been proved by indirect model-theoretic methods.

Clearly our application is one to which technology such as Mace4 and Prover9 is well suited. It is nonetheless a ringing tribute to the late Bill McCune that the accessibility and ease of use of these tools have enabled two naive users to get valuable results with very little effort.

Acknowledgments. We are grateful to the referees for pointers to the literature and for many other helpful suggestions; to Roger Bishop Jones for commenting on a draft of the chapter; to Geoff Sutcliffe for including our problem set in the TPTP Problem Library and for running the problems on a selection of provers; and to Bob Veroff for helping us understand Prover9 performance.

References

1. Anantharaman, S., Bonacina, M.P.: An Application of Automated Equational Reasoning to Many-Valued Logic. In: Okada, M., Kaplan, S. (eds.) CTRS 1990. LNCS, vol. 516, pp. 156–161. Springer, Heidelberg (1991)
2. Arthan, R., Oliva, P.: Hoops, coops and the algebraic semantics of continuous logic (2012), http://arXiv.org/abs/1212.2887v1
3. Ben Yaacov, I., Pedersen, A.P.: A proof of completeness for continuous first-order logic (2009), http://arxiv.org/0903.4051
4. Bierman, G.M.: On intuitionistic linear logic. PhD thesis, University of Cambridge Computer Laboratory (December 1993)
5. Blok, W.J., Ferreirim, I.M.A.: On the structure of hoops. Algebra Universalis 43(2-3), 233–257 (2000)
6. Bonacina, M.P.: Problems in Łukasiewicz logic. Newsletter of the Association for Automated Reasoning 18, 5–12 (1991), http://www.AARInc.org
7. Büchi, J.R., Owens, T.M.: Complemented monoids and hoops (1975) (Unpublished manuscript)
8. Chang, C.C.: Algebraic analysis of many valued logics. Trans. Amer. Math. Soc. 88, 467–490 (1958)
9. Chang, C.C.: A new proof of the completeness of the Łukasiewicz axioms. Trans. Amer. Math. Soc. 93, 74–80 (1959)
10. Girard, J.-Y.: Linear logic. Theoretical Computer Science 50(1), 1–102 (1987)
11. Hájek, P.: Metamathematics of Fuzzy Logic. Kluwer Academic Publishers (1998)
12. Harris, K., Fitelson, B.: Distributivity in $L_{\aleph_0}$ and other sentential logics. J. Autom. Reasoning 27, 141–156 (2001)

13. Hay, L.S.: Axiomatization of the infinite-valued predicate calculus. Journal of Symbolic Logic 28, 77–86 (1963)
14. Henson, C.W., Iovino, J.: Ultraproducts in analysis. In: Analysis and Logic. London Mathematical Society Lecture Notes, vol. 262, pp. 1–113. Cambridge University Press (2002)
15. Köhler, P.: Brouwerian semilattices. Trans. Amer. Math. Soc. 268, 103–126 (1981)
16. Łukasiewicz, J., Tarski, A.: Untersuchungen über den Aussagenkalkül. C. R. Soc. Sc. Varsovie 23, 30–50 (1930)
17. McCune, W.: Prover9 and Mace4 (2005-2010), http://www.cs.unm.edu/~mccune/prover9/
18. Raftery, J.G.: On the variety generated by involutive pocrims. Rep. Math. Logic 42, 71–86 (2007)
19. Slaney, J.K.: More proofs of an axiom of Łukasiewicz. J. Autom. Reasoning 29, 59–66 (2002)
20. Solovay, R., Arthan, R.D., Harrison, J.: Some new results on decidability for elementary algebra and geometry. Ann. Pure Appl. Logic 163(12), 1765–1802 (2012)
21. Sutcliffe, G.: The TPTP Problem Library and Associated Infrastructure: The FOF and CNF Parts, v3.5.0. Journal of Automated Reasoning 43(4), 337–362 (2009)
22. Veroff, R., Spinks, M.: On a homomorphism property of hoops. Bulletin of the Section of Logic 33(3), 135–142 (2004)
23. Wos, L.: New challenge problem in sentential calculus. Newsletter of the Association for Automated Reasoning 16, 7–8 (1990), http://www.AARInc.org

Appendix

Formal proof of Theorem 3 as output by Prover9:

```
1 x >= y & y >= z -> x >= z # label(non_clause).  [assumption].
2 x >= y & y >= x -> x = y # label(non_clause).  [assumption].
3 x + z >= y <-> z >= x ==> y # label(non_clause).  [assumption].
4 x >= y -> x + z >= y + z # label(non_clause).  [assumption].
5 x >= y -> y ==> z >= x ==> z # label(non_clause).  [assumption].
6 x >= y -> z ==> x >= z ==> y # label(non_clause).  [assumption].
7 y = y ==> x & z = z ==> x -> y = z
                    # label(non_clause) # label(goal).  [goal].
8 (x + y) + z = x + (y + z).  [assumption].
9 x + y = y + x.  [assumption].
10 x + 0 = x.  [assumption].
11 x >= x.  [assumption].
12 -(x >= y) | -(y >= z) | x >= z.  [clausify(1)].
13 -(x >= y) | -(y >= x) | y = x.  [clausify(2)].
14 -(x + y >= z) | y >= x ==> z.  [clausify(3)].
15 x + y >= z | -(y >= x ==> z).  [clausify(3)].
16 x >= 0.  [assumption].
17 -(x >= y) | x + z >= y + z.  [clausify(4)].
18 -(x >= y) | y ==> z >= x ==> z.  [clausify(5)].
19 -(x >= y) | z ==> x >= z ==> y.  [clausify(6)].
20 x + (x ==> y) = y + (y ==> x).  [assumption].
21 c1 ==> c2 = c1.  [deny(7)].
22 c3 ==> c2 = c3.  [deny(7)].
23 c3 != c1.  [deny(7)].
24 x + (y + z) = y + (x + z).  [para(9(a,1),8(a,1,1)),rewrite([8(2)])].
27 0 + x = x.  [para(10(a,1),9(a,1)),flip(a)].
28 x >= y ==> (y + x).  [hyper(14,a,11,a)].
30 -(x + y >= z) | x >= y ==> z.  [para(9(a,1),14(a,1))].
31 -(x >= y) | 0 >= x ==> y.  [para(10(a,1),14(a,1))].
32 x + (x ==> y) >= y.  [hyper(15,b,11,a)].
```

```
33 x >= y ==> 0.   [hyper(14,a,16,a)].
34 x + y >= y.   [hyper(17,a,16,a),rewrite([27(3)])].
35 0 ==> x >= y ==> x.   [hyper(18,a,16,a)].
36 x + ((x ==> y) + z) = y + ((y ==> x) + z).
                        [para(20(a,1),8(a,1,1)),rewrite([8(3)])].
41 c3 + x >= c2 | -(x >= c3).   [para(22(a,1),15(b,2))].
43 -(x + (y + z) >= u) | x + z >= y ==> u.   [para(24(a,1),14(a,1))].
46 0 ==> x = x + (x ==> 0).   [para(27(a,1),20(a,1))].
52 x ==> 0 = 0.   [hyper(13,a,16,a,b,33,a),flip(a)].
53 0 ==> x = x.   [back_rewrite(46),rewrite([52(4),10(4)])].
54 x >= y ==> x.   [back_rewrite(35),rewrite([53(2)])].
55 x ==> (y + z) >= x ==> z.   [hyper(19,a,34,a)].
70 x >= y ==> (x + y).   [para(9(a,1),28(a,2,2))].
81 c2 >= c1.   [para(21(a,1),54(a,2))].
82 c2 >= c3.   [para(22(a,1),54(a,2))].
86 x + c2 >= c1.   [hyper(12,a,34,a,b,81,a)].
89 x ==> c2 >= x ==> c3.   [hyper(19,a,82,a)].
127 x >= c2 ==> c1.   [hyper(30,a,86,a)].
171 c2 ==> c1 = 0.   [hyper(13,a,16,a,b,127,a),flip(a)].
180 c1 + c1 = c2.
        [para(171(a,1),20(a,1,2)),rewrite([9(3),27(3),21(5)]),flip(a)].
205 c1 + (x + c1) = x + c2.   [para(180(a,1),8(a,2,2)),rewrite([9(4)])].
271 x + ((x ==> y) + ((y ==> x) ==> z)) = y + (z + (z ==> (y ==> x))).
                        [para(20(a,1),36(a,1,2)),flip(a)].
275 (x ==> y) + z >= x ==> (y + ((y ==> x) + z)).   [para(36(a,1),28(a,2,2))].
418 c1 >= c1 ==> c3.   [para(21(a,1),89(a,1))].
419 0 >= c1 ==> (c1 ==> c3).   [hyper(31,a,418,a)].
609 c3 + (x + (x ==> c3)) >= c2.   [para(41,b,32,a)].
895 c3 ==> (x + c2) >= c3.   [para(22(a,1),55(a,2))].
996 c1 ==> (c1 ==> c3) = 0.   [hyper(13,a,16,a,b,419,a),flip(a)].
5220 c3 ==> (c1 + (x + c1)) >= c3.   [para(205(a,2),895(a,1,2))].
10398 c3 + (x ==> c3) >= x ==> c2.   [hyper(43,a,609,a)].
16713 c3 + ((c3 ==> c1) + ((c1 ==> c3) ==> c1)) = c2.
        [para(996(a,1),271(a,2,2,2)),rewrite([9(15),27(15),180(14)])].
20059 c1 + (c3 ==> c1) >= c3.   [hyper(12,a,275,a,b,5220,a),rewrite([9(5)])].
20066 c3 ==> c1 >= c1 ==> c3.   [hyper(14,a,20059,a)].
20564 c3 + (c1 ==> c3) >= c1.   [para(21(a,1),10398(a,2))].
20570 c1 ==> c3 >= c3 ==> c1.   [hyper(14,a,20564,a)].
20614 c3 ==> c1 = c1 ==> c3.   [hyper(13,a,20066,a,b,20570,a),flip(a)].
20625 c1 + c3 = c2.
        [back_rewrite(16713),rewrite([20614(4),20(10),996(7),9(4),27(4),9(3)])].
20634 c3 >= c1.   [para(20625(a,1),28(a,2,2)),rewrite([21(4)])].
20637 c1 >= c3.   [para(20625(a,1),70(a,2,2)),rewrite([22(4)])].
20793 -(c1 >= c3).   [ur(13,b,20634,a,c,23,a)].
20794 $F.   [resolve(20793,a,20637,a)].
```

Gibbard's Collapse Theorem for the Indicative Conditional: An Axiomatic Approach

Branden Fitelson

1   Background: Gibbard's (Informal) Argument

Gibbard [2] presents an argument to the effect that any conditional satisfying certain principles must be equivalent to the material (*viz.*, classical) conditional. Here is one rendition of Gibbard's (informal) argument.

Let $\supset$ be the classical material conditional, and let $\rightsquigarrow$ be the indicative conditional. Suppose that the indicative satisfies the *import-export law*. That is, suppose

(IE) $A \rightsquigarrow (B \rightsquigarrow C)$ is *logically equivalent* to $(A \& B) \rightsquigarrow C$.

If $\rightsquigarrow$ satisfies (IE), then (i) is equivalent to (ii).

(i) $(A \supset C) \rightsquigarrow (A \rightsquigarrow C)$.
(ii) $((A \supset C) \& A) \rightsquigarrow C$.

Substitutivity of logical equivalents then implies that (ii) [and $\therefore$ (i)] is equivalent to (iii).

(iii) $(A \& C) \rightsquigarrow C$.

So, if (iii) is a logical truth (as Gibbard supposes), then (i) and (ii) are too. Finally, suppose the indicative is at least as strong as the material conditional. That is, suppose $P \rightsquigarrow Q$ *entails* $P \supset Q$. Then, (i) entails (iv).

(iv) $(A \supset C) \supset (A \rightsquigarrow C)$.

Hence, (iv) is (also) a logical truth. Thus, $A \supset C$ entails $A \rightsquigarrow C$. Therefore, in general, $p \rightsquigarrow q$ entails $p \supset q$ and $p \supset q$ entails $p \rightsquigarrow q$. That is, in general, $\rightsquigarrow$ and $\supset$ are logically equivalent.   *QED.*

In this note, I present a formal axiomatization of the (theoretic and meta-theoretic) assumptions, which I take to be *essential* to the Gibbardian collapse phenomenon. This will lead to a formal proof of what I will call *Gibbard's collapse theorem*. Our formal treatment will reveal that collapse to the *classical, material* conditional is *not* inevitable. In fact, when one looks more closely at the assumptions involved (*essentially*) in proving Gibbard's collapse theorem, one realizes that *both classical and intuitionistic* interpretations of the indicative conditional are compatible with Gibbard's collapse phenomenon. This non-classical aspect of Gibbardian collapse is hidden by traditional presentations, which tend to (implicitly) presuppose various classical (theoretic and meta-theoretic) principles that are inessential to the theorem.

M.P. Bonacina and M.E. Stickel (Eds.): McCune Festschrift, LNAI 7788, pp. 181–188, 2013.

2   Axiomatization of the Gibbardian Collapse Phenomenon

Let $\mathscr{L}$ be a sentential (object) language containing atoms 'A', 'B', ..., and two *logical* connectives '&' and '$\rightarrow$'. In addition to these two *logical* connectives, $\mathscr{L}$ will also contain another binary connective '$\rightsquigarrow$', which is intended to be interpreted as the English indicative. In the meta-language for $\mathscr{L}$, we will have two meta-linguistic relations: $\Vdash$ and $\vdash$. '$\Vdash$' will denote a binary relation between individual sentences in $\mathscr{L}$. Specifically, '$\Vdash$' will be interpreted as the *single premise deducibility* (or *entailment*) relation of $\mathscr{L}$. '$\vdash$' will denote a monadic property of sentences of $\mathscr{L}$. Specifically, '$\vdash$' will be interpreted as the property of *theoremhood* (or *logical truth*) in $\mathscr{L}$. We will not presuppose anything about the relationship between '$\Vdash$' and '$\vdash$'. Rather, we will state explicitly all assumptions about these meta-theoretic relations that will be required (essentially) for Gibbard's collapse theorem. More precisely, I will state eight (8) independent axioms for $\rightarrow$, $\rightsquigarrow$, &, $\Vdash$, and $\vdash$, which will be jointly sufficient for (and severally essential for the proof of) Gibbard's collapse theorem.

First, two preliminary remarks: (a) the "if...then" (which I'll sometimes abbreviate as "$\Rightarrow$") and "and" in the meta-meta-language of $\mathscr{L}$ will be assumed throughout to be *classical*, and (b) the eight axioms are *schematic* (*i.e.*, they are to be interpreted as allowing *any instances* that can be formed with sentences of $\mathscr{L}$). With those caveats in mind, here are the eight (8) axioms that will form the basis of my formalization of Gibbard's collapse theorem.

(1) $\vdash (p \,\&\, q) \rightarrow p$.
(2) $\vdash p \rightarrow (q \rightarrow r)$ if and only if $\vdash (p \,\&\, q) \rightarrow r$.
(3) $\vdash p \rightsquigarrow (q \rightsquigarrow r)$ if and only if $\vdash (p \,\&\, q) \rightsquigarrow r$.
(4) If $\vdash p \rightsquigarrow q$, then $\vdash p \rightarrow q$.
(5) $\vdash (p \,\&\, q) \rightsquigarrow q$.
(6) If $p \Vdash q$ and $p \Vdash r$, then $p \Vdash q \,\&\, r$.
(7) If $\vdash p \rightarrow q$, then $p \Vdash q$.
(8) If $p \Vdash q$ and $q \Vdash p$, then p and q are *inter-substitutable* (in the context of $\rightsquigarrow$ theorems).[1]

Before stating our collapse theorem, I will make a few remarks about the axioms (1)–(8). Axiom (1) is a (left) conjunction-elimination *axiom* for $\langle \rightarrow, \& \rangle$. This is valid in every conditional logic I can think of. Axioms (2) and (3) are *import-export* rules for $\langle \rightarrow, \& \rangle$-*theorems* of $\mathscr{L}$ and $\langle \rightsquigarrow, \& \rangle$-*theorems* of $\mathscr{L}$, respectively. They say that import-export is *validity preserving* for both conditionals in $\mathscr{L}$. This is one of the more controversial axioms in the list. It is valid in many logics (*e.g.*, both classical and intuitionistic logic), but it also fails in many logics (*e.g.*, various substructural logics). Axiom (4) says that the indicative

---

[1] More precisely, we only need a *special case* of inter-substitutivity (in $\rightsquigarrow$ theorems). Let $p \approx q =_{df}$ for all r, $\vdash p \rightsquigarrow r$ iff $\vdash q \rightsquigarrow r$. Then, all we need to assume in (8) is: If $p \Vdash q$ and $q \Vdash p$, then $p \approx q$. This is made clearer in the proofs found in the Appendix.

conditional is "at least as strong as" the logical conditional — but *only* in the sense that if an indicative conditional is a *theorem* of $\mathcal{L}$, then the corresponding logical conditional is *also* a *theorem* of $\mathcal{L}$. Axiom (5) is a (right) conjunction-elimination *axiom* for $\langle \leadsto, \& \rangle$. Like (1), this is a universally valid principle for conditional logics. Axiom (6) is a form of the *conjunction introduction rule*. This is valid for just about any (single-premise) entailment relation I can think of. Axiom (7) says (informally) that if a logical conditional is a logical truth (theorem), then its antecedent entails its consequent. This is one direction of the deduction theorem for the logical conditional.[2] It holds in many logical systems (including both classical and intuitionistic logic). Axiom (8) is the assumption of inter-substitutivity of logical equivalents (in indicatives). This axiom is valid in many (positive) logics, including both intuitionistic and classical logic. Finally, axioms (1)–(8) are *independent*. And, they *suffice* to ensure that the indicative conditional collapses to the logical conditional. To wit, the following theorem.[3]

Theorem. Axioms (1)–(8) are independent, and they jointly entail the following *collapse* of $\leadsto$ to $\rightarrow$

(9) $p \rightarrow q \Vdash p \leadsto q$ and $p \leadsto q \Vdash p \rightarrow q$.

Before closing this section, two crucial remarks about our formal Gibbardian collapse theorem are in order.

Remark 1. Axioms (1)–(8) do *not* entail collapse of $\leadsto$ to $\supset$. That is to say, collapse of the indicative to the *classical, material* conditional does *not* follow from (1)–(8), and is therefore *inessential* to the (core) Gibbaridan collapse phenomenon. More precisely, we can show that (1)–(8) are compatible with

(10) $\nvdash ((p \leadsto q) \leadsto p) \leadsto p$ and $\nvdash ((p \rightarrow q) \rightarrow p) \rightarrow p$.

That is, *Peirce's Law* is not guaranteed by (1)–(8) to be a theorem (for either $\leadsto$ or $\rightarrow$).

Remark 2. Axioms (1)–(8) *do* entail that the indicative conditional must collapse to a logical conditional that is *at least as strong as the intuitionistic conditional*. This follows from the fact that (1)–(8) entail the following three additional theorems:

(11) If $\vdash p$ and $\vdash p \leadsto q$, then $\vdash q$. [And, if $\vdash p$ and $\vdash p \rightarrow q$, then $\vdash q$.]
(12) $\vdash p \leadsto (q \leadsto p)$. [And, $\vdash p \rightarrow (q \rightarrow p)$.]
(13) $\vdash (p \leadsto (q \leadsto r)) \leadsto ((p \leadsto q) \leadsto (p \leadsto r))$. [And, $\vdash (p \rightarrow (q \rightarrow r)) \rightarrow ((p \rightarrow q) \rightarrow (p \rightarrow r))$.]

It is well-known that (11)–(13) suffice to derive all theorems of intuitionistic implication. Therefore, (1)–(8) entail that all theorems of intuitionistic

---

[2] The other direction of the deduction theorem for $\rightarrow$ also follows from axioms (1)–(8), but I will not give a proof of that here.

[3] In the Appendix, I provide a proof of this theorem (and the other technical results) of the paper).

implication are theorems of both the indicative and the logical conditional. So, while the Gibbardian collapse phenomenon is compatible with a non-classical conditional, it does entail collapse to something that is no weaker (in terms of its theorems) than the intuitionistic conditional.

3 Concluding Remarks

We have given a rigorous formal rendition of the assumptions that we think are essential to a Gibbardian collapse theorem. This has revealed that the collapse phenomenon is not essentially classical in nature. But, it has also revealed that collapse to a conditional at least as strong as the intuitionistic conditional is essential to the phenomenon. This means that anyone who thinks that the indicative conditional does not have (at least) the logical strength of the intuitionistic conditional (*i.e.*, that the indicative lacks some theorems that the intuitionistic conditional has) is going to have to reject some of our axioms (1)–(8). The only axioms that seem plausibly deniable (to me — in the context of a sentential logic containing only conditionals and conjunctions) are axioms (2) and (3). These are the *import-export* laws, and they seem to be the most suspect of the bunch. I find it difficult to see how any of the other axioms could (plausibly) be denied (but I won't argue for that claim here). The two main purposes of this note have been (a) to reveal the non-classical nature of the (essence of the) Gibbardian collapse phenomenon, and (b) to make clear precisely what theoretic and meta-thoeretic assumptions underlie Gibbardian collapse.

4 Appendix: Proofs of Theorems

4.1 Proofs of the Independence of Our Axioms (1)–(8)

First, I must prove that the axioms (1)–(8) are *independent*. I will do so by providing eight countermodels.[4]

Independence of (1). We must show $\{(2),(3),(4),(5),(6),(7),(8)\} \not\Rightarrow (1)$. Here is a model on which (2)–(8) are **T**, but (1) is **F**.

&	0	1	2
0	0	1	2
1	0	1	2
2	2	2	2

→	0	1	2
0	1	1	2
1	2	1	2
2	1	1	1

⤳	0	1	2
0	1	1	2
1	2	1	2
2	1	1	1

⊩	0	1	2
0	**T**	**T**	**F**
1	**F**	**T**	**F**
2	**T**	**T**	**T**

⊢	0	1	2
	F	**T**	**F**

---

[4] All models and proofs in this Appendix were discovered and verified with the aid of the automated reasoning programs **paradox** [1], **vampire** [5], **prover9/mace4** [4], and **otter** [3]. All models are *smallest possible*, but there may be more elegant proofs of the theorems (I tried to find the simplest proofs I could, using various techniques for finding elegant proofs [7]).

Note: $\vdash (p \,\&\, q) \to p$ is **F** on this model, when $p := 0$ and $q := 1$.   □

Independence of (2). We must show $\{(1),(3),(4),(5),(6),(7),(8)\} \not\Rightarrow (2)$. Here is a model on which $\{(1),(3),(4),(5),(6),(7),(8)\}$ are **T**, but (2) is **F**.

&	0	1	2
0	0	1	0
1	1	1	1
2	0	1	2

→	0	1	2
0	2	0	2
1	2	2	2
2	0	0	2

⤳	0	1	2
0	2	1	2
1	2	2	2
2	0	1	2

⊩	0	1	2
0	T	F	T
1	T	T	T
2	F	F	T

⊢	0	1	2
	F	F	T

Note: $\vdash p \to (q \to r) \Rightarrow \; \vdash (p \,\&\, q) \to r$ is **F** on this model, when $p, q := 0$ and $r := 1$.   □

Independence of (3). We must show $\{(1),(2),(4),(5),(6),(7),(8)\} \not\Rightarrow (3)$. Here is a model on which $\{(1),(2),(4),(5),(6),(7),(8)\}$ are **T**, but (3) is **F**.

&	0	1	2
0	0	1	0
1	1	1	1
2	0	1	2

→	0	1	2
0	2	1	2
1	2	2	2
2	0	1	2

⤳	0	1	2
0	2	0	2
1	2	2	2
2	0	0	2

⊩	0	1	2
0	T	F	T
1	T	T	T
2	F	F	T

⊢	0	1	2
	F	F	T

Note: $\vdash p \rightsquigarrow (q \rightsquigarrow r) \Rightarrow \; \vdash (p \,\&\, q) \rightsquigarrow r$ is **F** on this model, when $p, q := 0$ and $r := 1$.   □

Independence of (4). We must show $\{(1),(2),(3),(5),(6),(7),(8)\} \not\Rightarrow (4)$. Here is a model on which $\{(1),(2),(3),(5),(6),(7),(8)\}$ are **T**, but (4) is **F**.

&	0	1
0	0	1
1	1	1

→	0	1
0	0	1
1	0	0

⤳	0	1
0	0	0
1	0	0

⊩	0	1
0	T	F
1	T	T

⊢	0	1
	T	F

Note: $\vdash p \rightsquigarrow q \Rightarrow \; \vdash p \to q$ is **F** on this model, when $p := 0$ and $q := 1$.   □

Independence of (5). We must show $\{(1),(2),(3),(4),(6),(7),(8)\} \not\Rightarrow (5)$. Here is a model on which $\{(1),(2),(3),(4),(6),(7),(8)\}$ are **T**, but (5) is **F**.

&	0	1
0	0	1
1	1	1

→	0	1
0	0	1
1	0	0

⤳	0	1
0	1	1
1	1	1

⊩	0	1
0	T	F
1	T	T

⊢	0	1
	T	F

Note: $\vdash (p \,\&\, q) \rightsquigarrow q$ is **F** on this model, when $p, q := 0$.   □

Independence of (6). We must show $\{(1),(2),(3),(4),(5),(7),(8)\} \not\Rightarrow (6)$. Here is a model on which $\{(1),(2),(3),(4),(5),(7),(8)\}$ are **T**, but (6) is **F**.

&	0	1	2
0	1	1	1
1	1	1	1
2	0	1	2

→	0	1	2
0	2	0	2
1	2	2	2
2	0	0	2

⤳	0	1	2
0	2	0	2
1	2	2	2
2	0	0	2

⊩	0	1	2
0	T	F	T
1	T	T	T
2	F	F	T

⊢	0	1	2
	F	F	T

Note: $(p \Vdash q$ and $p \Vdash r) \Rightarrow p \Vdash q \,\&\, r$ is **F** on this model, when $p, q, r := 0$. □

Independence of (7). We must show $\{(1),(2),(3),(4),(5),(6),(8)\} \not\Rightarrow (7)$. Here is a model on which $\{(1),(2),(3),(4),(5),(6),(8)\}$ are **T**, but (7) is **F**.

&	0	1
0	0	0
1	0	0

→	0	1
0	0	0
1	0	0

⤳	0	1
0	0	0
1	0	0

⊪	0	1
0	F	F
1	F	F

⊢	0	1
	T	F

Note: $\vdash p \to q \Rightarrow p \Vdash q$ is **F** on this model, when $p, q := 0$.   □

Independence of (8). We must show $\{(1),(2),(3),(4),(5),(6),(7)\} \not\Rightarrow (8)$. Here is a model on which $\{(1),(2),(3),(4),(5),(6),(7)\}$ are **T**, but (8) is **F**.

&	0	1
0	0	0
1	0	1

→	0	1
0	1	1
1	0	1

⤳	0	1
0	1	1
1	0	1

⊪	0	1
0	T	T
1	T	T

⊢	0	1
	T	F

Recall (see *fn.* 1) that the precise content of axiom (8) is the following: (8) If $p \Vdash q$ and $q \Vdash p$, then $p \approx q$, where $p \approx q$ just in case, for all r, $\vdash p \rightsquigarrow r$ iff $\vdash q \rightsquigarrow r$.

Thus, a counterexample to (8) must involve a model containing a triple $\{p, q, r\}$ such that both $p \Vdash q$ and $q \Vdash p$ are **T**, but $\vdash p \rightsquigarrow r \Leftrightarrow \vdash q \rightsquigarrow r$ is **F**. This is just such a model, where $p, r := 0$ and $q := 1$.   □

4.2   Proof of Our (Intuitionistic) Collapse Theorem

The following is a (unified) proof of the following two central theorems reported in the main text: (a) our Gibbardian collapse theorem (9), and (b) claim (13) for the indicative conditional. Claims (11) and (12) have easy proofs from (1)–(8), so I omit those here.[5] In this proof, I will present all axioms (and steps) in *clausal form*, and the only rule of inference I will use is *hyper-resolution*.[6]

---

[5] It can also be shown that *all* of our axioms (1)–(8) are *essential* to *any* proof of the collapse theorem. Moreover, it can be shown that axiom (2) is not needed to prove (13) for the indicative conditional, and Axiom (3) is not needed to prove of (13) for the logical conditional [but, in both cases, the remaining axioms *are* essential for proving (13)]. Obtaining (direct) axiomatic proofs of (13) was non-trivial. I thank Bob Veroff (and his *proof sketches* technique [6]) for his invaluable assistance in obtaining (direct) axiomatic proofs of (13) for both conditionals.

[6] The proof given here was discovered using McCune's theorem-prover `Otter` [3], and it was verified using his more recent `prover9` [4]. The requisite substitution instances are generally not too difficult to figure out for each hyper-resolution step. I omit those details, but they can be generated using McCune's `prooftrans` program [4]. Finally, I have posted two input files for exploring and verifying the results reported here. First, `http://fitelson.org/gibbard_fof.in` is a `tptp/fof` syntax input file for exploring the Gibardian collapse phenomenon (this file should work with most theorem-provers/model-finders that are available today). Second, `http://fitelson.org/gibbard_prover9.in` is a `prover9` input file, which allows for easy verification of the main proof of claims (9) and (13) reported below.

1. $\vdash (A \,\&\, B) \to A$ · Axiom (1)
2. $\vdash (A \,\&\, B) \rightsquigarrow B$ · Axiom (5)
3. $\not\vdash A \to (B \to C) \lor \vdash (A \,\&\, B) \to C.$ · · · · Axiom (2).
4. $\vdash A \to (B \to C) \lor \not\vdash (A \,\&\, B) \to C.$ · · · · Axiom (2).
5. $\not\vdash A \rightsquigarrow (B \rightsquigarrow C) \lor \vdash (A \,\&\, B) \rightsquigarrow C.$ · · · · Axiom (3).
6. $\vdash A \rightsquigarrow (B \rightsquigarrow C) \lor \not\vdash (A \,\&\, B) \rightsquigarrow C.$ · · · · Axiom (3).
7. $\not\vdash A \rightsquigarrow B \lor \vdash A \to B.$ · · · · · · · · · · · Axiom (4).
8. $A \not\Vdash B \lor A \not\Vdash C \lor A \Vdash B \,\&\, C.$ · · · · · · · Axiom (6).
9. $\not\vdash A \to B \lor A \Vdash B.$ · · · · · · · · · · · · · · · · · Axiom (7).
10. $A \not\Vdash B \lor B \not\Vdash A \lor \not\vdash A \rightsquigarrow C \lor \vdash B \rightsquigarrow C.$ · · Axiom (8).

---

11. $\vdash (((A \to B) \,\&\, C) \,\&\, A) \to B.$ · · · · · · · · · · · · · 3, 1.
12. $\vdash ((A \,\&\, (B \rightsquigarrow C)) \,\&\, B) \rightsquigarrow C.$ · · · · · · · · · · · 5, 2.
13. $\vdash (A \,\&\, B) \to B.$ · 7, 2.
14. $A \,\&\, B \Vdash A.$ · 9, 1.
15. $((A \to B) \,\&\, C) \,\&\, A \Vdash B.$ · · · · · · · · · · · · · · · · 9, 11.
16. $\vdash ((A \,\&\, (B \rightsquigarrow C)) \,\&\, B) \to C.$ · · · · · · · · · · 7, 12.
17. $A \,\&\, B \Vdash B.$ · 9, 13.
18. $(A \,\&\, (B \rightsquigarrow C)) \,\&\, B \Vdash C.$ · · · · · · · · · · · · · · · 9, 16.
19. $A \,\&\, B \Vdash A \,\&\, B.$ · 8, 14, 17.
20. $A \,\&\, B \Vdash B \,\&\, A.$ · 8, 17, 14.
21. $(A \,\&\, (B \rightsquigarrow C)) \,\&\, B \Vdash ((A \,\&\, (B \rightsquigarrow C)) \,\&\, B) \,\&\, C.$ · 8, 19, 18.
22. $A \,\&\, B \Vdash (A \,\&\, B) \,\&\, B.$ · · · · · · · · · · · · · · · · · · 8, 19, 17.
23. $((A \to B) \,\&\, C) \,\&\, A \Vdash (((A \to B) \,\&\, C) \,\&\, A) \,\&\, B.$ · 8, 19, 15.
24. $\vdash (A \,\&\, B) \rightsquigarrow A.$ · 10, 20, 20, 2.
25. $\vdash (((A \to B) \,\&\, C) \,\&\, A) \rightsquigarrow B.$ · · · · · · · · · · 10, 14, 23, 2.
26. $\vdash (((A \rightsquigarrow B) \,\&\, C) \,\&\, A) \rightsquigarrow B.$ · · · · · · · · · · 5, 24.
27. $\vdash ((A \to B) \,\&\, A) \rightsquigarrow B.$ · · · · · · · · · · · · · · · 10, 14, 22, 25.
28. $\vdash ((A \rightsquigarrow B) \,\&\, A) \rightsquigarrow B.$ · · · · · · · · · · · · · · · 10, 14, 22, 26.
29. $\vdash ((((A \rightsquigarrow (B \rightsquigarrow C)) \,\&\, D) \,\&\, A) \,\&\, B) \rightsquigarrow C.$ · · · 5, 26.
30. $\vdash (A \to B) \rightsquigarrow (A \rightsquigarrow B).$ · · · · · · · · · · · · · · 6, 27.
31. $\vdash ((A \rightsquigarrow B) \,\&\, A) \to B.$ · · · · · · · · · · · · · · · · 7, 28.
32. $\vdash (((A \rightsquigarrow (B \rightsquigarrow C)) \,\&\, (A \rightsquigarrow B)) \,\&\, A) \rightsquigarrow C.$ · · · 10, 14, 21, 29.
33. $\vdash (A \to B) \to (A \rightsquigarrow B).$ · · · · · · · · · · · · · · · 7, 30.
34. $\vdash (A \rightsquigarrow B) \to (A \to B).$ · · · · · · · · · · · · · · · 4, 31.
35. $\vdash ((A \rightsquigarrow (B \rightsquigarrow C)) \,\&\, (A \rightsquigarrow B)) \rightsquigarrow (A \rightsquigarrow C).$ · 6, 32.
36. $A \to B \Vdash A \rightsquigarrow B$ · 9, 33.
37. $A \rightsquigarrow B \Vdash A \to B.$ · 9, 34.
38. $\vdash (A \rightsquigarrow (B \rightsquigarrow C)) \rightsquigarrow ((A \rightsquigarrow B) \rightsquigarrow (A \rightsquigarrow C)).$ · 6, 35.   □

4.3   Counterexample to Peirce's Law for $\rightsquigarrow$ and $\to$

Here is a model on which (1)–(8) are all **T**, but $\vdash ((p \rightsquigarrow q) \rightsquigarrow p) \rightsquigarrow p$ and $\vdash ((p \to q) \to p) \to p$ are **F**.

&	0	1	2
0	0	1	0
1	1	1	1
2	0	1	2

→	0	1	2
0	2	1	2
1	2	2	2
2	0	1	2

⤳	0	1	2
0	2	1	2
1	2	2	2
2	0	1	2

⊩	0	1	2
0	T	F	T
1	T	T	T
2	F	F	T

⊢	0	1	2
	F	F	T

Note: ⊢ $((p \rightsquigarrow q) \rightsquigarrow p) \rightsquigarrow p$ and ⊢ $((p \rightarrow q) \rightarrow p) \rightarrow p$ are both **F** on this model, when $p := 0$ and $q := 1$.   □

References

1. Claessen, K., Sorensson, N.: New techniques that improve mace-style finite model finding. In: Proceedings of the CADE-19 Workshop: Model Computation—Principles, Algorithms, Applications (2003)
2. Gibbard, A.: Two recent theories of conditionals. In: Harper, W.L., Stalnaker, R., Pearce, G. (eds.) Ifs: Conditionals, Belief, Decision, Chance and Time. Reidel (1981)
3. Kalman, J.: Automated Reasoning with otter. Rinton Press (2001)
4. McCune, W.: Prover9 and mace4, http://www.cs.unm.edu/ mccune/mace4/
5. Riazanov, A., Voronkov, A.: The design and implementation of vampire. AI Communications (2002)
6. Veroff, R.: Solving open questions and other challenge problems using proof sketches. Journal of Automated Reasoning (2001)
7. Wos, L.: Automating the search for elegant proofs. Journal of Automated Reasoning (1998)

Geometric Quantifier Elimination Heuristics for Automatically Generating Octagonal and Max-plus Invariants*

Deepak Kapur[1], Zhihai Zhang[2], Matthias Horbach[1],
Hengjun Zhao[3], Qi Lu[1], and ThanhVu Nguyen[1]

[1] Department of Computer Science, University of New Mexico,
Albuquerque, NM, USA
[2] School of Mathematical Sciences, Peking University,
Beijing, China
[3] Institute of Software, Chinese Academy of Sciences,
Beijing, China

Abstract. Geometric heuristics for the quantifier elimination approach presented by Kapur (2004) are investigated to automatically derive loop invariants expressing weakly relational numerical properties (such as $l \leq x \leq h$ or $l \leq \pm x \pm y \leq h$) for imperative programs. Such properties have been successfully used to analyze commercial software consisting of hundreds of thousands of lines of code (using for example, the Astrée tool based on abstract interpretation framework proposed by Cousot and his group). The main attraction of the proposed approach is its much lower complexity in contrast to the abstract interpretation approach ($O(n^2)$ in contrast to $O(n^4)$, where n is the number of variables) with the ability to still generate invariants of comparable strength. This approach has been generalized to consider disjunctive invariants of the similar form, expressed using maximum function (such as $\max(x + a, y + b, z + c, d) \leq \max(x + e, y + f, z + g, h)$), thus enabling automatic generation of a subclass of disjunctive invariants for imperative programs as well.

1 Introduction

In [23,22], Kapur proposed an approach based on quantifier elimination for generating program invariants in general and loop invariants in particular. Depending upon the formulas of interest to serve as invariants at a program location, parametric formulas are identified such that when those parameters are fully instantiated, the results are the desired invariants. As an example, if a goal is to discover linear inequalities as the invariants, then the corresponding parametric form is $a \cdot x + b \cdot y + c \cdot z \leq d$, where x, y, z are program variables and a, b, c, d are parameters. Notice that the parametrized form is not a linear inequality because of

---

* Partially supported by NSF grants CCF-0729097 and CNS-0905222, by a fellowship from the Postdoc Program of the German Academic Exchange Service (DAAD), and by EXACTA and the China Scholarship Council.

M.P. Bonacina and M.E. Stickel (Eds.): McCune Festschrift, LNAI 7788, pp. 189–228, 2013.
© Springer-Verlag Berlin Heidelberg 2013

presence of terms like $a \cdot x, b \cdot y, c \cdot z$, but rather a nonlinear (quadratic) inequality. If the goal is to find quadratic inequalities (such as ellipsoid inequalities), then the associated parametric form is still a nonlinear inequality, albeit of degree 3. Once a parametric form for invariants of interest is identified, then Kapur's approach involves generating verification conditions using the parametrized formulas for each distinct program path and eliminating program variables from the verification conditions using quantifier elimination to produce constraints on parameters. For any given parameter value satisfying the resulting constraints, the verification conditions instantiated with these parameter values then become valid, implying that the corresponding instantiation of the parametrized invariant is indeed a program invariant.

The main contribution of the paper is the development of an efficient geometric local heuristic for a restricted version of quantifier elimination over a subset of parametrized linear formulas so as to generate octagonal invariants and max-plus invariants for programs. The quantifier elimination problem of interest is of the form

$$\forall x_1, \ldots, x_n \ \Phi(p_1, \ldots, p_m, x_1, \ldots, x_n),$$

where $p_1, \ldots, p_m$ are parameters, $x_1, \ldots, x_n$ are program variables, and Φ is a verification condition, a quantifier-free formula over $p_1, \ldots, p_m$ and $x_1, \ldots, x_n$. The objective is to generate a nontrivial quantifier-free formula over $p_1, \ldots, p_m$ that implies $\forall x_1, \ldots, x_n \ \Phi$. We focus on two types of parametric formulas: (i) a parametric formula obtained using a conjunction of atomic formulas of the form $l \leq x, x \leq h, l \leq x+y, x+y \leq h, l \leq x-y$, and $x-y \leq h$, where x, y are program variables and l, h are parameters, and (ii) a parametric formula obtained using a limited form of disjunctions of conjunctions of formulas of the form $l \leq x, x \leq h$, $l \leq x - y$, and $x - y \leq h$, which are equivalent to a pure conjunction of formulas of the form $\max(x_1 + a_1, x_2 + a_2, a_0) \geq \max(x_1 + b_1, x_2 + b_2, \ldots, x_k + b_k, b_0)$, where a_i, b_j are parameters. Such invariants have been found to be very effective in detecting bugs in commercial software for flight control and related embedded systems for memory violation [1] and numerical errors using tools such as Astrée [9].

Given that quantifier elimination is in general computationally a highly expensive operation (either undecidable or doubly exponential) and furthermore, outputs generated by complete quantifier elimination algorithms are huge, we address both of these problems by exploring a local incomplete quantifier-elimination geometric heuristic which considers formulas with constant number of variables (typically two variables) as well as which is geometrically based resulting in manageable outputs by focusing on relevant cases. As shown later in the paper, this quantifier elimination heuristic results in generating program invariants of strength and quality comparable to those obtained using the methods based on the abstract interpretation framework as discussed in [25], but with a much lower asymptotic complexity—$O(n^2)$ in the number of program variables in comparison to $O(n^4)$ for algorithms based on the abstract interpretation approach.

A fascinating aspect of our approach is that for octagonal invariants, since the parametric form is fixed and determined by the number of program variables, it is possible to develop local heuristics focusing on quantifier elimination to a pair of distinct variables. By analyzing different kinds of assignment statement, we have developed an approach for quantifier elimination using table look-ups based on the presence (or absence) of parameter-free atomic formulas (corresponding to various sides of octagons) appearing in a program path. Using these tables, it is possible to identify how these atomic formulas in a program path restrict octagonal invariants for various kinds of assignments on program variables. Parameter constraints generated by the quantifier-elimination heuristic can also be decomposed into subsets of constraints on at most four parameters, resulting in very efficient algorithms for a family of invariants of different quality and even generating the strongest possible invariants.

In our analysis, a formula over an arbitrary number of variables is decomposed into subformulas on a fixed number of variables. As a result, efficient heuristics can be designed exploiting the structure of these subformulas. This paper reports a few such heuristics we have developed; many more are still possible and are being explored.

A major advantage of the proposed approach is that it is highly parallelizable, which is especially good for scalability, since most of its steps can be done in parallel:

- Analysis for different program paths can be done in parallel.
- Table look-up for each of the tests to generate constraints on parameters can be done in parallel.
- Parameter constraints can be analyzed in parallel by decomposing them into blocks of constraints on a fixed number of variables.
- Generation of the strongest invariant after computing maximum lower bounds and minimum upper bounds on parameters can also be derived in parallel.

The sequential bottleneck in the analysis is the derivation of implicit tests from the tests appearing in a program path.

The paper is organized as follows: In the next subsection, we briefly review related work on the generation of octagonal and max plus invariants. This is followed by a high level comparison of the quantifier elimination approach and fixed point approaches for generating program invariants, with a particular focus on the abstract interpretation approach. We discuss the strength and quality of invariants generated by these approaches. Section 2 focuses on octagonal invariants. Section 3 reports our preliminary investigations for generating disjunctive invariants expressed using a conjunction of max-plus constraints. Section 4 briefly discusses future work.

1.1   Related Work

Quantifier elimination approaches for static program analysis have been investigated in many different ways, particularly for generating linear inequalities based on Farkas's lemma [26], using linear constraints and skeletons [15,16], program synthesis [14,30,28], termination of programs using linear and nonlinear ranking functions [7,32,31], as well as analysis of hybrid systems [29,21].

The most popular approach for automatically generating invariants is using the abstract interpretation framework pioneered by Cousot and Cousot [8]. This research direction has resulted in very powerful tools, Astrée and its descendants, which have been used in finding bugs in commercial software, and a related set of experimental freely available tools (including Interproc [20], the tool we use in this paper for comparative purposes because it was designed for similar programs). One of the main reasons for the success of the abstract interpretation approach on real large numerical software is a collection of efficient algorithms designed for various operations for different abstract domains. Two Ph.D. theses by Miné [25] and Allamigeon [1] are the closest to the results presented in this paper. Miné's thesis focused on *weakly relational numerical* abstract domains where program invariants are specified using octagonal constraints (or a subset of octagonal constraints). Allamigeon's thesis considered max-plus invariants. As pointed out by Miné, the use of linear inequalities as an abstract domain proposed in [10] does not scale because of the exponential complexity of algorithms needed to perform abstract domain operations on convex polyhedra including conversion back and forth between their representation as a conjunction of linear inequalities and the generator (frame) representation.

Octagonal constraints (also called unit two variable per inequality or UTVPI constraints) have been extensively investigated. Octagonal constraints are also interesting to study from a complexity perspective and are a good compromise between interval constraints and linear constraints. Linear constraint analysis over the rationals ($\mathbb{Q}$) and reals ($\mathbb{R}$), while of polynomial complexity, has been found in practice to be inefficient and slow, especially when the number of variables grows [25,9], since it must be used repeatedly in an abstract interpretation framework. Often, we are however interested in cases when program variables take integer values bound by computer arithmetic. If program variables are restricted to take integer values (which is especially the case for expressions serving as array indices and memory references), then octagonal constraints are among the most expressive fragments of linear (Presburger) arithmetic over the integers with a polynomial time complexity. It is well known that extending linear constraints to have three variables even with unit coefficients (i.e., ranging over $\{-1, 0, 1\}$) makes checking their satisfiability over the integers NP-complete [19,27]; similarly, restricting linear arithmetic constraints to be just over two variables, but allowing non-unit integer coefficients of the variables also leads to the satisfiability check over the integers being NP-complete. The Floyd-Warshall algorithm, which is typically used to analyze the difference bound matrices representation of octagonal constraints (see [25]), must be extended for computing integral closure if variables are over the integers; see [3], where an $O(n^3)$ algorithm for computing the tight closure of octagonal constraints over the integers is presented that exploits integrality of constraints.

Max-plus constraints, i.e. constraints of the form $\max(x_1+a_1, \ldots, x_n+a_n, c) \leq \max(x_1+b_1, \ldots, x_n+b_n, d)$, have been investigated in [4,6], and are an active area of research in combinatorics. They were first used for program analysis by Allamigeon et al. [2], who realized their value as an abstract domain that can express certain nonconvex sets, the so-called max-plus polyhedra, without the need for heuristics on how to manage the number of disjunctive components. Like classical convex polyhedra, (bounded) max-plus polyhedra can be equivalently represented by a set of constraints or by a set of extremal points. As in the classical case, the conversion between these representations is notoriously expensive. E.g. finding extremal points, which are usually called generators, is exponential in the number of constraints. Even a single inequation in n dimensions can give rise to quadratically many generators. The algorithms by Allamigeon et al. work purely on generators instead of constraints. In our quantifier-based approach, we will also restrict our attention to max-plus polyhedra represented by sets of generators.

1.2   On the Quality of Invariants Generated Using Quantifier Elimination

We briefly compare the quantifier elimination approach for generating inductive invariants as proposed in [23,22] to other approaches based on fixed point algorithms, in particular the abstract interpretation framework pioneered by [8,10,25,1].

An invariant at any location in a program captures a superset of the states reached (also called reachable states) whenever program control passes through that particular location. The strongest possible invariant at any location is thus simply a disjunction specifying that the state at that location is one of these reachable states. If a location is visited finitely many times, then this disjunction is a formula as long as a single reachable state can be precisely characterized by a formula in a first-order theory that is expressive enough (provided the set of initial states of the program can be described in such a way). However, if a location is visited infinitely often in the case of a nonterminating program, then this set of infinitely many states must be specified by a finite formula in a richer language with interpreted function symbols; this may often be an approximation in the sense that the set of states by which such a formula is satisfied is typically a superset of the reachable states at the location. For every program path through the location, the effect of the statements on that path preserve the reachable set of states at the location. Given a set S of states, a formula ϕ is the strongest in a theory (such as Presburger arithmetic, Tarski's theory of real closed fields, the theory of polynomial equalities over an algebraically closed field, etc.) specifying S iff there is no other formula γ not equivalent in the theory to ϕ such that (i) $\gamma \implies \phi$ and (ii) γ is satisfied by every state in the set. Typically, formulas used for specifying states are quantifier-free as quantified formulas are difficult to analyze.

In the abstract interpretation approach, concrete states are abstracted to abstract states using an abstraction function and abstract states are specified using the elements of an abstract domain, which is a lattice. In this context, each program variable, instead of taking a concrete value, may take an abstract value [8]. Concrete states and abstract states in their setting are related using a Galois connection. Examples of abstract values commonly used are the values of a variable being in an interval (zonal constraints), or in addition to being an interval, sum and difference of two different variables is also in an interval (octagonal constraint) [25], or more complex constraints on the values of variables including max-plus constraints [1] and linear constraints [10]; other domains have also been explored. An invariant is expressed as an abstract element or a set thereof and is computed by a terminating fixed point computation using a suitably defined widening operator (and narrowing operators); many heuristics have been proposed to improve the quality of the invariants computed using this approach [11,5].

Elements in an abstract domain typically can be represented as a conjunction of atomic formulas over a suitable theory. For example, conjunctions of interval constraints, octagonal constraints, and max plus constraints can all be written in a small fragment of quantifier-free Presburger arithmetic, whereas a general conjunction of linear constraints uses full quantifier-free Presburger arithmetic (but without disjunction and negation).

As stated above, the quantifier elimination approach for generating invariants at a program location hypothesizes invariants as formulas of a certain form, which can be parametrized. The intuition behind this approach is that for some parameter values, the resulting instantiated formula characterizes a superset of reachable states at the program location. In other words, when parameters in such a formula are fully instantiated, the resulting formula is in (a fragment of) the language used for writing the formulas, even though the parametrized formula may be in a richer language. In the case of octagonal constraints, a parametrized formula is a conjunction of atomic formulas of the form $l \leq e$ and $e \leq h$, where l, h are parameters, and e is a variable, the difference of two variables or the sum of two variables; in this case, both the the parametrized formula as well as its instantiation are in Presburger arithmetic (a parametrized atomic formula is expressed using at most three variables whereas a nonparametrized atomic formula is expressed using at most two variables). However, for linear constraints, a parametrized formula is not in the language of Presburger arithmetic since the parametrized formula could be $ax + by + cz \leq d$, where a, b, c, d are parameters, but its instantiations are in Presburger arithmetic.

Verification conditions corresponding to paths through the given program location are then generated and program variables are eliminated from the verification conditions, giving rise to constraints on parameters. If there is an invariant of the hypothesized shape associated with the program location, then the quantifier elimination approach would find such an invariant assuming that the method for quantifier elimination is complete. Furthermore, if all the solutions of constraints on parameters resulting from a complete quantifier elimination method can be finitely described, then this approach produces the strongest invariant of such shape, implying that it is stronger than the invariant generated by any other approach including the abstract interpretation framework. Even if the quantifier elimination method is incomplete (but sound), then its result would lead to constraints on parameters such that all parameter values that satisfy these constraints on parameters, will result in an invariant. In this sense, the quantifier elimination approach is the most general method for computing invariants of programs.

2   Octagonal Invariants

2.1   Overview

In this section, we propose a method based on quantifier elimination for automatically generating program invariants which are conjunctions of octagonal constraints over the integers as atomic formulas. Such an atomic formula is a lower bound on a program variable x, an upper bound on a program variable x, a lower bound on $x + y$, where x, y are program variables, an upper bound on $x + y$, and similarly, a lower bound on $x - y$ or an upper bound on $x - y$. The reader would notice that a negation of any of the above atomic formulas over the integers can also be easily expressed. An expression x, $x + y$ or $x - y$ need not have a lower bound or an upper bound; to allow such possibilities, it is convenient to extend $\mathbb{Z}$, the domain of integers, to include both $-\infty$ and $+\infty$ with the usual semantics (see the Appendix about how various arithmetic operations and ordering relations are extended to consider $-\infty$ and $+\infty$). This also ensures that a trivial invariant will always be generated by the proposed method, in which these expressions have $-\infty$ as the lower bound and $+\infty$ as the upper bound, much like the trivial invariant true.

Let us assume that invariants (which we will have to compute) are associated with sufficiently many program locations (usually it suffices to associate an invariant with every loop and the entry and exit of every procedure/method). Verification conditions are then generated for every possible program path among pairs of such invariants. In the case of nested loops, invariants must be associated with every loop.

Assuming n program variables $x_1, \cdots x_n$ appearing along a program path, an invariant $I(X)$ expressed as a conjunction of octagonal constraints is of the form:

$$I(X) = \wedge_{1 \leq i < j \leq n} octa(x_i, x_j),$$

where X is $\{x_1, \ldots, x_n\}$, $octa(x_i, x_j)$ is a conjunction of atomic formulas discussed above and expressed using program variables $x_i, x_j, i \neq j$. A typical verification condition $\Phi(X)$ corresponding to a program path in a loop expressed using such invariants is:

$$(I(X) \wedge T(X)) \implies I(X'),$$

where X' contains the new values of the variables in X after all the assignments along the path, and $T(X)$ is a conjunction of all the loop tests and branch conditions along the branch. Without any loss of generality we can and will assume that $T(X)$ is a conjunction of atomic formulas, as otherwise the verification condition can be split into a conjunction of several verification conditions, with each being considered separately. As an example, if a loop condition is $T(X) = T^1(X) \vee T^2(X)$, then the above subformula can be split into:

$$(I(X) \wedge T^1(X)) \implies I(X') \wedge$$
$$(I(X) \wedge T^2(X)) \implies I(X') .$$

It is also assumed in the analysis below that all branches indeed participate in determining the program behavior, i.e. there is no dead branch which is never executed for the initial states under consideration. Considering dead branches can unnecessarily weaken the invariants generated using the quantifier elimination approach by imposing unnecessary constraints on parameters.[1]

Assume a different parametrized loop invariant at the entry of every loop (and every function and procedure, if any, in a program). Given the fixed structure of octagonal constraints, this is relatively easy. The formula $octa(x_i, x_j)$ can be fully parametrized with 8 parameters, one parameter each for lower bound and upper bound for each of the two variables x_i, x_j, for the sum expression $x_i + x_j$ and the difference expression $x_i - x_j$. It is of the following form:

$$octa(x_i, x_j) \triangleq (l_1 \leq x_i - x_j \leq u_1 \wedge l_2 \leq x_i + x_j \leq u_2 \wedge l_3 \leq x_i \leq u_3 \wedge l_4 \leq x_j \leq u_4).^2$$

So there are $\frac{n \cdot (n-1)}{2}$ pairs of variables, and there are total $2n \cdot (n - 1) + 2n = 2n^2$ parameters for each loop invariant assuming all the variables are needed to specify the strongest possible loop invariant. We also have the constraints that $l_i \leq u_i, i = 1, 2, 3, 4$.

For generating invariants, all program paths must be considered. The initial state of a program, expressed by a precondition, as well as other initialization assignments to program variables, may impose additional constraints on parameters.

To ensure that the verification condition generated from any program path also has the same types of atomic formulas, it is assumed that tests (for a loop as

---

[1] This is a weakness of the quantifier elimination approaches in contrast to other approaches where dead code gets automatically omitted in the analysis. Incomplete but fast dead code detectors are however a standard component of the static analysis performed in state of the art integrated program development environment including ECLIPSE (JAVA/C^{++}) and Microsoft Visual Studio, and can be switched on in the GNU Compiler Collection.

In our current implementation of the quantifier elimination approach, many dead branches are detected during the generation of the verification conditions, which tend to have trivially false antecedents for inexecutable paths.

[2] As the reader would have noticed, the closure of these constraints imposes a relationship among various parameters; for instance, lower and upper bounds on x_i, x_j can be deduced from the lower and upper bounds on $x_i - x_j$ and $x_i + x_j$. However, the most generic octagon still requires 8 parameters.

well as in a conditional statement) are of the same form. And assignment statements are of the form x := x+A, x := −x+A, and x := A, where A is a constant. Otherwise, tests and assignments must be approximated:

- An unsupported assignment is approximated to take an unknown value, i.e. for the variable x being assigned and any other variable y: $-\infty \leq x \leq +\infty$, $-\infty \leq x - y \leq +\infty$ and $-\infty \leq x + y \leq +\infty$.
- An unsupported test of a loop can be approximated to be both `true` and `false`, i.e. the loop can be arbitrarily continued or left after each iteration.
- An unsupported test of a conditional can also be approximated to be both `true` and `false`, i.e. both branches can always be executed.

2.2   Program Analysis Using Octagonal Invariants

We now present our method for generating program invariants with octagonal constraints.

0. Associate a parametrized octagonal invariant with every loop entry as well as with the entry and exit of every function/procedure.
1. Within a function/procedure, for every program path, generate a verification condition from program invariants at every loop entry. This can be done by standard methods like the computation of weakest preconditions for each path using Hoare logic [18].
2. If the resulting verification condition cannot be expressed such that all atomic formulas are octagonal constraints and all assignments are of one of the supported forms, approximate them as detailed above (Section 2.1). This approximation is standard in program analysis. In this paper, we assume for simplicity that we do not have to perform any approximations for tests or assignments.
3. Eliminate the quantifiers from each verification condition. This results in a set of constraints on the parameters of the involved invariants. To keep the quantifier elimination procedure fast (quadratic in the number of program variables), the verification condition is approximated using a geometric heuristic (Section 2.5).
4. Take the union of all the constraint sets thus generated. Every parameter value that satisfies the constraints leads to an invariant. To accommodate program variables having no lower or upper bounds, parameters are allowed to have $-\infty$ and $+\infty$ as possible values. Because of this, an octagonal invariant is always generated, with the trivial invariant being the one where $-\infty$ and $+\infty$ serve as lower and upper bounds for every arithmetic expression $(x, x + y, x - y)$.

The remainder of this section is mainly devoted to the development of an efficient way to perform the quantifier elimination in Step 3: In Subsection 2.3, we will discuss ways to make the tests appearing in the verification condition leaner or richer, Subsections 2.4 and 2.5 contain an explanation of the rationale behind our method, and in Sections 2.6–2.8, we will show how to perform the actual quantifier elimination efficiently using a series of simple table look-ups.

Our approach does not involve any direct fixed point computation. The analysis is done only once for every program branch, in contrast to the abstract interpretation approach which requires the analysis to be done multiple times, depending upon the nature of the widening operator used for a particular abstract domain to ensure the termination of the fixed point computation. Furthermore, much like traditional Floyd-Hoare analysis, derivation of invariants is done without making any assumption about the termination of programs, which is handled separately. As illustrated below, our approach can derive invariants of nonterminating programs as well and can thus be effective for nonterminating reactive programs as well.

2.3   Trivially Redundant and Implicit Conditions

The formula $T(X)$, which is a conjunction of test conditions along a program path, can contain trivially redundant constraints and it can imply additional constraints (including their unsatisfiability). By trivially redundant constraints, we mean multiple constraints on the lower bound (upper bound) of the same expression (such as $5 \leq x - y$ as well as $4 \leq x - y$) out of which the respective greatest lower bound (the respective least upper bound) only needs to be retained; such trivially redundant constraints can be removed in $O(m)$, where m is the number of such constraints. It can be shown easily that if any trivially redundant constraints were retained and subsequently used to generate parameter constraints from the tables discussed in the later subsections, such parameter constraints will also be trivially redundant, without affecting the loop invariants generated. Henceforth, we will always assume that trivially redundant constraints are removed, such that e.g. a set of octagonal constraints between two variables contains at most 8 constraints.

Since our method for quantifier elimination is driven by parameter-free constraints in a verification condition, it is useful to derive implicit constraints from $T(X)$. Checking for satisfiability as well as deriving additional constraints can require in the worst case $O(n^3)$ steps [3] due to the use of the cubic Floyd-Warshall normalization algorithm, where n is the number of program variables appearing in these constraints. As an illustration,

$$x + y \geq 1, \; z - y \geq 2, \; z \leq 1$$

gives $-y \geq 1$ from $z - y \geq 2 \wedge z \leq 1$, which with $x + y \geq 1$ results in $x \geq 2$.

By localizing the derivation of additional constraints by considering each pair of variables, constant time is needed; this implies that for $O(n^2)$ pairs of variables, the derivation of implicit constraints requires $O(n^2)$ steps. This preprocessing is performed for each pair of variables in random order, since when performed sequentially, the order can affect the output of the results. In the above illustration, picking y, z led to an additional constraint on y which with interaction with constraints on x, y led to an additional constraint on x. If instead the pair x, y

had been picked first, then the additional implicit constraint on x would have been missed by this localized closure. Heuristics can be developed to come up with a good order such that the localized closure produces results which are a good approximation of the global closure. As will be shown below, such localized closure of constraints will keep the complexity of the loop invariant generation method quadratic in the number of program variables.

2.4 Localized Quantifier Elimination

Our goal is to efficiently generate sufficient conditions on parameters so that the verification condition $\Phi(X)$ is satisfied by all parameter values satisfying these conditions (the soundness condition). Of course, it is most desirable to generate as strong an approximation as possible to the quantifier-free formula on parameters equivalent to $\forall X\ \Phi(X)$. Let $\phi_{i,j}$ be the subformula of Φ expressed only using program variables x_i, x_j. E.g. if

$$\Phi = (x_1 \leq u^1 \wedge x_2 \leq u^2 \wedge x_2 \leq u^3) \implies (x_1 \leq 1 + u^1 \wedge x_2 \leq -l^2 \wedge x_2 \leq u^3) ,$$

then

$$\phi_{1,2} = (x_1 \leq u^1 \wedge x_2 \leq u^2) \implies (x_1 \leq 1 + u^1 \wedge x_2 \leq -l^2) .$$

Given the structure of Φ, it is easy to see that

$$[\wedge_{1 \leq i \neq j \leq n}(\forall x_i, x_j\ \phi_{i,j})] \equiv [\forall X\ \Phi(X)] .$$

The following theorem enables us to factor quantifier elimination of $\forall X\ \Phi(X)$ by considering subformulas $\forall x_i, x_j\ \phi_{i,j}$ corresponding to a single pair of distinct variables in Φ, generating sufficient conditions on parameters for the subformula (soundness requirement on quantifier elimination heuristic), and then doing a conjunction of such conditions on parameters for every subformula on every possible pairs of variables. The result is then a sufficient condition for the verification condition $\forall X\ \Phi(X)$.

Theorem 1. *Let $pc_{i,j}$ be a quantifier-free formula on parameters in $\phi_{i,j}$ such that for every possible parameter assignment σ, if σ satisfies $pc_{i,j}$, then σ satisfies $\forall x_i, x_j\ \phi_{i,j}$. Then any parameter assignment that satisfies $\wedge_{1 \leq i < j \leq n} pc_{i,j}$ also satisfies $\forall X\ \Phi(X)$.*

Proof. The proof follows from $\wedge_{1 \leq i < j \leq n}(\forall x_i, x_j\ \phi(x_i, x_j))$ being equivalent to $\forall X\ \Phi(X)$. □

The above theorem enables localizing quantifier elimination from a formula of arbitrary size to a formula of fixed size: the size of $\forall x_i, x_j\ \phi(x_i, x_j)$ is determined by the parameter-free part, which is a conjunction of tests along a program path; other than this subformula, the hypothesis is a conjunction of 8 atomic formulas with parameters and the conclusion is also a conjunction of 8 atomic formulas with parameters. For such a formula, quantifier elimination can be performed in constant time. In contrast, the worst case complexity of a complete quantifier elimination for linear constraints is exponential in the number of quantifiers alternations and doubly exponential in the number of quantified variables [24]. Below, we will focus on the subformula $\phi_{i,j}$ in the verification condition $\Phi(X)$ and discuss quantifier elimination of x_i, x_j from $\phi_{i,j}$. To make the presentation free of subscripts, we will replace x_i, x_j by x, y, and henceforth call this verification condition $\phi(x, y)$.

2.5   Geometric View of Quantifier Elimination

The subformula $\phi(x, y)$ is of the form

$$(octa(x, y) \wedge T(x, y)) \Rightarrow octa(x', y') ,$$

where x' and y' are the values of x and y after all the assignments on a program path have been executed and $T(x, y)$ is the conjunction of all the atomic formulas obtained by the localized closure of parameter-free octagonal constraints on x, y, resulting from the loop tests and branch conditions in the program path (after being appropriately modified due to assignment statements between any two tests).

As discussed above, the parametrized invariant $octa(x, y)$ specifies an octagon with 8 sides corresponding to each of the atomic constraints in $octa(x, y)$. In order to ensure that the verification condition $\phi(x, y)$ consists only of octagonal constraints, there can only be four different possibilities about the cumulative effect of all assignments of x, y along a program path (the fourth possibility is covered by the case 3 below by switching x and y).

Possibility 1 x := −x+A and y := −y+B,
Possibility 2 x := x+A and y := y+B,
Possibility 3 x := −x+A and y := y+B,

where A, B are constants.[3] Each of these possibilities gives rise to a transformed octagon $I(x', y')$; both the original (white) octagon $I(x, y)$ and the transformed (shaded) octagon $I(x', y')$ are depicted in Figures 1–3 on the following pages, corresponding to the three possibilities. The subformula $\phi(x, y)$ is:

$$\left(l_1 \leq x - y \leq u_1 \wedge l_2 \leq x + y \leq u_2 \wedge l_3 \leq x \leq u_3 \wedge l_4 \leq y \leq u_4\right) \wedge T(x, y)$$
$$\implies \left(l_1 \leq x' - y' \leq u_1 \wedge l_2 \leq x' + y' \leq u_2 \wedge l_3 \leq x' \leq u_3 \wedge l_4 \leq y' \leq u_4\right) ,$$

with $l_1, u_1, l_2, u_2, l_3, u_3, l_4, u_4$ as parameters. The goal is to eliminate program variables x, y from $\phi(x, y)$ and efficiently generate the strongest possible approximation of an equivalent quantifier-free formula on the parameters l_1, l_2, l_3, l_4 and u_1, u_2, u_3, u_4 to $\forall x, y \ \phi(x, y)$. The main requirement on the result is that it is a sound under-approximation in the sense that we may miss some valid invariants, but the method does not yield any formulas that are invalid invariants.

---

[3] Constant assignments of the form x := A or y := B are handled similarly. Their discussion is omitted due to lack of space.

It should be first observed that the hypothesis in $\phi(x, y)$ is a conjunction of a parametrized octagon $octa(x, y)$ and a (partial) concrete octagon $T(x, y)$; thus it corresponds to the intersection of these octagons. The conclusion $octa(x', y')$ is also a parametrized octagon; it is a displaced version of the octagon in the hypothesis. Constraints on parameters $l_1, l_2, l_3, l_4, u_1, u_2, u_3, u_4$ that ensure that the intersection octagon $octa(x, y) \land T(x, y)$ being contained in $octa(x', y')$ is a good approximation of a quantifier-free formula equivalent to $\forall x, y \; \phi(x, y)$.

In each subsection corresponding to one of the possibilities, using local geometric analysis, we consider one by one, every side of the concrete octagon (and hence every possible lower bound and upper bound on $x - y, x + y, x, y$ in T) to see how it rules out the portion of the parametrized octagon $octa(x, y)$ not included in $octa(x', y')$. This is ensured in two parts. Conditions on A, B must be identified for each of the above three possibilities such that there is an overlap between the transformed octagon and the original octagon; often, this overlap can be ensured by making a few of the sides to be at $-\infty$ (or $+\infty$) (the case of when all sides have to be unbounded, leads to a trivial invariant, similar to the invariant **true** for any loop).

Depending upon the presence or absence of a side in the concrete octagon defined by T (i.e., a bound in T), parameter values are constrained by A or B. As an example, in Figure 1, if there is no upper bound constraint on $x - y$ in T, then $u_1 \leq (-l_1 + \Delta_1)$ (where $\Delta_1 = A - B$) ensures that the $x - y$ side of the original octagon is contained in the corresponding inverted side in the displaced octagon; we can also see this from $(x - y \leq u_1) \Rightarrow (l_1 \leq -x + y + \Delta_1)$ which is equivalent to $(x - y \leq u_1) \Rightarrow (x - y \leq -l_1 + \Delta_1)$. In the presence of an upper bound constraint of the form $x - y \leq a$ in T, both $a \leq u_1$ and $a \leq (-l_1 + \Delta_1)$ can be used to prune the original octagon.

These constraints on parameter values are derived below once and for all, and a table is constructed corresponding to each of the above three possibilities (see tables in Figures 1, 2, 3). For each possible bound, there is a table entry depending upon whether that bound is present or absent in T.[4] To generate a quantifier-free formula pc for the above verification condition $\phi(x, y)$, it suffices to take a conjunction of all table entries corresponding to the absence or presence of the bounds for $x, y, x - y$, and $x + y$ in T, where each table entry specifies a constraint on a parameter.

For example, when the signs of both variables are reversed in an assignment x := −x+A, y := −y+B (Figure 1) and a constraint $x \leq 5$ is present, the constraint $e \leq A - l_3$ is generated. When the signs of both variables are reversed in an assignment x := x+A, y := y+B (Figure 2) and no constraint of the form $x \leq e$ is present, either the constraint $u_3 = +\infty$ is generated (if $A > 0$) or no constraint is generated (if $A \not> 0$).

Below we show the derivation for the entries of the table in Figure 1. The tables in Figures 2 and 3 can be constructed analogously. For possibility 1, the assignment is x := −x+A, y := −y+B; for possibility 2, x := x+A, y := y+B; for possibility 3, x := −x+A, y := y+B. As should be evident from the table entries,

---

[4] This is why implicit constraints from T become relevant.

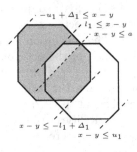

constraint	present	absent
$x - y \leq a$	$a \leq \Delta_1 - l_1$	$u_1 \leq \Delta_1 - l_1$
$x - y \geq b$	$\Delta_1 - u_1 \leq b$	$\Delta_1 - u_1 \leq l_1$
$x + y \leq c$	$c \leq \Delta_2 - l_2$	$u_2 \leq \Delta_2 - l_2$
$x + y \geq d$	$\Delta_2 - u_2 \leq d$	$\Delta_2 - u_2 \leq l_2$
$x \leq e$	$e \leq A - l_3$	$u_3 \leq A - l_3$
$x \geq f$	$A - u_3 \leq f$	$A - u_3 \leq l_3$
$y \leq g$	$g \leq B - l_4$	$u_4 \leq B - l_4$
$y \geq h$	$B - u_4 \leq h$	$B - u_4 \leq l_4$

Fig. 1. Signs of x and y are reversed: Constraints on Parameters

given two constraints on an expression $(x, x + y, x - y)$ such that one of them is trivially redundant (e.g., $x - y \leq 4$ and $x - y \leq 10$), the corresponding entry in a table to the trivially redundant constraint is also redundant.

2.6 Reversal of the Signs of Both Variables

The verification condition for this case is:

$$(I(x, y) \wedge T(x, y)) \Rightarrow I(-x + A, -y + B),$$

which is

$$\begin{pmatrix} l_1 \leq x - y \leq u_1 \wedge l_2 \leq x + y \leq u_2 \\ \wedge\, l_3 \leq x \leq u_3 \wedge l_4 \leq y \leq u_4 \end{pmatrix} \wedge T(x, y)$$
$$\Rightarrow \begin{pmatrix} \Delta_1 - u_1 \leq x - y \leq \Delta_1 - l_1 \wedge \Delta_2 - u_2 \leq x + y \leq \Delta_2 - l_2 \\ \wedge\, A - u_3 \leq x \leq A - l_3 \wedge B - u_4 \leq y \leq B - l_4 \end{pmatrix},$$

where $\Delta_1 = A - B$ and $\Delta_2 = A + B$.

For $x - y$, the original octagon has l_1 and u_1 as lower and upper bound, respectively, whereas the transformed octagon has $-u_1 + \Delta_1$ and $-l_1 + \Delta_1$ as the lower and upper bounds. In the absence of any bounds on $x - y$ in T, $u_1 \leq \Delta_1 - l_1$ and $\Delta_1 - u_1 \leq l_1$ have to hold for the transformed octagon to include the original octagonal.

A concrete upper bound a on $x - y$ in T however changes the constraint on the parameter l_1: For the transformed octagon to include the original octagonal, $\Delta_1 - l_1$ has to be greater than or equal to one of the upper bounds u_1 and a of the original, corresponding to a constraint $a \leq u_1 \vee a \leq \Delta_1 - l_1$. Using such disjunctive constraints would directly lead to a combinatorial explosion of the analysis. Instead, we decided to only admit $a \leq \Delta_1 - l_1$ to the table, a safe over-approximation which reflects that in practice, tests the programmer specifies are actually relevant to the semantics of the program. Similarly, a concrete lower bound b on $x - y$ in T changes the constraint for u_1 to $\Delta_1 - u_1 \leq b$.

A similar analysis can be done for $x + y$: the hypothesis has $l_2 \leq x + y \leq u_2$ and the conclusion includes: $\Delta_2 - u_2 \leq x + y \leq \Delta_2 - l_2$. In the absence of any bound on $x + y$ in $T(x, y)$, the parameter constraints $\Delta_2 - u_2 \leq l_2$ and $u_2 \leq \Delta_2 - l_2$ will ensure that the corresponding side of the transformed octagon includes that of the original octagon. If T contains a concrete upper bound c on $x + y$, then $c \leq \Delta_2 - l_2$. If T has a concrete lower bound d, then $\Delta_2 - u_2 \leq d$ ensures that the transformed octagon includes the original octagon.

We will omit the analysis leading to entries corresponding to the presence (or absence) of concrete lower and upper bounds on program variables x, y, as it is essentially the same. The above analysis is presented in the table in Figure 1. Depending upon a program and cases, looking up the table generates constraints on parameters in constant time.

The following lemma states that the above quantifier elimination method is sound.

Lemma 2. *Given a test T in the subformula $\phi(x, y)$ in the verification condition, let pc be the conjunction of the parameter constraints corresponding to the presence (or absence) of each type of concrete constraint in T. For every assignment of parameter values satisfying pc, substitution of these values for the parameters makes $\forall x, y \; \phi(x, y)$ valid.*

Proof. Follows directly from the above computations.   □

It thus follows:

Theorem 3. *Any assignment of parameter values satisfying pc in the above lemma generates $octa(x, y)$ as the subformula on program variables x, y serving as the invariant for the associated program path.*

We illustrate how the above table can be used to generate invariants for a program.

Example 4. Consider the following program:

```
x := 2; y := 3;
while (x+y ≥ 0) do
    if (y ≥ 2) then
        y := -y+4; x := -x+3;
    else
        x := -x-3; y := -y+5;
```

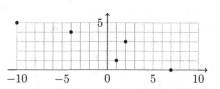

The actual state space for this program is depicted on the right. There are two branches in the loop body. For the first branch, the verification condition is:

$$\left(l_1 \leq x - y \leq u_1 \wedge l_2 \leq x + y \leq u_2 \wedge l_3 \leq x \leq u_3 \wedge l_4 \leq y \leq u_4 \right) \wedge T_1(x, y)$$
$$\Rightarrow \begin{pmatrix} -1 - u_1 \leq x - y \leq -1 - l_1 \wedge 7 - u_2 \leq x + y \leq 7 - l_2 \\ \wedge \, 3 - u_3 \leq x \leq 3 - l_3 \wedge 4 - u_4 \leq y \leq 4 - l_4 \end{pmatrix},$$

where $A = 3$, $B = 4$, $\Delta_1 = -1$, $\Delta_2 = 7$, and $T_1(x, y) = x + y \geq 0 \wedge y \geq 2$. We approximate this verification condition by separating the components containing common parameters:

$$((l_1 \leq x - y \leq u_1) \wedge (x + y \geq 0 \wedge y \geq 2)) \implies (-1 - u_1 \leq x - y \leq -1 - l_1)$$
$$((l_2 \leq x + y \leq u_2) \wedge (x + y \geq 0 \wedge y \geq 2)) \implies (7 - u_2 \leq x + y \leq 7 - l_2)$$
$$((l_3 \leq x \leq u_3) \wedge (x + y \geq 0 \wedge y \geq 2)) \implies (3 - u_3 \leq x \leq 3 - l_3)$$
$$((l_4 \leq y \leq u_4) \wedge (x + y \geq 0 \wedge y \geq 2)) \implies (4 - u_4 \leq y \leq 4 - l_4)$$

The following parameter constraints are generated for branch 1:

1. Given no concrete lower or upper bounds on $x - y$ in $T_1(x, y)$, the table in Figure 1 generates the parameter constraints $u_1 \leq -1 - l_1$ and $-1 - u_1 \leq l_1$.
2. Since $x + y \geq 0$ in $T_1(x, y)$, the table entry corresponding to it is $7 - u_2 \leq 0$. Since there is no concrete upper bound on $x + y$ specified in T_1, there is an additional constraint: $u_2 \leq 7 - l_2$.
3. Similarly, absence of concrete upper or lower bounds on x in $T_1(x, y)$ gives $l_3 + u_3 = 3$.
4. For the concrete lower bound $y \geq 2$ and absence of any concrete upper bound on y in $T_1(x, y)$ the table entries are $4 - u_4 \leq 2$ and $u_4 \leq 4 - l_4$.

Similarly, for the second branch, the verification condition is:

$$\left(l_1 \leq x - y \leq u_1 \wedge l_2 \leq x + y \leq u_2 \wedge l_3 \leq x \leq u_3 \wedge l_4 \leq y \leq u_4 \right) \wedge T_2(x, y)$$
$$\Rightarrow \left(\begin{array}{c} -u_1 - 8 \leq x - y \leq -l_1 - 8 \wedge -u_2 + 2 \leq x + y \leq -l_2 + 2 \\ \wedge -u_3 - 3 \leq x \leq -l_3 - 3 \wedge -u_4 + 5 \leq y \leq -l_4 + 5 \end{array} \right),$$

Here $\Delta_1 = -8$ and $\Delta_2 = 2$. The tests along the branch are $(y + x) \geq 0 \wedge y \leq 1$. Their closure is

$$T_2(x, y) = x + y \geq 0 \wedge y \leq 1 \wedge x \geq -1 \wedge x - y \geq -2 ,$$

because $x + y \geq 0 \wedge y \leq 1 \implies x \geq -1 \wedge x - y \geq -2$. The following parameter constraints are generated for branch 2:

1. The only concrete constraint on $x - y \geq -2$ in $T_2(x, y)$ generates the parameter constraints $-8 - u_1 \leq -2$ and $u_1 \leq -8 - l_1$.
2. Similarly, the only constraint on $x + y \geq 0$ in $T_2(x, y)$ produces the parameter constraints $2 - u_2 \leq 0$ and $u_2 \leq 2 - l_2$.
3. Similarly, $x \geq -1$ in $T_2(x, y)$ generates $-3 - u_3 \leq -1 \wedge u_3 \leq -3 - l_3$.
4. Finally, the constraint $y \leq 1$ in $T_2(x, y)$ generates the parameter constraints $-u_4 + 5 \leq l_4$ and $1 \leq -l_4 + 5$.

At the initial entry of the loop, $x = 2, y = 3$, which generates additional constraints on the parameters, like $l_3 \leq 2 \leq u_3$. Using these and the constraints computed from the two branches, the following parameter constraints are generated (after throwing away trivially redundant constraints on parameters):

$$l_1 \leq -1 \wedge u_1 \geq -1 \wedge l_1 + u_1 \geq -1 \wedge l_1 + u_1 \leq -8$$
$$\wedge l_2 \leq 5 \wedge u_2 \geq 7 \wedge l_2 + u_2 \leq 2$$
$$\wedge l_3 \leq 2 \wedge u_3 \geq 2 \wedge l_3 + u_3 = 3 \wedge l_3 + u_3 \leq -3$$
$$\wedge l_4 \leq 3 \wedge u_4 \geq 3 \wedge l_4 + u_4 \leq 4 \wedge l_4 + u_4 \geq 5.$$

Any values of l_i, u_is that satisfy the above constraints result in an invariant for the loop. However, our goal is to generate the strongest possible invariant.

It is easy to see that the above constraints can be decomposed into disjoint subsets of constraints: (i) constraints on l_1, u_1, (ii) constraints on l_2, u_2, (iii) constraints on l_3, u_3, and finally, (iv) constraints on l_4, u_4, implying that each disjoint subset can be analyzed by itself. This is so because the table entries only relate parameters $l_i, u_i, 1 \leq i \leq 4$. This structure of parameter constraints is exploited later to check for satisfiability of parameter constraints to generate invariants including the strongest possible invariant.

Consider all the inequalities about l_1 and u_1, as an illustration:

$$l_1 \leq -1 \wedge u_1 \geq -1 \wedge l_1 + u_1 = -1 \wedge l_1 + u_1 \leq -8.$$

As the reader would notice there are no integers that satisfy the above constraints, since $-1 > -8$. But recall that we extended numbers to include $-\infty$ and $+\infty$ to account for no lower bounds and no upper bounds, and that both $(+\infty) + (-\infty) \leq a$ and $(+\infty) + (-\infty) \geq a$ hold for every integer a. Taking those observations into account, we find that $l_1 = -\infty$ and $u_1 = +\infty$ is a solution of this system. Similarly, $l_3 = l_4 = -\infty$ and $u_3 = u_4 = +\infty$ is obtained.

Finally, all the inequalities relating to l_2 and u_2 are as follows

$$l_2 \leq 5 \wedge u_2 \geq 7 \wedge l_2 + u_2 \leq 2.$$

Since we want to get as strong an invariant as possible, l_2 should be as large as possible and u_2 should be as small as possible. Hence, $l_2 = -5 \wedge u_2 = 7$, giving us the invariant

$$-5 \leq x + y \leq 7.^5$$

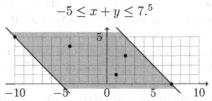

The reader should notice that we did not make use of the constraints $x \geq -1 \wedge x - y \geq -2$ obtained by closing the path constraints. So for this example, computing the closure did not change the constraints on parameters.

2.7   The Signs of Both Variables Remain Invariant

An analysis similar to the one discussed in the previous subsection is also done for the case when the effect of assignments on a pair of variables is of the form x := x+A and y := y+B. The verification condition $\phi(x, y)$ for this case is:

$$\left(l_1 \leq x - y \leq u_1 \wedge l_2 \leq x + y \leq u_2 \wedge l_3 \leq x \leq u_3 \wedge l_4 \leq y \leq u_4\right) \wedge T(x, y)$$
$$\Rightarrow \left(\begin{array}{c} l_1 - \Delta_1 \leq x - y \leq u_1 - \Delta_1 \wedge l_2 - \Delta_2 \leq x + y \leq u_2 - \Delta_2 \\ \wedge l_3 - A \leq x \leq u_3 - A \wedge l_4 - B \leq y \leq u_4 - B \end{array}\right),$$

---

[5] The Interproc tool produces the same invariant.

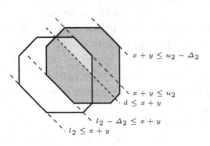

constraint	present	absent	side condition
$x - y \le a$	$u_1 \ge a + \Delta_1$	$u_1 = +\infty$	$\Delta_1 > 0$
$x - y \ge b$	$l_1 \le b + \Delta_1$	$l_1 = -\infty$	$\Delta_1 < 0$
$x + y \le c$	$u_2 \ge c + \Delta_2$	$u_2 = +\infty$	$\Delta_2 > 0$
$x + y \ge d$	$l_2 \le d + \Delta_2$	$l_2 = -\infty$	$\Delta_2 < 0$
$x \le e$	$u_3 \ge e + A$	$u_3 = +\infty$	$A > 0$
$x \ge f$	$l_3 \le f + A$	$l_3 = -\infty$	$A < 0$
$y \le g$	$u_4 \ge g + B$	$u_4 = +\infty$	$B > 0$
$y \ge h$	$l_4 \le h + B$	$l_4 = -\infty$	$B < 0$

Fig. 2. Signs of x and y do not change: Constraints on Parameters

where $\Delta_1 = A - B$ and $\Delta_2 = A + B$. Figure 2 contains the transformed octagon (shaded) and the original octagon (white) corresponding to the octagonal invariant, and along with concrete lower and upper bounds on $x + y$ possibly appearing in T.

There are two main differences between this table and the previous table. Firstly, each table entry imposes constraints on a single parameter corresponding to an octagonal side. Secondly, for some sides of octagons, the original octagon does not have to be pruned. For instance, in this case when $\Delta_1 \le 0$, an upper bound on $x - y$ does not make any difference; so there is no corresponding entry in the table.

We would like to point out similarities with the abstract interpretation approach, particularly how the widening operator is implicit in the quantifier elimination heuristic. Consider an entry in the table when $A > 0$. The condition $A > 0$ implies that if the path is executed several times, then x keeps increasing arbitrarily. In the absence of any upper bound on x in a test in the path, the value of x has no upper bound and u_3 has to be $+\infty$, similar to the widening operator of abstract interpretation. Similarly, if $\Delta_1 < 0$, then along that path, the value of $x - y$ keeps decreasing indefinitely unless again if there is a lower bound imposed by a test in the program path; the table entries correspond to both of these situations. We again illustrate the use of the above table to analyze the program below.

Example 5. Consider the following program:

```
x := 10; y := 0;
while (x−y ≥ 3) do
   if (x ≥ 5) then
      x := x−1;
   else
      y := y+1;
```

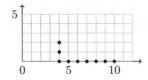

The state space for this program is depicted on the right. For initial values and the execution of the loop body, the following verification conditions are generated:

- Initial values: $l_1 \leq 10 \leq u_1 \wedge l_2 \leq 10 \leq u_2 \wedge l_3 \leq 10 \leq u_3 \wedge l_4 \leq 0 \leq u_4$.
- Branch 1:

$$\left(l_1 \leq x - y \leq u_1 \wedge l_2 \leq x + y \leq u_2 \wedge l_3 \leq x \leq u_3 \wedge l_4 \leq y \leq u_4\right) \wedge T_1(x, y)$$
$$\Rightarrow \left(\begin{array}{c} l_1 + 1 \leq x - y \leq u_1 + 1 \wedge l_2 + 1 \leq x + y \leq u_2 + 1 \\ \wedge\, l_3 + 1 \leq x \leq u_3 + 1 \wedge l_4 \leq y \leq u_4 \end{array}\right),$$

where $\Delta_1 = -1, \Delta_2 = -1$ and $T_1(x, y) = x - y \geq 3 \wedge x \geq 5$.
- Branch 2:

$$\left(l_1 \leq x - y \leq u_1 \wedge l_2 \leq x + y \leq u_2 \wedge l_3 \leq x \leq u_3 \wedge l_4 \leq y \leq u_4\right) \wedge T_2(x, y)$$
$$\Rightarrow \left(\begin{array}{c} l_1 + 1 \leq x - y \leq u_1 + 1 \wedge l_2 - 1 \leq x + y \leq u_2 - 1 \\ \wedge\, l_3 \leq x \leq u_3 \wedge l_4 - 1 \leq y \leq u_4 - 1 \end{array}\right),$$

where $\Delta_1 = -1, \Delta_2 = 1$ and $T_2(x, y) = x - y \geq 3 \wedge x \leq 4 \wedge y \leq 1 \wedge x + y \leq 5$.[6]

By combining the information from the initial condition with the constraints generated by the table in Figure 2, we get the following constraints on the parameters:

$$l_1 \leq 2 \wedge u_1 \geq 10 \wedge l_2 = -\infty \wedge u_2 \geq 10 \wedge l_3 \leq 4 \wedge u_3 \geq 10 \wedge l_4 \leq 0 \wedge u_4 \geq 2 .$$

To generate the strongest possible invariant, the l_is should be as large as possible, and u_is should be as small as possible. The parameter values $l_1 = 2, u_1 = 10$, $l_2 = -\infty, u_2 = 10, l_3 = 4, u_3 = 10, l_4 = 0$ and $u_4 = 2$ satisfy these requirements, giving the following invariant:[7]

$$2 \leq x - y \leq 10 \wedge x + y \leq 10 \wedge 4 \leq x \leq 10 \wedge 0 \leq y \leq 2 .$$

As the reader would notice, the above formula is not closed: we can easily obtain a lower bound of 4 on $x + y$ from lower bounds of x and y. This is despite the fact that parameter constraints above are closed. This suggests that even though the invariant thus generated is the strongest, its representation is not closed.

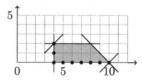

Note that computing the closure of the test conditions in the second branch strengthens the invariant generated using the proposed method. If the closure had not been computed, the implicit constraint $y \leq 1$ would not then

---

[6] The closure of $(x - y \geq 3 \wedge x \leq 4)$ gives $(y \leq 1 \wedge x + y \leq 5)$.

[7] Interproc outputs $2 \leq x - y \leq 10 \wedge 4 \leq x + y \leq 10 \wedge 4 \leq x \leq 10 \wedge 0 \leq y \leq \frac{7}{2}$ [sic].

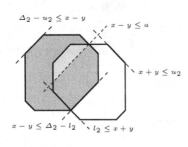

constraint	present	absent	side condition
$x - y \leq a$	$a \leq \Delta_2 - l_2$	$u_1 \leq \Delta_2 - l_2$	–
$x - y \geq b$	$\Delta_2 - u_2 \leq b$	$\Delta_2 - u_2 \leq l_1$	–
$x + y \leq c$	$c \leq \Delta_1 - l_1$	$u_2 \leq \Delta_1 - l_1$	–
$x + y \geq d$	$\Delta_1 - u_1 \leq d$	$\Delta_1 - u_1 \leq l_2$	–
$x \leq e$	$e \leq A - l_3$	$u_3 \leq A - l_3$	–
$x \geq f$	$A - u_3 \leq f$	$A - u_3 \leq l_3$	–
$y \leq g$	$u_4 \geq g + B$	$u_4 = +\infty$	$B > 0$
$y \geq h$	$l_4 \leq h + B$	$l_4 = -\infty$	$B < 0$

Fig. 3. Sign of only x is reversed in assignment: Constraints on Parameters

generate the parameter constraint $u_4 \geq 2$. This example illustrates that computing the closure of test conditions (even locally) can help in generating a stronger invariant.

2.8   The Sign of Exactly One Variable Is Reversed

As in the previous two subsections, a similar analysis can be done to investigate the effect of assignments of the form x := −x+A and y := y+B, where the sign of exactly one variable changes. The parametric verification condition in this case is:

$$\left(l_1 \leq x - y \leq u_1 \wedge l_2 \leq x + y \leq u_2 \wedge l_3 \leq x \leq u_3 \wedge l_4 \leq y \leq u_4\right) \wedge T(x, y)$$
$$\Rightarrow \left(\begin{array}{c} \Delta_1 - u_1 \leq x + y \leq \Delta_1 - l_1 \wedge \Delta_2 - u_2 \leq x - y \leq \Delta_2 - l_2 \\ \wedge\, A - u_3 \leq x \leq A - l_3 \wedge l_4 - B \leq y \leq u_4 - B \end{array}\right),$$

where $\Delta_1 = A - B$ and $\Delta_2 = A + B$. The associated table of generated constraints is given in Figure 3. The use of this table is once again illustrated using an example below.

Example 6. Let us consider the following loop:

```
x := 4; y := 6;
while (x+y ≥ 0)
   if (y ≥ 6) then
      x := -x; y := y-1;
   else
      x := x-1; y := -y;
```

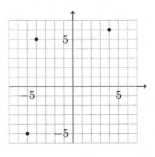

The state space for this program is again depicted on the right. For initial values and the execution of the loop body, the following verification conditions are generated:

– Initial condition:

$$l_1 \leq -2 \leq u_1 \wedge l_2 \leq 10 \leq u_2 \wedge l_3 \leq 4 \leq u_3 \wedge l_4 \leq 6 \leq u_4 .$$

- Branch 1:

$$\begin{pmatrix} l_1 \le x - y \le u_1 \wedge l_2 \le x + y \le u_2 \\ \wedge \, l_3 \le x \le u_3 \wedge l_4 \le y \le u_4 \end{pmatrix} \wedge T_1(x, y)$$

$$\Rightarrow \begin{pmatrix} 1 - u_1 \le x + y \le 1 - l_1 \wedge -1 - u_2 \le x - y \le -1 - l_2 \\ \wedge -u_3 \le x \le -l_3 \wedge 1 + l_4 \le y \le 1 + u_4 \end{pmatrix},$$

where $\Delta_1 = 1$, $\Delta_2 = -1$ and $T_1(x, y) = (x + y \ge 0 \wedge y \ge 6)$.

- Branch 2:

$$\begin{pmatrix} -u_1 \le y - x \le -l_1 \wedge l_2 \le y + x \le u_2 \\ \wedge \, l_4 \le y \le u_4 \wedge l_3 \le x \le u_3 \end{pmatrix} \wedge T_2(x, y)$$

$$\Rightarrow \begin{pmatrix} l_1 + 1 \le y + x \le u_1 + 1 \wedge -1 - u_2 \le y - x \le -1 - l_2 \\ \wedge \, 1 + l_3 \le x \le 1 + u_3 \wedge -u_4 \le y \le -l_4 \end{pmatrix},$$

where $T_2(x, y) = (x + y \ge 0 \wedge y \le 5 \wedge x \ge -5 \wedge x - y \ge -10)$.[8]

From the table, we get the following parameter constraints:

$$l_1 \le -2 \wedge u_2 \ge 10 \wedge -1 \le l_1 + u_2 \le 1$$
$$\wedge \, l_2 \le -11 \wedge u_1 \ge 1 \wedge l_2 + u_1 \le -1 \wedge -1 \le u_2 - u_1 \le 1$$
$$\wedge \, l_3 \le -6 \wedge u_3 \ge 4 \wedge u_3 + l_3 = 0$$
$$\wedge \, l_4 \le -5 \wedge u_4 \ge 6 \wedge u_4 + l_4 \ge 0.$$

Making the l_is as large as possible, and u_is as small as possible, we get:

$$l_1 = -9, u_1 = 9, l_2 = -11, u_2 = 10, l_3 = -6, u_3 = 6, l_4 = -5, u_4 = 6.$$

The corresponding invariant is:[9]

$$-9 \le x - y \le 9 \wedge -11 \le x + y \le 10 \wedge -6 \le x \le 6 \wedge -5 \le y \le 6 \,.$$

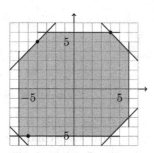

If the closure of the test conditions for the second branch is not computed, then the invariant generated is weaker since the parameter constraint due to the fact that the implicit constraint $x \ge -5$ is omitted.

---

[8] This is because the closure of the constraint $(x + y \ge 0 \wedge y \le 5)$ also includes $x \ge -5$ and $x - y \ge -10$.

[9] The Interproc tool's output is $y \ge -5$.

2.9   The Effects of Computing the Closure of the Tests and Choosing Different Table Entries

The examples in the previous subsections illustrated how implicit constraints derived from the test conditions in a program path can improve the strength of invariants by generating additional constraints on the parameters. However, that is not always the case. The following example is a slight adaptation of Example 6.

Example 7. Let us consider the following loop:

```
x := 4; y := 6;
while (x+y ≥ 0)
    if (y ≥ 6) then
        x := -x; y := y-1;
    else
        x := -x; y := -y;
```

The parameter constraints after computing the closure $x \geq -5 \wedge x - y \geq -10$ of $x + y \geq 0 \wedge y \leq 5$ in the second branch lead to the computation of the following strongest invariant:

$$-10 \leq x - y \leq 10 \wedge -11 \leq x + y \leq 10 \wedge -5 \leq x \leq 5 \wedge -5 \leq y \leq 6 .$$

In contrast, without generating implicit constraints, the strongest invariant generated is:

$$-9 \leq x - y \leq 9 \wedge -10 \leq x + y \leq 10 \wedge -4 \leq x \leq 4 \wedge -5 \leq y \leq 6 .$$

The reader would note that $-10 \leq x + y$ is not closed; normalization would transform this into $-9 \leq x + y$.

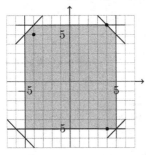
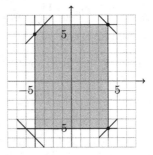

invariant with closure   invariant without closure

The optimal bound, based on the states, would be $-1 \leq x + y$. This effect stems from the way we approximate the verification condition: To derive, say, the upper bound for x in the presence of a sign-reversing assignment, we look at the part of a verification condition involving x (and no other variable). Depending on whether or not a test on x is present on the path, the condition may have one of two forms:

$$x \leq u_3 \qquad \Longrightarrow \quad x \leq A - l_3 \text{ or}$$
$$x \leq u_3 \wedge x \leq e \Longrightarrow x \leq A - l_3 .$$

The former implication is equivalent to $u_3 \leq A - l_3$, which is the corresponding table entry, and the latter is equivalent to $u_3 \leq A - l_3 \vee e \leq A - l_3$. Since the introduction of disjunctions in this way would adversely affect the complexity, our decision was to assume that the test $x \leq e$ carries more semantics and is most likely the more restrictive of the two, and thus to choose $e \leq A - l_3$ as the table entry. In the previous example, this choice did not lead to the optimal result. The effect of choosing the right table entry is even more evident in the following trivial example:

Example 8. Consider the following program:

```
x := 0;
while (x ≤ 10)
    x := 1-x;
```

The value of x jumps back and forth between 0 and 1 in each iteration and never comes close to violating the loop condition. The optimal octagonal invariant, which in one dimension is an interval, is $0 \leq x \leq 1$.

With the table entry $u_3 \leq A - l_3$ for the upper bound of x, we derive exactly this invariant. Our heuristic that the loop condition should impact the invariant is not optimal for this loop. So with the table entry $e \leq A - l_3$, we instead derive $-9 \leq x \leq 10$, which is exactly what also Interproc returns as the loop's invariant.

Below we discuss how the structure of parameter constraints generated by the above method can be exploited to develop fast methods to checking their satisfiability, generating an assignment to parameters to result in an invariant, as well as generating the strongest invariant.

The role of derived implicit constraints obtained from the closure operation and their relationship to different possible table entries needs further investigation. Conversely, it might also be useful to investigate dropping constraints which are otherwise implied by the remaining constraints.

2.10   Programs with Multiple Loops

The proposed approach extends to programs containing several loops, both nested and sequential. The verification conditions for such programs will often relate two invariants associated with different loops, i.e. they take a shape like

$$(I(x,y) \wedge T(x,y)) \implies J(x',y') \,,$$

where I and J are the two invariants that are involved. Following our quantifier elimination approach, the verification conditions can as before be approximated by constraints on the parameters using similar tables to the ones presented in Figures 1–3, which now contain constraints on the parameters of both invariants. Due to lack of space, we restrict ourselves to illustrating this extension on an example.

Example 9. Consider the following nonterminating program that contains two nested loops.

```
x := 0; y := 5;
while (true) do
    if (x < 10 and y = 5) then
        x := x+1;
    else if (x = 10 and y > 0) then
        y := y-1;
    else
        while (y < 5) do
            x := x-1; y := y+1;
```

In the depiction of this program's state space, states reachable at the outer loop are indicated by the continuous line and states reachable at the inner loop are indicated by the dotted line.

Let $I(x, y)$ be the invariant associated with the outer loop with parameters $l_1, l_2, l_3, l_4, u_1, u_2, u_3, u_4$. Similarly, let $J(x, y)$ be the invariant associated with the inner loop with parameters $l'_1, l'_2, l'_3, l'_4, u'_1, u'_2, u'_3, u'_4$. Besides verification conditions of the form discussed above, there are two verification conditions relating $I(x, y)$ to $J(x, y)$ and vice versa.

(i) $(J(x, y) \land \neg(y < 5)) \implies I(x, y)$, and
(ii) $(I(x, y) \land \neg(x < 10 \land y = 5) \land \neg(x = 10 \land y > 0)) \implies J(x, y)$.

These verification conditions relate parameters of the two loop invariants. From condition (i), for example, we can derive the constraints $l'_1 \leq l_1$ and $u_1 \leq u'_1$ and similarly for other parameters; because of the test $\neg(y < 5)$, or equivalently $\neg(y \leq 4)$, we get $l_4 \leq 5$. The loop invariants generated by the proposed approach are

$$I(x, y) = J(x, y) = \begin{pmatrix} x - y \leq 10 \land 5 \leq x + y \leq 15 \\ \land\, x \leq 10 \land 0 \leq y \leq 5 \end{pmatrix},$$

or after closure:

$$I(x, y) = J(x, y) = \begin{pmatrix} -5 \leq x - y \leq 10 \land 5 \leq x + y \leq 15 \\ \land\, 0 \leq x \leq 10 \land 0 \leq y \leq 5 \end{pmatrix}.$$

This is indeed the strongest octagonal invariant for this program.

2.11   Strongest Invariants and Complexity

As the reader would have noticed, every entry in each of the three tables is also an octagonal constraint on at most two parameters. Constraints on parameters imposed due to initialization are also octagonal. Further, there are constraints $l_i \leq u_i$ relating a lower bound with the associated upper bound. Most importantly, even though there are $2n^2$ parameters in a parametrized invariant, parameter interaction is localized. Constraints can be decomposed into disjoint parts based on parameters appearing in them, so that parameters appearing in one subset of constraints do not overlap with parameters appearing in other subsets of constraints. Consequently, a large formula expressed using many parameters can be decomposed into smaller subformulas expressed using a few parameters. And each of these subformulas can be processed separately.

We illustrate this using the parameter constraints generated in Example 6:

$$l_1 \leq -2 \wedge u_1 \geq 1 \wedge l_2 \leq -11 \wedge u_2 \geq 10 \wedge l_3 \leq -6 \wedge u_3 \geq 4$$
$$\wedge\, l_4 \leq -5 \wedge u_4 \geq 6 \wedge -1 \leq l_1 + u_2 \leq 1 \wedge l_2 + u_1 \leq -1$$
$$\wedge\, u_3 + l_3 = 0 \wedge -1 \leq u_2 - u_1 \leq 1 \wedge u_4 + l_4 \geq 0.$$

The above constraints can be separated into three disjoint parts $S_1 \wedge S_2 \wedge S_3$, where

$$S_1 = (l_1 \leq -2 \wedge u_1 \geq 1 \wedge l_2 \leq -11 \wedge u_2 \geq 10 \wedge -1 \leq l_1 + u_2 \leq 1$$
$$\wedge\, l_2 + u_1 \leq -1 \wedge -1 \leq u_2 - u_1 \leq 1)\,,$$
$$S_2 = (l_3 \leq -6 \wedge u_3 \geq -4 \wedge u_3 + l_3 = 0)\,,$$
$$S_3 = (l_4 \leq -5 \wedge u_4 \geq 6 \wedge u_4 + l_4 \geq 0)\,.$$

Each of these subsystems can be solved separately, reducing the complexity of parameter constraint solving to $O(m)$, where m is the number of constraints, since the number of parameters is constant (2 or 4). The closure of S_1 gives

$$l_1 \leq -2 \wedge -u_1 \leq -9 \wedge l_2 \leq -11 \wedge -u_2 \leq -10 \wedge -l_1 - u_2 \leq 1$$
$$\wedge\, l_1 + u_2 \geq 1 \wedge l_2 + u_1 \leq -1 \wedge u_2 - u_1 \leq 1 \wedge l_2 + u_2 \leq 0 \wedge l_2 - l_1 \leq 1\,.$$

The maximum values of l_1 and l_2 are, respectively, $-9, -11$; the minimum values of u_1 and u_2 are 9 and 10, respectively. For S_2, its closure adds $l_3 \leq -6$, $-u_3 \leq -6$, and $l_3 + u_3 = 0$. The maximum value of l_3 is -6, the minimum value of u_3 is 6. For S_3, its closure is itself; the maximum value of l_4 is -5; the minimum value of u_4 is 6. These parameter values give the optimal invariant:

$$-9 \leq x - y \leq 9 \wedge -11 \leq x + y \leq 10 \wedge -6 \leq x \leq 6 \wedge -5 \leq y \leq 6.$$

As was stated above, the Interproc tool computes a much weaker invariant.

With these considerations in mind, it can be shown that program analysis using our geometric heuristic for quantifier elimination for octagons takes quadratic time in the number of program variables:

Theorem 10. *A constraint description of octagonal loop invariants for a program can be automatically derived using the geometric quantifier elimination heuristics proposed above in $O(k \cdot n^2 + p)$ steps, where n is the number of program variables, k is the number of program paths, and p is the size of the program's abstract syntax tree.*

Proof. There are two steps involved in automatically deriving an octagonal invariant, and the complexity of each step is analyzed below.

First, for every program path all tests performed must be collected as well as the cumulative effect of assignment statements on program variables must be computed to generate a verification condition. This can be done in $O(p)$ steps using a standard Hoare-style backward analysis.

For every pair of distinct program variables, there are at most 8 types of atomic formulas appearing as tests in a given program path (after removing redundant tests which again can be performed globally in $O(p)$ steps); the local closure of these tests is performed in time $O(n^2)$ for each path. From this, the verification condition for each path can be constructed in constant time. The generation of parameter constraints involves one table look-up for each lower and upper bound of the expressions $x, x + y, x - y$, where x and y are program variables. As mentioned before, there are $O(n^2)$ tests to consider for each path (present or absent) and thus also $O(n^2)$ table look-ups, each of which takes constant time. Given all k program paths, $O(k \cdot n^2)$ parameter constraints are generated in $O(k \cdot n^2)$ steps.   □

The result of the presented algorithm does not consist of an explicit list of invariants for the program's loops, but it implicitly describes the invariants by constraints on their parameters. Each of the usually infinitely many instantiations of the parameters satisfying the constraints describes one list of invariants. Among these, we can effectively extract a unique *strongest* list of invariants that implies all other described invariant lists and that is (globally) closed.

Lemma 11. *From the set of constraints for a program with a single loop, computed by our algorithm and described by constraints on the invariants' parameters, the unique strongest one can be automatically derived in $O(k \cdot n^2)$ steps, where n is the number of program variables and k is the number of program paths.*

Proof. Using the algorithm by Bagnara et al. [3], it is possible to generate the maximum value of each lower bound parameter l_i ($-\infty$ if none exists) as well as the minimum value of each upper bound parameter u_j (similarly $+\infty$ if none exists) as follows:

For every l_i that has a maximum value in $\mathbb{Z}$, there is an invariant in which l_i takes that maximum value allowed by the constraints; similarly, for every u_j that has a minimum value in $\mathbb{Z}$, there is also an invariant in which u_j takes the minimum value allowed by the constraints (as otherwise, it would be possible to further tighten the associated octagon using the closure operation). If there is no invariant where l_i or u_i has a value in $\mathbb{Z}$, they are always $-\infty$ or ∞, respectively. The conjunction of any two octagonal invariants thus generated is also an octagonal invariant. It thus follows that the invariant in which l_i's have the maximum of all their lower bounds and u_j's have the minimum of all their upper bounds gives an octagonal invariant that implies all other invariants described by the constraints.

The key to computing these optimal bounds efficiently lies in decomposing the parameter constraints into disjoint subsets of constant size that do not interact with each other:

From the above tables, it is clear that for any particular program variable x, its lower bound l and upper bound u can be related by constraints, but these parameters do not appear in relation with other parameters. This has also been discussed at the beginning of the current subsection. Thus for a particular program variable x, all constraints on l and u can be analyzed separately without having to consider other constraints. Their satisfiability as well as lower and upper bounds on these parameters can be computed in linear time of the number of constraints relating l and u, which is at most 2 per path, i.e. $O(k)$: Since there are only a constant number of parameters in these constraints (namely 2), the algorithm for tight integral closure in [3] takes time proportional to the number of constraints on these two parameters. There are n variables, so computing their optimal lower and upper bounds takes $O(k \cdot n)$.

Similarly, for each variable pair (x_i, x_j) the parameters expressing lower and upper bounds for $x_i - x_j$ and $x_i + x_j$ appear together in table entries, and they do not appear in relation with other parameters. So each of those parameter sets of constant size 4 can be analyzed independently. There are $O(n^2)$ variable pairs, so analogously to above, computing their optimal lower and upper bounds takes $O(k \cdot n^2)$.

Overall, the strongest invariant can be computed from a conjunction of parameter constraints in $O(k \cdot n^2)$ steps. By construction of the algorithm by Bagnara et al., every program state satisfying this invariant satisfies all invariants described by the initial parameter constraints. □

Theorem 12. *From the set of constraints for a program with no nesting of loops, computed by our algorithm and described by constraints on the invariants' parameters, the unique strongest one can be automatically derived in $O(k \cdot n^2)$ steps, where n is the number of program variables and k is the number of program paths.*

Proof. There are no nested loops, so the loops can be partially ordered by their order of execution. For each verification condition $(I(X) \land T(X)) \implies J(X')$ involving two different parameterized loop invariants I and J, the loop for I is then executed before the one for J.

This ordering is obviously well-founded. We can thus examine the loops one at a time, inductively assuming that strongest invariants have already been computed for all loops that are executed before the one under consideration. When we analyze each loop, we can without loss of generality assume that all constraints include only parameters for the current loop: Optimal parameters for the previous loops have already been instantiated and constraints for following loops do not impose any upper bounds on parameters l_i or lower bounds for parameters u_i of the current loop. We can also safely assume that during the generation phase, the constraints have already been assembled following the execution order, so we do not need to sort the constraints now to access the ones for the current loop.

Then the result follows directly from Lemma 11, since the global number of program paths corresponds to the sum of paths that are relevant for the individual loops.   □

Our complexity analysis contains a parameter whose significance we have mostly ignored so far: The number k of program paths. In principle, it is easy to construct a program with 2^p program paths, where p is the size of the program, by simply making every statement a conditional. This would mean that our analysis would not scale at all.

In practice, however, a program has a far less paths; see also [12,17] where techniques for managing program paths and associated verification conditions are discussed. Since our analysis is based on computing invariants that are satisfied at certain points in the program, the number of paths can also be kept in check by introducing invariants at additional program locations. These invariants provide a convenient way to trade off precision for complexity.

Example 13. Consider the following program stub:

```
x=0; y=0;
while (true) do
    if  ...  then  ...  ;
    if  ...  then  ...  ;
    // add invariant J here
    if  ...  then  ...  ;
    if  ...  then  ...  ;
```

To compute the invariant I for this loop, $2^4 = 16$ paths inside the loop have to be analyzed, depending on which of the 4 conditionals are used and which are not. If we introduce an additional invariant J after the second conditional, there are only 2^2 paths leading from I to J and another 2^2 paths from J to I, or 8 paths in total.

In general, adding an invariant after every conditional to directly merge the control flows would lead to an analysis resembling an abstract interpretation approach. As a side effect, such additional invariants allow for a convenient way to check assertions: Just add invariants at the locations of all assertions, analyze the program and check whether the computed invariants entail the corresponding assertions.

3   Invariant Generation with Max-plus Polyhedra

In many cases where programs contain loops with several execution paths, purely conjunctive invariants can only inadequately model the program semantics.

Example 14. The strongest invariant at the loop entry point in the program

```
x := 0; y := 5;
while (x < 10) do
  if (x < 5) then
    x := x+1;
  else
    x := x+1; y := y+1;
```

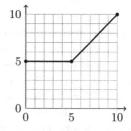

is $(0 \leq x \leq 5 \wedge y = 5) \vee (5 \leq x \leq 10 \wedge x = y)$. This invariant is depicted on the right. It consists of two lines and is clearly not convex. Hence it cannot be expressed as a conjunction of linear constraints, including octagonal constraints or even general polyhedra.

One way of allowing more expressive loop invariants including disjunctive invariants is to use formulas defining so-called *max-plus polyhedra*, as proposed by Allamigeon [1,6]. Such a formula allows disjunctions of a subset of octagonal constraints, particularly of constraints of the form $l_i \leq x_i \leq u_i$ and $l_{i,j} \leq x_i - x_j \leq u_{i,j}$. Atomic formulas of the form $a \leq x_i + x_j \leq b$ are not allowed in a strict max-plus setting. More specifically, the allowed disjunctions are those that can be written as

$$\max(x_1 + a_1, \ldots, x_n + a_n, c) \leq \max(x_1 + b_1, \ldots, x_n + b_n, d) ,$$

where a_i, b_i, c, d are integers or $-\infty$. For example,

$$\max(x + 0, y - \infty, -\infty) \leq \max(x - \infty, y + b, d)$$

would represent the disjunction $x \leq d \vee x - y \leq b$.

For a reader who is mainly interested in the high-level concepts, it suffices to note that sets of linear constraints formed in this way not with the standard addition and multiplication operators, but instead with max and $+$, can also be regarded as describing convex state spaces, which only have a different shape than for the standard operators.

Formally, the basis of max-plus polyhedra is formed by the *max-plus semiring* $(\mathbb{Z}_{\max}, \oplus, \otimes)$. The elements of $\mathbb{Z}_{\max}$ are those of $\mathbb{Z}$ and $-\infty$, and the semiring addition $\oplus$ and multiplication $\otimes$ are given by $x \oplus y = \max(x, y)$ and $x \otimes y = x+y$, with additive unit $\mathbb{0} = -\infty$ and multiplicative unit $\mathbb{1} = 0$. The usual order on $\mathbb{Z}$ extends to $\mathbb{Z}_{\max}$ by making $-\infty$ the least element. A *max-plus polyhedron* is a subset of $\mathbb{Z}_{\max}^n$ satisfying a finite set of linear max-plus inequalities of the form

$$(a_1 \otimes x_1) \oplus \cdots \oplus (a_n \otimes x_n) \oplus c \leq (b_1 \otimes x_1) \oplus \cdots \oplus (b_n \otimes x_n) \oplus d ,$$

where $a_i, b_i, c, d \in \mathbb{Z}_{\max}$.

Max-plus convex sets were introduced by Zimmermann [33], a general introduction can be found for example in Gaubert and Katz [13]. They are more expressive than Difference Bound Matrices: On the one hand, every invariant that can be represented by Difference Bound Matrices can also be represented by a max-plus polyhedron. To convert a Difference Bound Matrix into a max-plus polyhedron, each constraint $x_i - x_j \leq c$ is simply written as the equivalent max-plus constraint $0 \otimes x_i \leq c \otimes x_j$. On the other hand, a max-plus polyhedron can in general not be represented by a single Difference Bound Matrix. For example, the invariant from example 14 is a max-plus polyhedron but it cannot be represented by a Difference Bound Matrix, nor by any other other form of conjunctions of linear constraints.

Max-plus polyhedra are convex sets (with respect to the max-plus semiring). As such, they can be represented in multiple ways, just like traditional polyhedra. Of particular interest is the representation as the convex hull of a finite set of generator points (analogous to a frame representation of a convex polyhedron; cf. Allamigeon et al. [1] for details on how to convert between the representations). If $p_1, \ldots, p_n$ are points, we denote the max-plus polyhedron that is generated by these points as their convex hull by $\mathrm{co}(\{p_1, \ldots, p_n\})$. In contrast to standard polyhedra, the generator representation is arguably more intuitive for humans than the constraint representation for max-plus polyhedra (due to constraints like $\max(x, y+1, 2) \leq \max(x+2, y)$), and we will utilize it throughout most of this article. To keep the presentation even more intuitive, we will also restrict ourselves mainly to bounded polyhedra in $\mathbb{Z}_{\max}^2$, i.e. to programs with two variables whose range during program execution is bounded. As is customary when working with max-plus polyhedra, we write points in $\mathbb{Z}_{\max}^2$ as column vectors, i.e. the point with x-coordinate 0 and y-coordinate 5 is written as $\binom{0}{5}$.

Example 15. Consider the max-plus polyhedron with generators $\binom{0}{5}$ and $\binom{10}{10}$: As the convex hull of these two points, this max-plus polyhedron contains all points of the form $\alpha \otimes \binom{0}{5} \oplus \beta \otimes \binom{10}{10}$ with $\mathbb{0} \leq \alpha, \beta$ and $\alpha \oplus \beta = \mathbb{1}$. Expressed in standard arithmetic, these are the points of the form $\binom{\max(\alpha+0, \beta+10)}{\max(\alpha+5, \beta+10)}$ where $\max(\alpha, \beta) = 0$. It is easy to verify that this max-plus polyhedron is exactly the invariant from Example 14. A constraint representation of this max-plus polyhedron is

$$0 \leq x \leq 10 \qquad\qquad 5 \leq y \leq 10$$
$$-5 \leq x - y \leq 0 \qquad\qquad 0 \otimes y \leq (0 \otimes x) \oplus 5 \text{ (i.e. } y \leq \max(x, 5)) .$$

This representation is of course harder to understand for the human reader than the one from Example 14. However, it is a pure (max-plus) conjunction, which is the key to its fully automatic treatment.

To better understand the possible shapes of invariants, let us consider the case of a bounded two-dimensional max-plus polyhedron generated by two points $\binom{a_1}{b_1}$ and $\binom{a_2}{b_2}$. The possible shapes of such a max-plus polyhedron are depicted in Figure 4, where we assume, without loss of generality, $a_1 \leq a_2$.

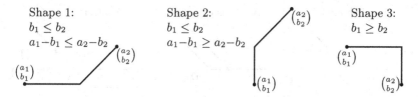

Fig. 4. Possible shapes of bounded two-dimensional max-plus polyhedra

In general, also in higher dimensions, two generators are always connected by lines that run parallel, perpendicular, or at a 45 degree angle with all coordinate axes. A polynomial with multiple generators will consist of all these connections as well as the area that is surrounded by the connec-

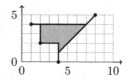

tions. Such an example with four generators (marked by dots) can be seen to the right. Knowledge about these shapes will be used in the next section to develop a quantifier elimination heuristics based on table lookups, comparable to the one for octagonal constraints.

3.1 Invariant Generation

Similar to the procedure for octagonal constraints described in the previous section, we derive loop invariants described by max-plus polyhedra as follows: We start with a parametrized invariant given as a max-plus polyhedron M with a predetermined number of parametrized generators. Each path through the loop gives rise to a verification condition $M \cap \Gamma \subseteq M'$, where Γ is the polyhedron defined by the branch and loop tests along the path and M' arises from M as usual by the translation induced by the executed assignments. This verification condition can equivalently be expressed as $(M \setminus M') \cap \Gamma = \emptyset$. We then look up a representation of $M \setminus M'$ from a precomputed table that describes $M \setminus M'$ as a set of max-plus polyhedra and express the requirement that Γ cuts off all of $M \setminus M'$ to convert the verification condition into a set of ground constraints on the parameters of M. The verification condition for the loop entry is also encoded by such constraints.

Every solution of the constraints then corresponds to a valid loop invariant. As a heuristic to select a strong invariant, we choose a solution of the constraints that minimizes the distance between the generators: We maximize a_1 and b_1 and minimize a_2 and b_2, just as we maximized the l_i and minimized the u_i before. We will show in Theorem 18 that for two generators, this results in an optimal solution.

Example 16. To find a loop invariant for Example 14, we nondeterministically guess that the invariant is a max-plus polyhedron $M = \mathrm{co}(\{\binom{a_1}{b_1}, \binom{a_2}{b_2}\})$ of shape 1, as depicted below, and look for suitable instantiations of the parameters a_i and b_i. The loop contains two execution paths, depending on whether or not the condition x<5 is satisfied.

Assumed shape of M: Analysis of path 1:   Analysis of path 2:

The loop invariant M must be satisfied when the loop is entered. We guess that the initial value is situated on the horizontal branch of M. This results in the constraints $a_1 \leq 0 \leq a_2 - (b_2 - b_1)$ and $b_1 = 5$.

Furthermore, the loop invariant must be maintained during the execution of the loop. Consider the first path with the assignment x := x+1. The verification condition for this branch is

$$\binom{x}{y} \in co(\{\binom{a_1}{b_1}, \binom{a_2}{b_2}\}) \wedge (x < 10 \wedge x < 5) \implies \binom{x+1}{y} \in co(\{\binom{a_1}{b_1}, \binom{a_2}{b_2}\})$$

or equivalently

$$\binom{x}{y} \in co(\{\binom{a_1}{b_1}, \binom{a_2}{b_2}\}) \wedge (x < 10 \wedge x < 5) \implies \binom{x}{y} \in co(\{\binom{a_1-1}{b_1}, \binom{a_2-1}{b_2}\}) .$$

So if M_1' is the max-plus polyhedron generated by $\binom{a_1-1}{b_1}$ and $\binom{a_2-1}{b_2}$, this verification condition can be rewritten, following the considerations above, as

$$(M \setminus M_1') \cap \{x \mid x < 10 \wedge x < 5\} = \emptyset .$$

The relation between M and M_1' is depicted in the middle of the above figure. For M to be an invariant, it is necessary that every point in $M \setminus M_1'$, i.e. every point in M right of $\binom{a_2-(b_2-b_1)-1}{b_1}$, is cut off by the constraints $x < 10 \wedge x < 5$ (or equivalently $x \leq 4$) along the path. This directly implies $a_2 - (b_2 - b_1) - 1 \geq 4$.

For the second path with the assignment x := x+1; y := y+1, let M_2' be the max-plus polyhedron generated by $\binom{a_1-1}{b_1-1}$ and $\binom{a_2-1}{b_2-1}$. For M to be an invariant, it is necessary that every point in $M \setminus M_2'$, i.e. every point in M left of $\binom{a_2-(b_2-b_1)}{b_1}$ or right of $\binom{a_2-1}{b_2-1}$, is cut off by the constraints $x < 10 \wedge x \geq 5$ (or equivalently $x \leq 9 \wedge x \geq 5$) along the path. This implies $a_2 - (b_2 - b_1) \leq 5$ and $a_2 - 1 \geq 9$.[10]

All these constraints can be combined to $a_1 \leq 0$, $b_1 = 5$, $a_2 \geq 10$, and $a_2 - b_2 = 5$. Every instantiation of the parameters that satisfies these constraints leads to a valid loop invariant. To find the strongest invariant, we maximize a_1 and b_1 and minimize a_2 and b_2. This yields the max-plus polyhedron generated by the two points $\binom{0}{5}$ and $\binom{10}{10}$.

Had we guessed shape 2 or 3 for M, the resulting set of constraints would have been unsatisfiable, indicating that the loop does not have an invariant that is expressible as one of these shapes.

In general, multiple tests in Γ may be used to cut off a component of $M \setminus M'$. Similar to our approach to quantifier elimination for octagons, we approximate

---

[10] The two other possibilities $a_1 \geq 10$ and $a_2 < 5$ that would result in all of M being cut off would later contradict the loop entry condition.

assignment statements	max-plus polyhedra to be cut off
$A = 0, B = 0$	–
$A > 0, B = 0, \ A \leq \Delta_a - \Delta_b$	$co(\{\binom{a_2 - \Delta_b - A + 1}{b_1}, \binom{a_2}{b_2}\})$
$A > 0, B = A, A \leq \Delta_b$	$co(\{\binom{a_1}{b_1}, \binom{a_2 - \Delta_b - 1}{b_1}\})$, if $\Delta_a \neq \Delta_b$
	$co(\{\binom{a_2 - A + 1}{b_2 - A + 1}, \binom{a_2}{b_2}\})$
$A > B, B > 0, \Delta \leq \Delta_a - \Delta_b, B \leq \Delta_b$	$co(\{\binom{a_1}{b_1}, \binom{a_2 - \Delta_b - \Delta - 1}{b_1}\})$, if $\Delta_b + \Delta \neq \Delta_a$
	$co(\{\binom{a_2}{b_2}, \binom{a_2 - \Delta_b - \Delta + 1}{b_1}\})$
$A < 0, B = 0, \ \|A\| \leq \Delta_a - \Delta_b$	$co(\{\binom{a_1}{b_1}, \binom{a_1 - A - 1}{b_1}\})$
	$co(\{\binom{a_2 - \Delta_b + 1}{b_1 + 1}, \binom{a_2}{b_2}\})$, if $\Delta_b \neq 0$
$A < 0, B = A, \|A\| \leq \Delta_b$	$co(\{\binom{a_1}{b_1}, \binom{a_2 - \Delta_b - A - 1}{b_1 - A - 1}\})$
$A < B, B < 0, \Delta \leq \Delta_a - \Delta_b, \|B\| < \Delta_b$	$co(\{\binom{a_1}{b_1}, \binom{a_2 - \Delta_b - B - 1}{b_1 - B - 1}\})$
	$co(\{\binom{a_2}{b_2}, \binom{a_2 - \Delta_b - B + 1}{b_1 - B + 1}\})$, if $B + \Delta_b \neq 0$
all other cases	$co(\{\binom{a_1}{b_1}, \binom{a_2}{b_2}\})$

Fig. 5. Max-plus polyhedra to be cut off for shape 1

the verification conditions by assuming instead that only one of the tests in Γ may be used for an individual component. Due to this approximation, exactly which parts of a polyhedron have to be cut off by the constraints along a given path can be precomputed, just like for octagonal constraints. This gives rise to a finite table and reduces the analysis of each path in a loop to a table look-up. Such tables for the three shapes from Figure 4 are presented in Figures 5–7. The tables link assignments x := x+A; y := y+B along a path with the respective max-plus polyhedra that must be cut off. Throughout the tables, we use the abbreviations $\Delta_a = a_2 - a_1$, $\Delta_b = b_2 - b_1$, and $\Delta = A - B$.

Theorem 17. *The loop verification condition for an invariant of shape 1–3 and a path in the loop with assignments* x := x+A; y := y+B *and path condition* Γ *is equivalent to the condition that* $\Gamma \cap M = \emptyset$ *for each max-plus polyhedron* M *given in the respective entry of Figures 5–7.*

Proof. The proofs of correctness for all of the various table entries are very similar. As a typical representative, we show how to derive the entries for shape 1 in the case $A = B > 0$. Let M be the assumed (parametrized) invariant and let M^A result from translating M by $-A$ along both axes. Furthermore, let Γ be the set of all points that satisfy the constraints along the given path.

To obtain an invariant, M has to be chosen such that every point in $M \cap \Gamma$ is also in M^A. Considering a generic point $p = \binom{\alpha_1 + a_1}{\alpha_1 + b_1} \oplus \binom{\alpha_2 + a_2}{\alpha_2 + b_2}$ with $\alpha_1 \oplus \alpha_2 = 1 = 0$ in M, this means that the parameters a_i, b_i must be such that if $p \in \Gamma$, then p can be written as $p = \binom{\beta_1 + a_1 - A}{\beta_1 + b_1 - A} \oplus \binom{\beta_2 + a_2 - A}{\beta_2 + b_2 - A}$ with $\beta_1 \oplus \beta_2 = 1 = 0$.

assignment statements	max-plus polyhedra to be cut off
$A = 0, B = 0$	$-$
$A = 0, B > 0, \ B \le \Delta_b - \Delta_a$	$co(\{\binom{a_1}{b_2-\Delta_a-B+1}, \binom{a_2}{b_2}\})$
$A = 0, B < 0, \ |B| \le \Delta_b - \Delta_a$	$co(\{\binom{a_1}{b_1}, \binom{a_1}{b_1-B-1}\})$ $co(\{\binom{a_1+1}{b_2-\Delta_a+1}, \binom{a_2}{b_2}\})$, if $a_1 \ne a_2$
$A > 0, B = A, \ A \le \Delta_a$	$co(\{\binom{a_1}{b_1}, \binom{a_1}{b_2-\Delta_a-1}\})$, if $\Delta_a \ne \Delta_b$ $co(\{\binom{a_2}{b_2}, \binom{a_2-A+1}{b_2-A+1}\})$
$A > 0, B > A, \ A \le \Delta_a, \Delta \ge \Delta_b - \Delta_a$	$co(\{\binom{a_1}{b_1}, \binom{a_1}{b_2-\Delta_a+\Delta-1}\})$ $co(\{\binom{a_2}{b_2}, \binom{a_1}{b_2-\Delta_a+\Delta+1}\})$, if $\Delta_a - \Delta \ne \Delta_b$
$A < 0, B = A, \ |A| \le \Delta_a$	$co(\{\binom{a_1}{b_1}, \binom{a_1-A-1}{b_2-\Delta_a-A-1}\})$
$A < 0, B < A, \ |A| \le \Delta_a, \Delta \le \Delta_b - \Delta_a$	$co(\{\binom{a_1}{b_1}, \binom{a_1-A-1}{b_2-\Delta_a-A-1}\})$ $co(\{\binom{a_2}{b_2}, \binom{a_1-A+1}{b_2-\Delta_a-A+1}\})$, if $\Delta_b + A \ne 0$
all other cases	$co(\{\binom{a_1}{b_1}, \binom{a_2}{b_2}\})$

Fig. 6. Max-plus polyhedra to be cut off for shape 2

assignment statements	max-plus polyhedra to be cut off
$A = 0, B = 0$	$-$
$A = 0, B > 0, \ B \le -\Delta_b$	$co(\{\binom{a_1}{b_1}, \binom{a_2}{b_1-B+1}\})$
$A = 0, B < 0, \ B \ge \Delta_b$	$co(\{\binom{a_1}{b_1}, \binom{a_2-1}{b_1}\})$, if $a_1 \ne a_2$ $co(\{\binom{a_2}{b_2}, \binom{a_2}{b_2-B-1}\})$
$A > 0, B = 0, \ A \le \Delta_a$	$co(\{\binom{a_2}{b_2}, \binom{a_2-A+1}{b_1}\})$
$A > 0, B < 0, \ A \le \Delta_a, B \ge \Delta_b$	$co(\{\binom{a_1}{b_1}, \binom{a_2-A-1}{b_1}\})$, if $A \ne \Delta_a$ $co(\{\binom{a_2}{b_2}, \binom{a_2-A+1}{b_1}\})$
$A < 0, B = 0, \ |A| \le \Delta_a$	$co(\{\binom{a_1}{b_1}, \binom{a_1-A-1}{b_1}\})$ $co(\{\binom{a_2}{b_2}, \binom{a_2}{b_1-1}\})$, if $b_1 \ne b_2$.
$A < 0, B > 0, \ |A| \le \Delta_a, B \le -\Delta_b$	$co(\{\binom{a_1}{b_1}, \binom{a_2}{b_1-B+1}\})$ $co(\{\binom{a_2}{b_2}, \binom{a_2}{b_1-B-1}\})$, if $\Delta_b + B \ne 0$.
all other cases	$co(\{\binom{a_1}{b_1}, \binom{a_2}{b_2}\})$

Fig. 7. Max-plus polyhedra to be cut off for shape 3

- Case $-(b_2 - b_1) \leq \alpha_2 \leq -A$. Then $\alpha_1 = 0$. Choose $\beta_1 = 0, \beta_2 = \alpha_2 + A$ to show that $p \in M^A$.

$$(\beta_1 + a_1 - A) \oplus (\beta_2 + a_2 - A) = (a_1 - A) \oplus (\alpha_2 + a_2) = \alpha_2 + a_2$$
$$\text{(because } \alpha_2 + a_2 \geq b_1 - b_2 + a_2 \geq a_1 > a_1 - A \text{ due to the shape.)}$$
$$(\beta_1 + b_1 - A) \oplus (\beta_2 + b_2 - A) = (b_1 - A) \oplus (\alpha_2 + b_2) = \alpha_2 + b_2$$
$$\text{(because } \alpha_2 + b_2 \geq b_1 > b_1 - A)$$

So points that fall into this case are in the invariant and do not restrict the choice of the parameters.

- Case $\alpha_2 > -A$. If we assume $p \in M^A$, then we have the following contradiction:

$$b_2 - A \overset{\alpha_2 > -A}{<} \alpha_2 + b_2 \overset{\text{def } \oplus}{\leq} (\alpha_1 + b_1) \oplus (\alpha_2 + b_2)$$

$$\overset{p \in M^A}{=} (\beta_1 + b_1 - A) \oplus (\beta_2 + b_2 - A) \leq (b_1 - A) \oplus (b_2 - A) \overset{b_1 \leq b_2}{\leq} b_2 - A.$$

So $p \notin M^A$, which implies $p \notin \Gamma$.

- Case $\alpha_2 < -(b_2 - b_1)$. Then $\alpha_1 = 0$. Again assume $p \in M^A$. A look at the y-coordinate reveals:

$$b_1 = b_1 + ((\overset{=0}{\overbrace{\alpha_1}} + \overset{=0}{\overbrace{b_1 - b_1}}) \oplus (\overset{<0}{\overbrace{\alpha_2 + b_2 - b_1}}))$$

$$= (\alpha_1 + b_1) \oplus (\alpha_2 + b_2) \overset{p \in M^A}{=} (\beta_1 + b_1 - A) \oplus (\beta_2 + b_2 - A)$$

Because $\beta_1 + b_1 - A < b_1$, this means that $b_1 = \beta_2 + b_2 - A$, or $\beta_2 = b_1 - b_2 + A$. So $\beta_2 = -(b_2 - b_1) - A < 0$, i.e. $\beta_1 = 0$ because of $\beta_1 \oplus \beta_2 = 0$. For the x-coordinate, this leads to the following contradiction:

$$a_1 \overset{\text{def } \oplus}{\leq} a_1 \oplus (\alpha_2 + a_2) = (\overset{=0}{\overbrace{\alpha_1}} + a_1) \oplus (\alpha_2 + a_2)$$

$$\overset{p \in M^A}{=} (\beta_1 + a_1 - A) \oplus (\beta_2 + a_2 - A) = (a_1 - A) \oplus (\overset{<a_1}{\overbrace{b_1 - b_2 + a_2}}) \overset{\text{shape}}{<} a_1$$

This means that the polyhedron has to degenerate to a line or $p \notin \Gamma$.

Overall, we have the following result:

- Γ must cut off the top right part of the polyhedron (case 2).
- If $b_1 - b_2 + A > 0$, Γ must cut off the left or lower arm. Otherwise either $a_1 = b_1 - b_2 + a_2$, i.e. the whole polyhedron consists only of one arm, or Γ must cut off the left or lower arm (case 3). $\qquad \square$

As a heuristic to reduce the computational cost, we usually assume that each component of $M \setminus M'$ is cut off by a single constraint.

Note that Γ can contain any path constraint that the employed constraint solver can handle. In particular, Γ is not restricted to a conjunction of constraints of the form $\pm x_i \le c$ or $x_i \pm x_j \le c$. E.g. in Example 14 and 16 the loop condition $3x+2y < 50$ would have led to a similar analysis.

For two generators, our heuristics of minimizing the distance between the generators leads to an optimal invariant.

Theorem 18. *Among a set of bounded max-plus polyhedra with two generators $\binom{a_1}{b_1}$ and $\binom{a_2}{b_2}$, the state spaces described by those max-plus polyhedra that mini-mize $|a_2 - a_1| + |b_2 - b_1|$ are minimal.*

Proof. Since $|a_2 - a_1|$ and $|b_2 - b_1|$ are (discrete) nonnegative integers, the minimum is assumed, even if we consider an infinite set of bounded max-plus polyhedra.

A polyhedron of shape 1 contains $(|a_2-a_1|-|b_2-b_1|)+|b_2-b_1|+1 = |a_2-a_1|+1$ points (over $\mathbb{Z}$), independently of $|b_2 - b_1|$. Analogously, a polyhedron of shape 2 contains $(|b_2-b_1|-|a_2-a_1|)+|a_2-a_1|+1 = |b_2-b_1|+1$ points, independently of $|a_2 - a_1|$, and one of shape 3 contains $|a_2 - a_1| + |b_2 - b_1| + 1$ points. So whatever the shape of the polyhedron, the described state space is minimal whenever $|a_2 - a_1| + |b_2 - b_1|$ is minimal. □

In the case of more than two generators, the sit-uation is more complicated, and indeed increas-ing the Euclidean distance between two generators may result in a smaller polyhedron. In practice, however, the constraints imposed by the invariant generation process usually circumvent this situa-

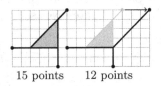

15 points   12 points

tion. The reason is that, for example in the depicted situation, generators of two invariants can only differ by translations along the one-dimensional components leading to them. So the leftmost generator can only be moved horizontally, the lower one only vertically and the top one only along the diagonal. If generators are moved in a different way, the result cannot be an invariant any more.

Example 19. As a simple example whose invariant cannot be expressed with two generators, consider the following program:

```
x := 0;  y := 0;
while (true) do
    if (y = 0 and x < 10) then
        x := x+1;
    else if (x ≥ 10 and y < 5) then
        y := y+1;
    else x := x-1; y := y-1;
```

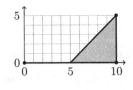

The minimal max-plus invariant for this loop is generated by the points $\binom{0}{0}$, $\binom{10}{0}$, and $\binom{10}{5}$, that can be found with an analysis similar to the one shown above. Note that the program is not terminating, and that the max-plus invariant consists of two components of different codimension, namely the line $0 \le x \le 5 \wedge y = 0$ and the triangle $x \le 10 \wedge y \ge 0 \wedge x - y \ge 5$. The true optimal invariant of the program is the boundary of the one computed.

Example 20. Let us look again at the program from Example 9, which is also nonterminating and contains a nested loop:

```
x := 0; y := 5;
while (true) do
    if (x < 10 and y = 5) then
        x := x+1;
    else if (x = 10 and y > 0) then
        y := y−1;
    else
        while (y < 5) do
            x := x−1; y := y+1;
```

The minimal max-plus invariant $J(x, y)$ for the inner loop is the polyhedron generated by $\binom{5}{0}$, $\binom{5}{5}$, and $\binom{10}{0}$, i.e. the square $5 \leq x \leq 10 \wedge 0 \leq y \leq 5$. The resulting minimal max-plus invariant I for the outer loop is generated by $\binom{0}{5}$ and $\binom{10}{0}$, i.e. it is the union of two (one-dimensional) lines. This invariant is also the true optimal invariant of the loop.

In contrast, the octagonal invariant derived in Example 9 is a (two-dimensional) polygon. Indeed, every invariant for the outer loop that is expressed as a conjunction of traditional linear constraints is convex in the classical sense and therefore must contain at least the whole convex hull of I.

4 Concluding Remarks and Future Work

We have presented a new approach for investigating quantifier elimination of a subclass of formulas in Presburger arithmetic that exploits the structure of verification conditions generated from programs as well as the (lack of) interaction among different variables appearing in a verification condition. This approach has led to a geometric local method which is of quadratic complexity, much lower than the complexity of invariant generation using other approaches. Furthermore, the approach is amenable to development of additional heuristics that can additionally exploit the structure of verification conditions. The invariants generated by the proposed approach are of comparable strength to the invariants generated by Miné's method. To further improve the heuristics and assess the mentioned scalability issues arising from large path numbers, we are implementing the proposed approach and will experiment with it on a large class of benchmarks.

As should be evident from the discussion in the previous section, the work on max plus invariants is somewhat preliminary and still under progress. Of course, we are implementing the derivation of max plus constraints as well and, more importantly, investigating heuristics to make the approach scalable. While the method can easily be extended to more generators to specify more and more expressive disjunctive invariants, the complexity of quantifier elimination heuristics increases with the number of generators, due to the large number of possible shapes. We are currently exploring an alternative approach that uses a representation of max plus invariants by constraints instead of generators.

References

1. Allamigeon, X.: Static analysis of memory manipulations by abstract interpretation–Algorithmics of tropical polyhedra, and application to abstract interpretation. PhD thesis, Ecole Polytechnique, Palaiseau, France (2009)
2. Allamigeon, X., Gaubert, S., Goubault, É.: Inferring Min and Max Invariants Using Max-Plus Polyhedra. In: Alpuente, M., Vidal, G. (eds.) SAS 2008. LNCS, vol. 5079, pp. 189–204. Springer, Heidelberg (2008)
3. Bagnara, R., Hill, P.M., Zaffanella, E.: An Improved Tight Closure Algorithm for Integer Octagonal Constraints. In: Logozzo, F., Peled, D.A., Zuck, L.D. (eds.) VMCAI 2008. LNCS, vol. 4905, pp. 8–21. Springer, Heidelberg (2008)
4. Bezem, M., Nieuwenhuis, R., Rodriguez-Carbonell, E.: Hard problems in max-algebra, control theory, hypergraphs and other areas. Information Processing Letters 110(4), 133–138 (2010)
5. Blanchet, B., Cousot, P., Cousot, R., Feret, J., Mauborgne, L., Miné, A., Monniaux, D., Rival, X.: Design and Implementation of a Special-Purpose Static Program Analyzer for Safety-Critical Real-Time Embedded Software. In: Mogensen, T.Æ., Schmidt, D.A., Sudborough, I.H. (eds.) The Essence of Computation. LNCS, vol. 2566, pp. 85–108. Springer, Heidelberg (2002)
6. Butkovič, P.: Max-Linear Systems: Theory and Algorithms. Springer Monographs in Mathematics. Springer (2010)
7. Chawdhary, A., Cook, B., Gulwani, S., Sagiv, M., Yang, H.: Ranking Abstractions. In: Drossopoulou, S. (ed.) ESOP 2008. LNCS, vol. 4960, pp. 148–162. Springer, Heidelberg (2008)
8. Cousot, P., Cousot, R.: Abstract Interpretation: a Unified Lattice Model for Static Analysis of Programs by Construction or Approximation of Fixpoints. In: Conference Record of the Fourth Annual ACM SIGPLAN-SIGACT Symposium on Principles of Programming Languages, pp. 238–252. ACM Press, New York (1977)
9. Cousot, P., Cousot, R., Feret, J., Mauborgne, L., Miné, A., Monniaux, D., Rival, X.: The Astrée analyzer. In: Programming Languages and Systems, p. 140 (2005)
10. Cousot, P., Halbwachs, N.: Automatic Discovery of Linear Restraints among Variables of a Program. In: Conference Record of the Fifth Annual ACM SIGPLAN-SIGACT Symposium on Principles of Programming Languages, Tucson, Arizona, pp. 84–97. ACM Press, New York (1978)
11. Cousot, P., Cousot, R.: Abstract interpretation frameworks. Journal of Logic and Computation 2, 511–547 (1992)
12. Flanagan, C., Saxe, J.B.: Avoiding exponential explosion: generating compact verification conditions. In: POPL, pp. 193–205 (2001)
13. Gaubert, S., Katz, R.: Max-Plus Convex Geometry. In: Schmidt, R.A. (ed.) RelMiCS/AKA 2006. LNCS, vol. 4136, pp. 192–206. Springer, Heidelberg (2006)
14. Gulwani, S., Jha, S., Tiwari, A., Venkatesan, R.: Synthesis of loop-free programs. In: Proceedings of the 32nd ACM SIGPLAN Conference on Programming Language Design and Implementation, pp. 62–73. ACM (2011)
15. Gulwani, S., Srivastava, S., Venkatesan, R.: Program analysis as constraint solving. In: PLDI, pp. 281–292 (2008)
16. Gulwani, S., Tiwari, A.: Constraint-Based Approach for Analysis of Hybrid Systems. In: Gupta, A., Malik, S. (eds.) CAV 2008. LNCS, vol. 5123, pp. 190–203. Springer, Heidelberg (2008)
17. Henzinger, T.A., Jhala, R., Majumdar, R., McMillan, K.L.: Abstractions from proofs. In: Jones, N.D., Leroy, X. (eds.) POPL, pp. 232–244. ACM (2004)

18. Hoare, C.A.R.: An axiomatic basis for computer programming. Communications of the ACM 12(10), 576–580 (1969)
19. Jaffar, J., Maher, M., Stuckey, P., Yap, R.: Beyond Finite Domains. In: Borning, A. (ed.) PPCP 1994. LNCS, vol. 874, pp. 86–94. Springer, Heidelberg (1994)
20. Jeannet, B., Argoud, M., Lalire, G.: The Interproc interprocedural analyzer
21. Jha, S., Seshia, S.A., Tiwari, A.: Synthesis of optimal switching logic for hybrid systems. In: EMSOFT 2011, pp. 107–116. ACM (2011)
22. Kapur, D.: A quantifier-elimination based heuristic for automatically generating inductive assertions for programs. Journal of Systems Science and Complexity 19(3), 307–330 (2006)
23. Kapur, D.: Automatically generating loop invariants using quantifier elimination—preliminary report. Technical report, University of New Mexico, Albuquerque, NM, USA (2004)
24. Loos, R., Weispfenning, V.: Applying linear quantifier elimination. Computer Journal (1993)
25. Miné, A.: Weakly relational numerical abstract domains. These de doctorat en informatique, École polytechnique, Palaiseau, France (2004)
26. Sankaranarayanan, S., Sipma, H.B., Manna, Z.: Non-linear Loop Invariant Generation using Gröbner Bases. In: Symp. on Principles of Programming Languages (2004)
27. Sheini, H.M., Sakallah, K.A.: A Scalable Method for Solving Satisfiability of Integer Linear Arithmetic Logic. In: Bacchus, F., Walsh, T. (eds.) SAT 2005. LNCS, vol. 3569, pp. 241–256. Springer, Heidelberg (2005)
28. Srivastava, S., Gulwani, S., Foster, J.S.: From program verification to program synthesis. In: POPL, pp. 313–326 (2010)
29. Sturm, T., Tiwari, A.: Verification and synthesis using real quantifier elimination. In: ISSAC 2011, pp. 329–336. ACM (2011)
30. Taly, A., Gulwani, S., Tiwari, A.: Synthesizing switching logic using constraint solving. STTT 13(6), 519–535 (2011)
31. Xia, B., Zhang, Z.: Termination of linear programs with nonlinear constraints. J. Symb. Comput. 45(11), 1234–1249 (2010)
32. Yang, L., Zhou, C., Zhan, N., Xia, B.: Recent advances in program verification through computer algebra. Frontiers of Computer Science in China 4(1), 1–16 (2010)
33. Zimmermann, K.: A general separation theorem in extremal algebras. Ekonomia Matematyczna Obzory 13, 179–201 (1977)

A   Arithmetic of Infinity

It is convenient to introduce extend $\mathbb{Z}$ with $-\infty$ and $+\infty$, to stand for no lower bound and no higher bound, respectively, for an arithmetic expression. Below, arithmetic operators and ordering relations used in this paper are extended to work on $-\infty$ and $+\infty$.

Addition

$$\forall a\ a \neq +\infty \implies a + (-\infty) = -\infty$$
$$\forall b\ b \neq -\infty \implies b + (+\infty) = +\infty$$
$$(+\infty) + (-\infty) \text{ is undefined}$$

Negation and Subtraction

$$-(-\infty) = +\infty,\ -(+\infty) = -\infty$$
$$\forall a, b\ a - b = a + (-b)$$

Orderings

$$\forall a\ (a \neq +\infty) \implies (a < +\infty)$$
$$\forall a\ (a \neq -\infty) \implies (a > -\infty)$$
$$\forall a\ (a \in \mathbb{Z}) \implies ((+\infty) + (-\infty) > a)$$
$$\forall a\ (a \in \mathbb{Z}) \implies ((+\infty) + (-\infty) < a)$$

In particular, the relations $\leq$ and $>$ are *not* complementary when infinities are involved, and neither are $\geq$ and $<$.

Moreover, the equivalence $a + b \leq a + c \iff b \leq c$ that is regularly used to simplify formulas over $\mathbb{Z}$ is *not* valid, for example $+\infty + 1 \leq +\infty + 0$, but $1 \not\leq 0$

Toward a Procedure for Data Mining Proofs

Zachary Ernst[1] and Seth Kurtenbach[2]

[1] Department of Philosophy, University of Missouri, USA
[2] Department of Computer Science, University of Missouri, USA

Abstract. In this paper, we report results of an experiment to use the mutual information criterion to automatically select formulas to guide the search for proofs using McCune's Prover9 system. The formulas were selected from the TPTP library of problems for theorem-provers.

1 Introduction

One assumption that guided Bill McCune's development of the automated reasoning program Otter [4] and its successor Prover9 [5] is that there is no single 'best' algorithm for discovering proofs. For McCune, a consequence of this fact is that automated reasoning systems should be flexible enough to allow the user to experiment with different search strategies. A strategy that is successful for one problem may be useless for another, even when there is no obvious qualitative difference between the two. Therefore, the user of an automated reasoning system must be able to adjust the behavior of the system along a number of different dimensions.

A large number of strategies for guiding Otter and Prover9 have been suggested over the years. However, a disproportionately large number of significant results discovered with McCune's systems have been achieved through the use of one strategy in particular, the so-called 'proof sketches' or 'hints' strategy. This strategy, due to Robert Veroff [9], allows the user to input a set of formulas or clauses, which are evocatively called 'hints'. As the theorem-prover traverses the search space of derivable clauses, each derived clause is tested against each hint. If a match is discovered, the program prioritizes the matching clause, thereby shifting its focus to the part of the search space containing consequences of the hint.

Of course, for someone who intends to use this strategy, the immediate problem is determining which hints are likely to be useful. Common heuristics for the selection of hints include taking steps from proofs of the same conclusion in a different domain, or proofs of related theorems in the same domain. For example, if a user were searching for a proof in semigroups, it would be reasonable to begin by proving the theorem in group theory, which is stronger and may admit of a relatively simple proof of the theorem. The steps from the proof in group theory could be used to guide the search for the theorem in semigroups. Despite the fact that many of the proof steps from the group theory proof might be unreachable in semigroups, it is possible that several key steps in the proof may

M.P. Bonacina and M.E. Stickel (Eds.): McCune Festschrift, LNAI 7788, pp. 229–239, 2013.

be reachable. When the system discovers a derivation of one of those steps, it is reasonable to suspect that it may play an important role in proving the desired theorem. The proof sketch strategy allows the theorem-prover to leverage that information, often enabling the user to discover quite difficult proofs.

For users of Otter and Prover9, it can be quite surprising just how effective the proof sketches strategy can be. Furthermore, as a user discovers a larger number of proofs, a positive feedback loop takes hold. With a larger number of proofs to draw upon, the user can take advantage of a growing library of hints, thereby accelerating the discovery of new theorems.

Unfortunately, the flip side of this situation is that the unavailability of an existing library of hints creates a barrier to entry for new users of the automated reasoning system. It is too common for a new user of the system to become frustrated when a proof is difficult to find; and a new user is unlikely to possess a suitable list of hints to guide the search. When we also consider the sometimes daunting learning curve associated with these systems, it is not surprising that automated reasoning systems have not penetrated deeper into mainstream mathematical research.

This paper addresses this problem from a user's perspective. We outline an approach for data-mining an existing library of theorem proving problems – the TPTP library [8] – for clauses that are likely to be useful hints. Given a large database of problems and proofs, we recast the task of selecting hints as a data-mining problem. In particular, we ask, 'given the premises and conclusion of a desired proof, which formulas from the TPTP library are likely to be useful hints?' The approach we advocate for answering this question is to apply an information-theoretic criterion to select formulas that are related to the premises and conclusion. Specifically, we apply the so-called 'mutual information criterion' to order the formulas according to their relevance to the problem. Hints are selected by choosing the formulas that score highest according the mutual information criterion.

The method we outline is novel. To our knowledge, the closest related approaches are ones that use information-theoretic measures or probability functions to select lemmata for inclusion in the problem description, or entropy measures to guide a theorem-prover through the search space [3,2,6]. What these methods have in common is that they apply various measures to identify useful clauses generated in the course of the proof search itself. In contrast, our aim here is to use an information-theoretic measure to leverage information gleaned from successful proof searches for different problems. Our hope is that this difference will enable researchers to 'bootstrap' a theorem prover's capability by building upon past successes, hopefully creating a positive feedback loop that will increase the power of theorem provers. The experimental results we report should be understood as providing a preliminary 'proof of concept'. We are far from having a general procedure or production-ready implementation of the procedure we outline here. However, our results are surprisingly positive and warrant further investigation.

Section 2 briefly reviews the proof sketches strategy. In Section 3, we provide background about the mutual information criterion and motivate its use for the present application. Section 4 outlines the method we used for testing the criterion with the TPTP library, and reports our results. We conclude with some observations and suggestions for lowering the barriers to entry for new users of McCune's systems.

2   Proof Sketches

In this section, we provide preliminary background on the proof sketches technique. First, we set some notation. We will use lower-case Greek letters $\alpha, \beta, \gamma, \cdots$ to represent first-order formulas. The set of variables occurring in α will be written VAR(α). The set of all well-formed first-order formulas will be $\mathcal{L}$. We omit all quantifiers, assuming that each free variable in α is universally quantified. If two formulas α and β are syntactically identical, we we will write this fact as $\alpha \equiv \beta$.

A *substitution* μ is a function mapping VAR(α) into $\mathcal{L}$. Accordingly, we will write $\mu\alpha$ to denote the result of replacing each occurrence of each variable $x \in$ VAR(α) with $\mu(x)$ in α. For example, suppose $\mu(x) = f(y)$ and $\mu(z) = c$. Then if α is $g(x, z)$ then $\mu\alpha$ is $g(f(y), c)$. Note that we may use an alphabetic variant of α if VAR(α) includes variables that are in the image of μ. We will say that α *subsumes* β if there exists some substitution μ such that $\mu\alpha \equiv \beta$. Of course, every formula trivially subsumes itself.

The idea behind proof sketches is simple and intuitively appealing. We provide a set of formulas to the theorem prover, which we call *hints*. When, in the course of its search, the prover finds a clause that subsumes or is subsumed by a hint, the clause is given priority in the search. In the main loop of Prover9 (and its predecessor Otter), each derived clause is assigned a 'weight', which may be any integer. Clauses with a lower weight are typically chosen earlier than clauses with a higher weight. Accordingly, clauses that subsume or are subsumed by a hint are assigned a lower weight, and are thereby prioritized in the search.

As we mentioned above, a large number of published results that use the Prover9 system have been discovered through the use of carefully selected hints. Of course, the challenge in employing the proof sketch strategy is selecting the appropriate hints to use to guide the theorem prover. Practitioners who have domain-specific expertise may have a good idea what system(s) or specific problem(s) are related to the current problem. For example, if a theorem has been proven in a relatively strong system such as groups, then it may be worthwhile to use the steps of that proof as hints when searching for a proof of the same conclusion in semigroups. Although it would be unlikely that the proofs are identical in the two systems, it often turns out that some of the hints match formulas that are derivable in the weaker system; frequently, these turn out to be the keys for finding the theorem.

However, it is not always obvious, even to an expert, which clauses are the best candidates for useful hints. Furthermore, new users are unlikely to possess related proofs, even if they have a good idea of which proofs would be most likely to be useful in the present problem. Therefore, it is highly desirable to have a more principled procedure for selecting hints, one that does not rely on having domain-specific expertise or a personal library of successful proofs.

3   Applications of Mutual Information

3.1   Mutual Information

Mutual information is a measure from Claude Shannon's information theory [7]. As its name indicates, it measures the information shared between two objects, events, variables, etc. In this section we provide background on mutual information and explain how it has been fruitfully applied in the fields of bioinformatics and computational linguistics. We think its fruitful application in those fields helps motivate its application to the present problem.

The mutual information $M(X; Y)$ of random variables X and Y, quantifies the degree to which having information about the elements of X raises the level of information about Y, and vice-versa. If X and Y are independent, then their mutual information is zero bits. If they are dependent, then their mutual information will be some $n > 0$ bits. If X and Y are probabilistically independent (that is, $P(X|Y) = P(X)$, and vice-versa), then their mutual information is simply the uncertainty of one of them.

The uncertainty of a variable in information theory is the amount of information content lost by not knowing the variable's value. This is its entropy. Thus, if a variable has high entropy, it contains a large amount of information. One may also think of mutual information as the amount by which Y's entropy decreases when we are provided information about X. Their mutual information is the information they share, so if $M(X; Y)$ is high, then learning information about X decreases Y's entropy by a large amount.

Formally, mutual information of $x \in X$ and $y \in Y$ is:

$$M(x; y) = P(x, y) \log_2 \frac{P(x, y)}{P(x)P(y)} \tag{1}$$

where $P(x, y)$ is the probability that x and y are both true. Accordingly, the mutual information over the sets X and Y is just:

$$M(X; Y) = \sum_{x \in X} \sum_{y \in Y} M(x; y)$$

According to Church and Hanks, the mutual information measure is motivated by comparing the quantity $P(x, y)$ with the two quantities $P(x)$ and $P(y)$. Information is highest when there is a genuine correlation between x and y, so that if $P(x, y) > P(x)P(y)$, then $M(x; y)$ will be high [1].

3.2   Applications of Mutual Information

Mutual information has already been put to use in a variety of fields, with quite positive results. The motivation here is that, despite obvious differences in these applications, the tasks to which mutual information have already been applied are similar enough to the present problem to warrant investigation. Here, we briefly mention some of those applications.

In bioinformatics, one task is to organize and make sense of the vast amounts of data generated through years of published research. Reading through all of the literature and noticing correlations available only from a literature-wide perspective is an impossible task for a human. Data-mining with mutual information measures makes this possible, and provides avenues for discoveries that would otherwise elude researchers. For example, a human researcher may wonder if there is a connection between chili peppers and neurotoxicity. A program can easily sift through the medical literature and note the frequencies of items, like *capsaicin* (the chemical that makes chili peppers hot) and the items that co-occur with *capsaicin*. It discovers a correlation between *capsaicin* and *neurogenic inflammation*, and the computed mutual information of the two items turns out to be quite high, (35.33 bits) [10]. This discovery permits the inference that large quantities of capsaicin, perhaps due to eating too many chili peppers, may cause neurogenic inflammation. Of course, this inference must then be tested through more direct experimental methods, but it is the novel use of mutual information that indicates that such experiments are potentially valuable.

To use mutual information, the researcher designates an item of interest. In the above case, *capsaicin* is the item of interest. Next, the electronic literature is scanned to obtain the base-rate of the item of interest, and the rate at which it co-occurs with some other item of interest, whose base-rate is likewise computed. With these numbers in hand, the mutual information measure is easily obtained, and the researcher will have some evidence of a correlation, or not. This method dramatically cuts down the amount of time and energy previously required of a human researcher analyzing the literature for correlations.

In [1] Kenneth Ward Church and Patrick Hanks propose a mutual information measure for word association, which they call the *association ratio*. In doing so, they note five practical applications, including enhancing the retrieval of texts from large databases, and enhancing lexicographers' abilities to research usage norms. We briefly explain their application of mutual information, then turn to our own, drawing on analogies to these applications.

Prior to Church and Hanks, lexical research was a costly and inefficient endeavor. The most prominent measure of word association was the result of a 1964 subjective method, in which thousands of subjects were given a list of two-hundred words, and were asked to write a word after each on the list. Church and Hanks revolutionized the discipline by providing an objective measure that takes a database of texts, composed of millions of words, and measuring the mutual information of each word. This provides researchers a cheap, efficient, and objective method for measuring associations among words.

A lexicographer attempts to identify the prominent usage norms of a word. Take, for example, the word *save*. To identify usage norms, a lexicographer sifts through a large corpus of example sentences using the word of interest, and relies on her own ability to discern relevant patters. In British learners' dictionaries, *save* is identified as commonly preceding *from*. This likely matches a native English speaker's intuitions, but based on those same intuitions, can one think of other words associated with *from*? A lexicographer is interested in identifying such a list, and ranking the associated words according to strength of association. This is a difficult task for a human mind, but for a computer equipped with a mutual information measure, it is a simple task.

Church and Hanks, using the association ratio, identified 911 verbs also associated with *from*, and discovered that *saved...from* ranks 319th. The number one ranked association was *refrain...from*. Additionally, their investigation reveals seven verbs, among the twenty-seven verbs most associated with *from*, that are absent from a prominent learner's dictionary. Their inclusion would improve the dictionary's coverage.

We use mutual information among formulas to determine how helpful a given formula is likely to be for arriving at a proof. The analogies between this proposed use of mutual information and those above is simple. The formulas are like the medical or linguistic items. Our database is composed of previously obtained proofs, whose formulas have rewritten in a canonical form in order to standardize notation. We use mutual information to scan the database of proofs and identify correlated formulas, just as is done in bioinformatics and computational linguistics. Then, in new problems for the theorem prover, a list of hints is constructed of formulas with high mutual information to those formulas given in the problem's assumptions and desired conclusion. Thus, we develop a systematic and objective method for identifying formulas that will most likely be useful for developing a proof.

4   Experimental Results

Our approach is to data-mine the TPTP ('Thousands of Problems for Theorem-Provers'), which is a set of publicly available problems suitable for testing theorem provers. For our purposes, the TPTP library has several advantages that enable a study such as this one:

1. The library is designed to be easily translated into the proper syntax for many theorem-provers, including Prover9.
2. The problems have been carefully curated and organized into categories such as ring theory, group theory, puzzles, lattices, etc.
3. It contains problems that range from simple to extremely difficult, and so it does a good job of measuring the relative virtues and drawbacks of various theorem-proving strategies.

Our initial task was simply to generate a number of proofs from the TPTP library, and use the proofs to measure the mutual information of their clauses.

For a given domain of problems in the TPTP, we run Prover9 for a few seconds, recording the output. If the output contains a proof, we extract the clauses from the proof, and calculate their mutual information. That is, for any particular formula α, we would set $P(\alpha)$ equal to the number of proofs in which α occurs, divided by the total number of proofs. Similarly, $P(\alpha, \beta)$ was defined as the frequency of proofs in which both α and β occur. In this way, we define the mutual information of α and β exactly as in Equation 1 above.

First, however, we used the mutual information measure to gauge the usefulness of an individual formula α by averaging $M(\alpha, \beta)$ for all β. The motivation for making this measurement was to test an idea that McCune was often intrigued by, namely, that there might be formulas that are especially likely to be useful in proofs within a particular area of mathematics or logic.[1] That is, McCune often wondered whether there was some principled way to automatically generate formulas for a given system, which are especially likely to play an important role for proofs in that system. This is in agreement with intuitions that (e.g.) idempotencies, commutativity, distributivity, and so on are often crucial to completing proofs within a system that admits of those rules.

Although this preliminary study cannot possibly come close to answering this rather open-ended and vague question, a brief survey of some high-information formulas for some systems is intriguing. For example, within the problems taken from Boolean algebra, the ten formulas having the highest information content are the following:

$$- - x = x \qquad\qquad x + y = y + x$$
$$x \cdot y = y \cdot x \qquad\qquad x + x = x$$
$$x \cdot x = x \qquad\qquad x + (x \cdot y) = x$$
$$x + (y \cdot x) = x \qquad\qquad x + -x = 0$$
$$(x \cdot y) + (-x \cdot y) = y \qquad (x \cdot y) + (x \cdot -y) = x$$

It is worth noting that these ten formulas are neither the shortest, nor the most frequent formulas appearing in proofs. For example, there are several premises among the Boolean algebra problems in the TPTP library that occur more frequently than many of the formulas with higher information. We would expect premises and the immediate results of premises to have *lower* information than many other formulas because the mutual information criterion effectively penalizes clauses that occur too frequently or too rarely. In this way, clauses with higher information will tend to be ones that occur fairly frequently in proofs, yet are not immediate or obvious consequences of common axiomatizations. Furthermore, these clauses are the ones that tend to cluster with other high-information formulas; that is, they are ones whose probability of occurring in a proof is positively correlated with other high-information formulas. We believe that in this way, the mutual information criterion goes a long way toward characterizing the formulas that we would expect to play a significant role in the discovery of proofs.

---

[1] To our knowledge, this possibility was first mentioned to McCune by Dolph Ulrich.

Of course, the important question is whether high-information formulas are of any practical value in discovering proofs. Because even the TPTP library is not large enough for large-scale data mining, our experiments should be understood as a preliminary 'proof of concept'. Nevertheless, there is a straightforward experimental approach that we implemented to test the value of the mutual information criterion for finding proofs:

1. We ran Prover9 with a two-second time limit over domains of problems.
2. We recorded all the formulas that were used in the proof search, including the formulas that appeared in successful proofs, and calculated the mutual information for each pair of formulas.
3. For each problem for which Prover9 did not find a proof, we generated a list of the one-hundred formulas with the highest mutual information score, relative to the formulas that appeared in the unsuccessful search. The high-information formulas were added to the input file as hints.
4. The search was run again, using the newly-added hints, and with the same two-second time limit.

Our hope, of course, was that when the hints were added, Prover9 would discover proofs that had evaded it during the first run. Here we report a few results.[2] We found that the addition of new hints by this method enabled Prover9 to discover a significant number of proofs it was unable to discover otherwise. Furthermore, the addition of these hints did not cause Prover9 to be unable to discover any proofs that it had found previously.

4.1   Boolean Algebra

The TPTP Boolean algebra domain contains 140 problems. Prover9 was able to find the proofs of 51 problems unaided, within the two-second time limit. By adding high-information formulas, Prover9 is able to find an additional 8 proofs within the same time limit.[3] These are non-trivial proofs. The shortest of the newly-discovered proofs has 52 steps and is 15 levels deep; the longest has 124 steps with level 29.[4] Of the new proofs, we found proofs of the following 'famous' properties:

Associativity of Boolean addition: $x + (y + z) = (x + y) + z$
DeMorgan's Law for Inverse and Product: $(x + y)^{-1} = x^{-1} \cdot y^{-1}$
DeMorgan's Law for Inverse and Sum: $(x^{-1} + y^{-1}) = (x \cdot y)^{-1}$

---

[2] For this experiment, we used a computer running Linux, with a Xeon processor and 16G of memory (although we did not nearly approach any memory limitations in these short runs). Version 5.4.0 of the TPTP library was used. We chose a two-second time limit because any new proofs generated by adding our hints to Prover9 would be more clearly attributable to the hints, rather than to small changes to the search space whose effects would magnify over time.

[3] They are: BOO007-2, BOO007-4, BOO014-2, BOO014-4, BOO015-2, BOO015-4, BOO023-1, and BOO028-1.

[4] The 'level' of a premise is 1, and the level of any other formula is the level of the highest-level parent, plus one. The level of the proof is the level of its conclusion.

We also tested the procedure by randomly selecting from the existing proofs the same number of formulas to use as hints. However, Prover9 was unable to find any additional proofs using random hints.

4.2 Field Theory

The TPTP library contains 279 problems from field theory, 91 of which can be proven by Prover9 in two seconds or less. When the clauses with the highest information scores are collected, we find that they seem to be quite different from the high information clauses from the Boolean algebra set. This may be because the TPTP field theory problems tend to use a different syntax for representation. Instead of expressing the value of an n-ary function as $f(x_1, x_2, \cdots, x_n) = y$, these problems use an $(n+1)$-ary relation $f(x_1, x_2, \cdots, x_n, y)$ without explicit use of equality. For clarity, we will use equality to represent clauses taken from this domain.

The high-information clauses from the field theory proofs seem to fall into two categories. First, we have ones that express basic facts about fields. For example, as we might expect, two high-information clauses are $0 + x = x$ and $1 \cdot x = x$. Similar clauses expressing facts about additive and multiplicative inverse can also be found in this set.

The second category is useful to Prover9 because they contain several disjunctions. Frequently, one of the difficulties in discovering a proof is that the theorem-prover generates too many formulas having a large number of disjuncts. Because Prover9 uses resolution as an inference rule, these formulas generate many other formulas; and it often turns out that formulas with many disjunctions take up an increasing share of the search space to the detriment of more useful lemmas. For this reason, the ability to find a small number of useful formulas with a large number of disjuncts would allow the user to put limits on the search space. This may be accomplished by using Prover9's weighting scheme to increase the weight of large disjunctions that have low information, while decreasing the weight of large disjunctions that have high information.

Some of the high information clauses are ones that express rules of inference that are repeatedly used to infer unit clauses (formulas without disjunctions). For example, the following two formulas have high average mutual information (we have rewritten the formulas as implications for clarity):

$$((w + v = x) \land (v + y = u) \land (w + u = z)) \to x + y = z$$
$$((x \cdot w = v) \land (w \cdot u = y) \land (v \cdot u = z)) \to x \cdot y = z$$

Intuitively these formulas allow us to combine several additions and multiplications over five variables and reduce them to a simple form with just three variables. A possible strategy (which we have not yet pursued) would be to include such formulas as lemmas, but otherwise impose sharp restrictions on the number of disjuncts allowed in a derived formula. This could have the effect of severely restricting the search space, while still permitting the use of more complex formulas that are likely to play an important role in generating proofs.

As before, we added formulas with the highest mutual information using the same criteria as with the Boolean algebra problems. Additional proofs were discovered for the following two theorems:[5]

$$x \geq y \rightarrow 2 \cdot x \geq 2 \cdot y$$
$$x \leq y \wedge z \leq w \rightarrow x + z \leq y + w$$

4.3 Group Theory

There are a total of 986 problems in the group theory category of the TPTP library. Unaided, Prover9 is able to find proofs for 486 of them within a two-second time limit.

The high-information formulas for the group theory problems are similar to those for the Boolean algebra problems (possibly because both sets tend to use equality to represent the formulas). In addition to the usual multiplication, inverse, and identity relations, the following high-information formulas also use so-called 'double division', where $x \odot y = x'y'$.

$$x'' = x \qquad x \odot (y \odot x) = y \quad xy = (x \odot y)'$$
$$(x \odot y) \odot x = y \quad xy = (y \odot x) \odot e \quad x' = x \odot e$$

$$x \odot y = y \odot x \qquad ex = x$$
$$e = x \odot x' \quad x \odot (x \odot e) = e$$

With the addition of hints automatically selected based on their information score, Prover9 is able to find an additional 19 proofs.[6] These include proofs of single axioms, some results about Moufang loops, and the associativity of product in one formulation of group theory. We believe that the relatively large number of new proofs in group theory is especially promising because the difficulty ratings for these problems tends to be much higher than for the other problem areas we tested.[7]

5   Conclusion

In this paper, we have reported the results of initial experiments to data-mine an existing library of problems to find hints for discovering new proofs. Of course, these small-scale experiments are only preliminary. However, we believe they are encouraging enough to warrant further testing on a larger scale, which is now underway.

There are at least three potential benefits to the approach we advocate. First, we aim to extend the reach of automated theorem-provers such as Prover9 by

---

[5] FLD060-4 and FLD061-4.
[6] GRP014-1 GRP057-1 GRP083-1 GRP167-1 GRP167-2 GRP202-1 GRP205-1 GRP261-1 GRP405-1 GRP413-1 GRP427-1 GRP428-1 GRP429-1 GRP438-1 GRP440-1 GRP502-1 GRP504-1 GRP644+1 GRP697-1.
[7] We are grateful to an anonymous referee for pointing this out.

leveraging information from previously discovered proofs. Ultimately, our aim is to enable the discovery of new, more complex and elusive theorems.

However, a merely technical advance is probably insufficient to reach that goal. Rather, we need to increase the size of the user base for systems like Prover9. Therefore, the second potential benefit is to increase the number of mathematicians and other scientists who use Prover9 by lowering the most significant barrier to entry. This barrier, as we have mentioned above, is that new users are often discouraged by the difficulty of selecting formulas to guide the system's search for difficult proofs. If a data-mining procedure like the one we have outlined is successful, then new users will be able to leverage existing work immediately, without having to construct that work from scratch.

But perhaps most importantly, we believe that in order for automated theorem-proving systems like Prover9 to have a greater impact, we must find a way to make their use more collaborative. Just as nobody expects a mathematician to independently re-discover all the lemmas necessary to prove a new theorem, we should not have to use a theorem-prover without the benefit of others' work. It is ironic that computers have enabled a much greater level of collaboration in virtually every research field, while collaboration among users of theorem-provers is as difficult as it ever was. Hopefully, a data-mining approach like the one we outline here is a first step toward developing a collaborative environment for automated theorem-proving.[8]

References

1. Church, K.W., Hanks, P.: Word Association Norms, Mutual Information, and Lexicography. Computational Linguistics 16(1)
2. Draeger, J.: Acquisition of Useful Lemma-Knowledge in Automated Reasoning. In: Giunchiglia, F. (ed.) AIMSA 1998. LNCS (LNAI), vol. 1480, pp. 230–239. Springer, Heidelberg (1998)
3. Draeger, J., Schulz, S., et al.: Improving the performance of automated theorem provers by redundancy-free lemmatization. In: Proceedings of the 14th Florida Artificial Intelligence Research Symposium, pp. 345–349 (2001)
4. McCune, W.: OTTER: 3.0 Reference Manual and Guide, Technical Report ANL-94/6. Argonne National Laboratory, Argonne, Illinois (1994)
5. McCune, W.: Prover9 Manual (2006), http://www.cs.unm.edu/mccune/prover9/manual
6. Schulz, S.: Information-based selection of abstraction levels. In: Proc. of the 14th FLAIRS, Key West, pp. 402–406 (2001)
7. Shannon, C., Weaver, E.: The Mathematical Theory of Communication. University of Chicago Press, Chicago (1949)
8. Sutcliffe, G.: The TPTP Problem Library and Associated Infrastructure: The FOF and CNF Parts, v3.5.0. Journal of Automated Reasoning 43(4), 337–362 (2009)
9. Veroff, R.: Using hints to increase the effectiveness of an automated reasoning program: Case studies. J. Automated Reason. 16(3), 223–239 (1996)
10. Wren, J.: Extending the Mutual Information Measure to Rank Inferred Literature Relationships. BMC Bioinformatics 5 (2004)

---

[8] We thank the participants of the ADAM workshops on automated deduction, especially Robert Veroff, as well as several anonymous referees.

Theorem Proving in Large Formal Mathematics as an Emerging AI Field

Josef Urban[1,*] and Jiří Vyskočil[2,**]

[1] Radboud University Nijmegen, The Netherlands
[2] Czech Technical University

Abstract. In the recent years, we have linked a large corpus of formal mathematics with automated theorem proving (ATP) tools, and started to develop combined AI/ATP systems working in this setting. In this paper we first relate this project to the earlier large-scale automated developments done by Quaife with McCune's Otter system, and to the discussions of the QED project about formalizing a significant part of mathematics. Then we summarize our adventure so far, argue that the QED dreams were right in anticipating the creation of a very interesting semantic AI field, and discuss its further research directions.

1 OTTER and QED

Twenty years ago, in 1992, Art Quaife's book *Automated Development of Fundamental Mathematical Theories* [30] was published. In the conclusion to his JAR paper [29] about the development of set theory Quaife cites Hilbert's *"No one shall be able to drive us from the paradise that Cantor created for us"*, and makes the following bold prediction:

> The time will come when such crushers as Riemann's hypothesis and Goldbach's conjecture will be fair game for automated reasoning programs. For those of us who arrange to stick around, endless fun awaits us in the automated development and eventual enrichment of the corpus of mathematics.

Quaife's experiments were done using an ATP system that has left so far perhaps the greatest influence on the field of Automated Theorem Proving: Bill McCune's Otter. Bill McCune's opinion on using Otter and similar Automated Reasoning methods for general mathematics was much more sober. The Otter manual [22] (right before acknowledging Quaife's work) states:

> Some of the first applications that come to mind when one hears "automated theorem proving" are number theory, calculus, and plane geometry, because these are some of the first areas in which math students try to

---

* Supported by The Netherlands Organization for Scientific Research (NWO) grants *Knowledge-based Automated Reasoning* and *MathWiki*.
** Supported by the Czech institutional grant MSM 6840770038.

M.P. Bonacina and M.E. Stickel (Eds.): McCune Festschrift, LNAI 7788, pp. 240–257, 2013.

prove theorems. Unfortunately, OTTER cannot do much in these areas: interesting number theory problems usually require induction, interesting calculus and analysis problems usually require higher-order functions, and the first-order axiomatizations of geometry are not practical.

Yet, Bill McCune was also a part of the QED[1] discussions about making a significant part of mathematics computer understandable, verified, and available for a number of applications. These discussions started around 1993, and gained considerable attention of the theorem proving and formalization communities, leading to the publication of the anonymous QED manifesto [1] and two QED workshops in 1994 and 1995. And ATP systems such as SPASS [50], E [33], and Vampire [32] based on the ideas that were first developed in Otter have now really used for several years to prove lemmas in general mathematical developments in the large ATP-translated libraries of Mizar [13] and Isabelle [25].

This paper summarizes our experience so far with the QED-inspired project of developing automated reasoning methods for large general computer understandable mathematics, particularly in the large Mizar Mathematical Library. [2] A bit as Art Quaife did, we believe (and try to argue below) that automated reasoning in general mathematics is one of the most exciting research fields, where a number of new and interesting topics for general AI research emerge today. We hope that the paper might be of some interest to those QED-dreamers who remember the great minds of the recently deceased Bill McCune, John McCarthy, and N.G. de Bruijn.

2 Why Link Large Formal Mathematics with AI/ATP Methods?

Formalization of Mathematics and Automated Theorem Proving are often considered to be two subfields of the field of Automated Reasoning. In the former, the focus is on developing methods for (human-assisted) computer verification of more and more complex parts of mathematics, while in ATP, the focus is on developing stronger and stronger methods for finding proofs automatically. The QED Manifesto has the following conservative opinion about the usefulness of automated theorem proving (and other AI) methods to a QED-like project:

It is the view of some of us that many people who could have easily contributed to project QED have been distracted away by the enticing lure of AI or AR. It can be agreed that the grand visions of AI or AR are much more interesting than a completed QED system while still believing that there is great aesthetic, philosophical, scientific, educational, and technological value in the construction of the QED system, regardless of whether its construction is or is not largely done 'by hand' or largely automatically.

---

[1] http://www.rbjones.com/rbjpub/logic/qedres00.htm
[2] www.mizar.org

Our opinion is that formalization and automation are two sides of the same coin. There are three kinds of benefits in linking formal proof assistants like Mizar and their libraries with the Automated Reasoning technology and particularly ATPs:

1. The obvious benefits for the proof assistants and their libraries. Automated Reasoning and AI methods can provide a number of tools and strong methods that can assist the formalization, provide advanced search and hint functions, and prove lemmas and theorems (semi-)automatically. The QED Manifesto says:

> The QED system we imagine will provide a means by which mathematicians and scientists can scan the entirety of mathematical knowledge for relevant results and, using tools of the QED system, build upon such results with reliability and confidence but without the need for minute comprehension of the details or even the ultimate foundations of the parts of the system upon which they build

2. The (a bit less obvious) benefits for the field of Automated Reasoning. Research in automated reasoning over very large formal libraries is painfully theoretical (and practically useless) until such libraries are really available for experiments. Mathematicians (and scientists, and other human "reasoners") typically know a lot of things about the domains of discourse, and use the knowledge in many ways that include many heuristic methods. It thus seems unrealistic (and limiting) to develop the automated reasoning tools solely for problems that contain only a few axioms, make little use of previously accumulated knowledge, and do not attempt to further accumulate and organize the body of knowledge. In his 1996 review of Quaife's book, Desmond Fearnley-Sander says:

> The real work in proving a deep theorem lies in the development of the theory that it belongs to and its relationships to other theories, the design of definitions and axioms, the selection of good inference rules, and the recognition and proof of more basic theorems. Currently, no resolution-based program, when faced with the stark problem of proving a hard theorem, can do all this. That is not surprising. No person can either. Remarks about standing on the shoulders of giants are not just false modesty...

3. The benefits for the field of general Artificial Intelligence. These benefits are perhaps the least mentioned ones, perhaps due to the frequent feeling of too many unfulfilled promises from general AI. However to the authors they appear to be the strongest long-term motivation for this kind of work. In short, the AI fields of *deductive reasoning* and *inductive reasoning* (represented by machine learning, data mining, knowledge discovery in databases, etc.) have so far benefited relatively little from each other's progress. This is an obvious deficiency in comparison with the human mind, which can both inductively suggest new ideas and problem solutions based on analogy, memory, statistical evidence, etc., and also confirm, adjust, and even significantly modify these ideas and problem solutions by deductive reasoning and explanation, based on the understanding of the world. Repositories of "human thought" that are both large (and thus allow inductive methods), and have precise and deep semantics (and thus allow deduction) should be a very useful component for cross-fertilization of these two fields. QED-like large formal mathematical libraries are currently the closest approximation to such a computer-understandable repository of "human thought" usable for these purposes. To be really usable, the libraries however again have to be presented in a form that is easy to understand for existing automated reasoning tools. The Fearnley-Sander's quote started above continues as:

> ... Great theorems require great theories and theories do not, it seems, emerge from thin air. Their creation requires sweat, knowledge, imagination, genius, collaboration and time. As yet there is not much serious collaboration of machines with one another, and we are only just beginning to see real symbiosis between people and machines in the exercise of rationality.

3   Why Mizar?

Mizar is a language for formalization of mathematics and a proof checker for the language, developed since 1974 [19] by the Mizar team led by Andrzej Trybulec. This system was chosen by the first author for experiments with automated reasoning tools because of its focus on building the large formal Mizar Mathematical Library (MML). This formalization effort was begun in 1989 by the Mizar team, and its main purpose is to verify a large body of mainstream mathematics in a way that is close and easily understandable to mathematicians, allowing them to build on this library with proofs from more and more advanced mathematical fields. In 2012 the library consisted of over 1100 formal articles containing more than 50,000 proved theorems and 10,000 definitions. The QED discussions often used Mizar and its library as a prototypical example for the project. The particular formalization goals have influenced:

- the choice of a relatively human-oriented formal language in Mizar
- the choice of the declarative Mizar proof style (Jaskowski-style natural deduction [17])
- the choice of first-order logic and set theory as unified common foundations for the whole library
- the focus on developing and using just one human-obvious first-order justification rule in Mizar
- and the focus on making the large library interconnected, usable for more advanced formalizations, and using consistent notation.

There have always been other systems and projects that are similar to Mizar in some of the above mentioned aspects. Building large and advanced formal libraries seems to be more and more common today in systems like Isabelle, Coq [7], and HOL Light [16]. An example of a recent major formalization project (in HOL Light) is Flyspeck [15], which required the formal verification of thousands of lemmas in general mathematics. In the work that is described here, Mizar thus should be considered as a suitable particular choice of a system for formalization of mathematics which uses relatively common and accessible foundations, and produces a large formal library written in a relatively simple and easy-to-understand style. Some of the systems described below actually already work also with other than Mizar data: for example, MaLARea [41,48] has already been successfully used for reasoning over problems from the large formal SUMO ontology [24], and for experiments with Isabelle/Sledgehammer problems [23].

4   MPTP: Translating Mizar for Automated Reasoning Tools

The Mizar's translation (MPTP - Mizar Problems for Theorem Proving) to pure first-order logic is described in detail in [38,39,42]. The translation has to deal with a number of Mizar extensions and practical issues related to the Mizar implementation, implementations of first-order ATP systems, and the most frequent uses of the translation system.

The first version (published in early 2003[3]) has been used for initial exploration of the usability of ATP systems on the Mizar Mathematical Library (MML). The first important number obtained was the 41% success rate of ATP re-proving of about 30000 MML theorems from selected Mizar theorems and definitions (precisely: other Mizar theorems and definitions mentioned in the Mizar proofs) taken from corresponding MML proofs.

No previous evidence about the feasibility and usefulness of ATP methods on a very large library like MML was available prior to the experiments done with MPTP 0.1[4], sometimes leading to overly pessimistic views on such a project. Therefore the goal of this first version was to relatively quickly achieve a "mostly-correctly" translated version of the whole MML that would allow us to measure and assess the potential of ATP methods for this large library. Many shortcuts and simplifications were therefore taken in this first MPTP version, for example direct encoding in the DFG [14] syntax used by the SPASS system, no proof export, incomplete export of some relatively rare constructs (*structure* types and *abstract terms*), etc.

Many of these simplifications however made further experiments with MPTP difficult or impossible, and also made the 41% success rate uncertain. The lack of proof structure prevented measurements of ATP success rate on all internal proof lemmas, and experiments with unfolding lemmas with their own proofs. Additionally, even if only several abstract terms were translated incorrectly, during such proof unfoldings they could spread much wider. Experiments like finding

---

[3] http://mizar.uwb.edu.pl/forum/archive/0303/msg00004.html

[4] A lot of work on MPTP has been inspired by the previous work done in the ILF project [12] on importing Mizar. However it seemed that the project had stopped before it could finish the export of the whole MML to ATP problems and provide some initial overall statistics of ATP success rate on MML.

new proofs, and cross-verification of Mizar proofs (described below) would suffer from constant doubt about the possible amount of error caused by the incorrectly translated parts of Mizar, and debugging would be very hard.

Therefore, after the encouraging initial experiments, a new version of MPTP started to be developed in 2005, requiring first a substantial re-implementation [40] of Mizar interfaces. This version consists of two layers (Mizar-extended TPTP format processed in Prolog, and a Mizar XML format) that are sufficiently flexible and have allowed a number of gradual additions of various functions over the past years. The experiments described below are typically done on this version (and its extensions).

5   Experiments and Projects Based on the MPTP

MPTP has so far been used for

- experiments with re-proving Mizar theorems and simple lemmas by ATPs from the theorems and definitions used in the corresponding Mizar proofs,
- experiments with automated proving of Mizar theorems with the whole Mizar library, i.e. the necessary axioms are selected automatically from MML,
- finding new ATP proofs that are simpler than the original Mizar proofs,
- ATP-based cross-verification of the Mizar proofs,
- ATP-based explanation of Mizar atomic inferences,
- inclusion of Mizar problems in the TPTP [37] problem library, and unified web presentation of Mizar together with the corresponding TPTP problems,
- creation of the MPTP $100 Challenges[5] for reasoning in large theories in 2006, creation of the MZR category of the CASC Large Theory Batch (LTB) competition [36] in 2008, and creation of the MPTP2078 [2] benchmark[6] in 2011 used for the CASC@Turing[7] competition in 2012,
- testbed for AI systems like MaLARea and MaLeCoP [49] targeted at reasoning in large theories and combining inductive techniques like machine learning with deductive reasoning.

5.1   Re-proving Experiments

As mentioned in Section 4, the initial large-scale experiment done with MPTP 0.1 indicated that 41% of the Mizar proofs can be automatically found by ATPs, if the users provide as axioms to the ATPs the same theorems and definitions which are used in the Mizar proofs, plus the corresponding background formulas (formulas implicitly used by Mizar, for example to implement type hierarchies). As already mentioned, this number was far from certain, e.g., out of the 27449 problems tried, 625 were shown to be counter-satisfiable (i.e., a model of the

---

[5] www.tptp.org/MPTPChallenge

[6] http://wiki.mizar.org/twiki/bin/view/Mizar/MpTP2078

[7] http://www.cs.miami.edu/~tptp/CASC/J6/Design.html

negated conjecture and axioms exists) in a relatively low time limit given to SPASS (pointing to various oversimplifications taken in MPTP 0.1). This experiment was therefore repeated with MPTP 0.2, however only with 12529 problems that come from articles that do not use internal arithmetical evaluations done by Mizar. These evaluations were not handled by MPTP 0.2 at the time of conducting these experiments, being the last (known) part of Mizar that could be blamed for possible ATP incompleteness. The E prover version 0.9 and SPASS version 2.1 were used for this experiment, with 20s time limit (due to limited resources). The results (reported in [42]) are given in Table 1. 39% of the 12529 theorems were proved by either SPASS or E, and no counter-satisfiability was found.

Table 1. Re-proving MPTP 0.2 theorems from non-numerical articles in 2005

description	proved	counter-satisfiable	timeout or memory out	total
E 0.9	4309	0	8220	12529
SPASS 2.1	3850	0	8679	12529
together	4854	0	7675	12529

These results have thus to a large extent confirmed the optimistic outlook of the first measurement in MPTP 0.1. In later experiments, this ATP performance has been steadily going up, see Table 2 for results from 2007 run with 60s time limit. This is a result of better pruning of redundant axioms in MPTP, and also of ATP development, which obviously was influenced by the inclusion of MPTP problems in the TPTP library, forming a significant part of the FOF problems in the CASC competition since 2006. The newer versions of E and SPASS solved in this increased time limit together 6500 problems, i.e. 52% of them all. With addition of Vampire and its customized Fampire[8] version (which alone solves 51% of the problems), the combined success rate went up to 7694 of these problems, i.e. to 61%. The caveat is that the methods for dealing with arithmetic are becoming stronger and stronger in Mizar. The MPTP problem creation for problems containing arithmetic is thus currently quite crude, and the ATP success rate on such problems will likely be significantly lower than on the nonarithmetical ones. Fortunately, various approaches have been recently started to endow theorem provers with better capabilities for arithmetic [28,18,4,8].

5.2 Finding New Proofs and the AI Aspects

MPTP 0.2 was also used to try to prove Mizar theorems without looking at their Mizar proofs, i.e., the choice of premises for each theorem was done automatically, and all previously proved theorems were eligible. Because giving ATPs thousands of axioms was usually hopeless in 2005, the axiom selection was done

---

[8] Fampire is a combination of Vampire with the FLOTTER [26] clausifier.

Table 2. Re-proving MPTP 0.2 theorems from non-numerical articles in 2007

description	proved	counter-satisfiable	timeout or memory out	total
E 0.999	5661	0	6868	12529
SPASS 2.2	5775	0	6754	12529
E+SPASS together	6500	-	-	12529
Vampire 8.1	5110	0	7419	12529
Vampire 9	5330	0	7119	12529
Fampire 9	6411	0	6118	12529
all together	7694	-	-	12529

by symbol-based machine learning from the previously available proofs using the SNoW [10] (Sparse Network of Winnows) system run in the naive Bayes learning mode. The ability of systems to handle many axioms has improved a lot since 2005: the CASC-LTB category and also the related work on the Sledgehammer [23] link between ATPs with Isabelle/HOL have sparked interest in ATP systems dealing efficiently with large numbers of axioms. See [45] for a brief overview of the large theory methods developed so far.

The results (reported in [42]) are given in Table 3. 2408 from the 12529 theorems were proved either by E 0.9 or SPASS 2.1 from the axioms selected by the machine learner, the combined success rate of this whole system was thus 19%.

Table 3. Proving MPTP 0.2 theorems with machine learning support in 2005

description	proved	counter-satisfiability	timeout or memory out	total
E 0.9	2167	0	10362	12529
SPASS 2.1	1543	0	10986	12529
together	2408	0	10121	12529

This experiment demonstrates a very real and quite unique benefit of large formal mathematical libraries for conducting novel integration of AI methods. As the machine learner is trained on previous proofs, it recommends relevant premises from the large library that (according to the past experience) should be useful for proving new conjectures. A variety of machine learning methods (neural nets, Bayes nets, decision trees, nearest neighbor, etc.) can be used for this, and their performance evaluated in the standard machine learning way, i.e., by looking at the actual axiom selection done by the human author in the Mizar proof, and comparing it with the selection suggested by the trained learner. However, what if the machine learner is sometimes more clever than the human, and suggests a completely different (and perhaps better) selection of premises, leading to a different proof? In such a case, the standard machine learning evaluation (i.e. comparison of the two sets of premises) will say that the two sets of premises differ too much, and thus the machine learner has failed. This is considered acceptable for machine learning, as in general, there is no deeper concept of *truth* available, there are just training and testing data. However in our domain we do have a method for showing that the trained learner was right (and possibly smarter than the human): we can run an ATP system on its axiom selection. If a proof is found, it provides a much stronger measure of correctness. Obviously, this is only true if we know that the translation from Mizar to TPTP was correct, i.e., conducting such experiments really requires that we take extra care to ensure that no oversimplifications were made in this translation.

In the above mentioned experiment, 329 of the 2408 (i.e. 14%) proofs found by ATPs were shorter (used less premises) than the original MML proof. An example of such proof shortening is discussed in [42], showing that the newly found proof is really valid. Instead of arguing from the first principles (definitions) like in the human proof, the combined inductive-deductive system was smart enough to find a combination of previously proved lemmas (properties) that justify the conjecture more quickly.

A similar newer evaluation is done on the whole MML in [3], comparing the original MML theory graph with the theory graph for the 9141 automatically found proofs. An illustrative example from there is theorem COMSEQ_3:40[9], proving the relation between the limit of a complex sequence and its real and imaginary parts:

Theorem 1. Let $(c_n) = (a_n + ib_n)$ be a convergent complex sequence. Then (a_n) and (b_n) converge and $\lim a_n = Re(\lim c_n)$ and $\lim b_n = Im(\lim c_n)$.

The convergence of (a_n) and (b_n) was done the same way by the human formalizer and the ATP. The human proof of the limit equations proceeds by looking at the definition of a complex limit, expanding the definitions, and proving that a and b satisfy the definition of the real limit (finding a suitable n for a given ϵ). The AI/ATP just notices that this kind of groundwork was already done in a "similar" case COMSEQ_3:39[10], which says that:

Theorem 2. If (a_n) and (b_n) are convergent, then $\lim c_n = \lim a_n + i \lim b_n$.

And it also notices the "similarity" (algebraic simplification) provided by theorem COMPLEX1:28[11]:

Theorem 3. $Re(a + ib) = a \wedge Im(a + ib) = b$

Such (automatically found) manipulations can be used (if noticed!) to avoid the "hard thinking" about the epsilons in the definitions.

---

[9] http://mizar.cs.ualberta.ca/ mptp/cgi-bin/browserefs.cgi?refs=t40_comseq_3

[10] http://mizar.cs.ualberta.ca/ mptp/cgi-bin/browserefs.cgi?refs=t39_comseq_3

[11] http://mizar.cs.ualberta.ca/ mptp/cgi-bin/browserefs.cgi?refs=t28_complex1

5.3   ATP-Based Explanation, Presentation, and Cross-verification of Mizar Proofs

While the whole proofs of Mizar theorems can be quite hard for ATP systems, re-proving the Mizar atomic justification steps (called *Simple Justifications* in Mizar; these are the atomic steps discharged by the very limited and fast Mizar refutational checker [51]) turns out to be quite easy for ATPs. The combinations of E and SPASS usually solve more than 95% of such problems, and with smarter automated methods for axiom selection 99.8% success rate (14 unsolved problems from 6765) was achieved in [43]. This makes it practical to use ATPs for explanation and presentation of the (not always easily understandable) Mizar simple justifications, and to construct larger systems for independent ATP-based cross-verification of (possibly very long) Mizar proofs. In [43] such a cross-verification system is presented, using the GDV [34] system which was extended to process Jaskowski-style natural deduction proofs that make frequent use of assumptions (suppositions). MPTP was used to translate Mizar proofs to this format, and GDV together with the E, SPASS, and MaLARea systems were used to automatically verify the structural correctness of proofs, and 99.8% of the proof steps needed for the 252 Mizar problems selected for the MPTP Challenge (see below). This provides the first practical method for independent verification of Mizar, and opens the possibility of importing Mizar proofs into other proof assistants. A web presentation allowing interaction with ATP systems and GDV verification of Mizar proofs [44] has been set up,[12] and an online service [47] integrating the ATP functionalities has been built.[13]

5.4   Use of MPTP for ATP Challenges and Competitions

The first MPTP problems were included in the TPTP library in 2006, and were already used for the 2006 CASC competition. In 2006, the MPTP $100 Challenges[14] were created and announced. This is a set of 252 related large-theory problems needed for one half (one of two implications) of the Mizar proof of the general topological Bolzano-Weierstrass theorem [5]. Unlike the CASC competition, the challenge had an overall time limit (252 * 5 minutes = 21 hours) for solving the problems, allowing complementary techniques like machine learning from previous solutions to be experimented with transparently within the overall time limit. The challenge was won a year later by the leanCoP [27] system, having already revealed several interesting approaches to ATP in large theories: goal-directed calculi like connection tableaux (used in leanCoP), model-based axiom selection (used e.g. in SRASS [35]), and machine learning of axiom relevance (used in MaLARea) from previous proofs and conjecture characterizations (where *axiom relevance* can be defined as the likelihood that an axiom will be needed for proving a particular conjecture). The MPTP Challenge problems

---

[12] http://www.tptp.org/MizarTPTP

[13] http://mws.cs.ru.nl/~mptp/MizAR.html, and
http://mizar.cs.ualberta.ca/~mptp/MizAR.html

[14] http://www.tptp.org/MPTPChallenge/

were again included in TPTP and used for the standard CASC competition in 2007. In 2008, the CASC-LTB (Large Theory Batch) category appeared for the first time with a similar setting like the MPTP Challenges, and additional large-theory problems from the SUMO and Cyc [31] ontologies. A set of 245 relatively hard Mizar problems was included for this purpose in TPTP, coming from the most advanced parts of the Mizar library. The problems come in four versions that contain different numbers of the previously available MML theorems and definitions as axioms. The largest versions thus contain over 50000 axioms. An updated larger version (MPTP2078) of the MPTP Challenge benchmark was developed in 2011 [2], consisting of 2078 interrelated problems in general topology, and making use of precise dependency analysis of the MML for constructing the easy versions of the problems. The MPTP2078 problems were used in the Mizar category of the 2012 CASC@Turing ATP competition. This category had similar rules as the MPTP Challenge, i.e., a large number (400) of related MPTP2078 problems is being solved in an overall time limit. Unlike in the MPTP Challenge, an additional training set of 1000 MPTP2078 problems (disjoint from the competition problems) was made available before the competition together with their Mizar and Vampire proofs. This way, more expensive training and tuning methods could be developed before the competition, however (as common in machine-learning competitions) the problem-solving knowledge extracted from the problems by such methods has to be sufficiently general to be useful on the disjoint set of competition problems.

5.5   Development of Larger AI Metasystems Like MaLARea and MaLeCoP on MPTP Data

In Section 5.2, it is explained how the notion of mathematical *truth* implemented through ATPs can improve the evaluation of learning systems working on large semantic knowledge bases like translated MML. This is however only one part of the AI fun made possible by such large libraries being available to ATPs. Another part is that the newly found proofs can be recycled, and again used for learning in such domains. This closed loop (see Figure 1) between using deductive methods to find proofs, and using inductive methods to learn from the existing proofs and suggest new proof directions, is the main idea behind the Machine Learner for Automated Reasoning (MaLARea [41,48]) metasystem, which turns out to have so far the best performance on large theory benchmarks like the MPTP Challenge and MPTP2078. There are many kinds of information that such an autonomous metasystem can try to use and learn. The second version of MaLARea already uses also structural and semantic features of formulas for their characterization and for improving the axiom selection.

MaLARea can work with arbitrary ATP backends (E and SPASS by default), however, the communication between learning and the ATP systems is *high-level*: The learned relevance is used to try to solve problems with varied limited numbers of the most relevant axioms. Successful runs provide additional data for learning (useful for solving related problems), while unsuccessful runs can yield countermodels, which can be re-used for semantic pre-selection and as additional

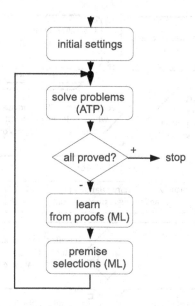

Fig. 1. The basic MaLARea loop

input features for learning. An advantage of such high-level approach is that it gives a generic inductive (learning)/deductive (ATP) metasystem to which any ATP can be easily plugged as a blackbox. Its disadvantage is that it does not attempt to use the learned knowledge for guiding the ATP search process once the axioms are selected.

Hence the logical next step done in the Machine Learning Connection Prover (MaLeCoP) prototype [49]: the learned knowledge is used for guiding the proof search mechanisms in a theorem prover (leanCoP in this case). MaLeCoP follows a general advising design that is as follows (see also Figure 2): The theorem prover (P) has a sufficiently fast communication channel to a general advisor (A) that accepts queries (proof state descriptions) and training data (characterization of the proof state[15] together with solutions[16] and failures) from the prover, processes them, and replies to the prover (advising, e.g., which clauses to choose). The advisor A also talks to external (in our case learning) system(s) (E). A translates the queries and information produced by P to the formalism used by a particular E, and translates E's guidance back to the formalism used by P. At suitable time, A also hands over the (suitably transformed) training data to E, so that E can update its knowledge of the world on which its advice is based.

MaLeCoP is a very recent work, which has so far revealed interesting issues in using detailed smart guidance in large theories. Even though naive Bayes is a

---

[15] instantiated, e.g., as the set of literals/symbols on the current branch.

[16] instantiated, e.g., as the description of clauses used at particular proof states.

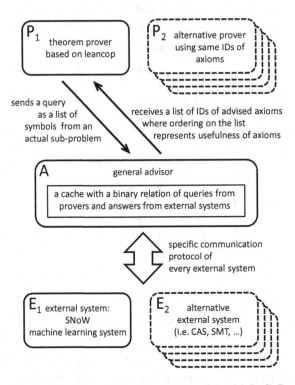

Fig. 2. The General Architecture used for MaLeCoP

comparatively fast learning and advising algorithm, in a large theory it turned out to be about 1000 times slower than a primitive tableaux extension step. So a number of strategies had to be defined that use the smart guidance only at the crucial points of the proof search. Even with such limits, the preliminary evaluation done on the MPTP Challenge already showed an average proof search shortening by a factor of 20 in terms of the number of tableaux inferences.

There are a number of development directions for knowledge-based AI/ATP architectures like MaLARea and MaLeCoP. Extracting lemmas from proofs and adding them to the set of available premises, creating new interesting conjectures and defining new useful notions, finding optimal strategies for problem classes, faster guiding of the internal ATP search, inventing policies for efficient governing of the overall inductive-deductive loop: all these are interesting AI tasks that become relevant in this large-theory setting, and that also seem to be relevant for the task of *doing mathematics* (and exact sciences) more automatically. A particularly interesting issue is the following.

Consistency of Knowledge and Its Transfer: An important research topic in the emerging AI approaches to large-theory automated reasoning is the issue of *consistency of knowledge and its transfer*. In an unpublished experiment with MaLARea in 2007 over a set of problems exported by an early Isabelle/Sledgehammer version, MaLARea quickly solved all of the problems, even though some of them were supposed to be hard. Larry Paulson tracked the problem to an (intentional) simplification in the first-order encoding of Isabelle/HOL types, which typically raises the overall ATP success rate (after checking in Isabelle the imported proofs) in comparison with heavier (but sound) type encodings that are also available in Isabelle/Sledgehammer. Once the inconsistency originating from the simplification was found by the guiding AI system, MaLARea focused on fully exploiting it even in problems where such inconsistency would be ignored by standard ATP search.

An opposite phenomenon happened recently in experiments with a clausal version of MaLARea. The CNF form introduces a large number of new skolem symbols that make similar problems and formulas look different after the reduction to clausal form (despite the fact that the skolemization attempts hard to use the same symbol whenever it can), and the AI guidance based on symbols and terms deteriorates. The same happens with the AI guidance based on finite counter-models generated by Mace [21] and Paradox [11]. The `clausefilter` utility from McCune's Library of Automated Deduction Routines (LADR) [20] is used by MaLARea to evaluate all available formulas in each newly found finite model, resulting in an additional semantic characterization of all formulas that often improves the premise selection. Disjoint skolem symbols however prevent a straightforward evaluation of many clauses in models that are found for differently named skolem functions. Such inability of the AI guidance to obtain and use the information about the similarity of the clauses results in about 100 fewer problems solved (700 vs. 800) in the first ten MaLARea iterations over the MPTP2078 benchmark.

Hence a trade-off: smaller pieces of knowledge (like clauses) allow better focus, but techniques like skolemization can destroy some explicit similarities useful for learning. Designing suitable representations and learning methods on top of the knowledge is therefore very significant for the final performance, while inconsistent representations can be fatal. Using CNF and its various alternatives and improvements has been a topic discussed many times in the ATP community (also for example by Quaife in his book). Here we note that it is not just the low-level ATP algorithms that are influenced by such choices of representation, but the problem extends to and significantly influences also the performance of high-level heuristic guidance methods in large theories.

6 Future QED-Like Directions

There is large amount of work to be done on practically all the projects mentioned above. The MPTP translation is by no means optimal (and especially proper encoding of arithmetic needs more experiments and work). Import of ATP proofs to Mizar practically does not exist (there is a basic translator taking Otter proof objects to Mizar, however this is very distant from the readable proofs in MML). With sufficiently strong ATP systems, the cross-verification of the whole MML could be attempted, and work on import of such detailed ATP proofs into other proof assistants could be started. The MPTP handling of second-order constructs is in some sense incomplete, and either a translation to (finitely axiomatized) Neumann-Bernays-Gödel (NBG) set theory (used by Quaife), or usage of higher-order ATPs like Satallax [9] and LEO-II [6] would be interesting from this point of view.

More challenges and interesting presentation tools can be developed, for example an ATP-enhanced wiki for Mizar is an interesting QED-like project that is now being worked on [46]. The heuristic and machine learning methods, and combined AI metasystems, have a very long way to go, some future directions are mentioned above. This is no longer only about mathematics: all kinds of more or less formal large knowledge bases are becoming available in other sciences, and automated reasoning could become one of the strongest methods for general reasoning in sciences when sufficient amount of formal knowledge exists. Strong ATP methods for large formal mathematics could also provide useful semantic filtering for larger systems for automatic formalization of mathematical papers. This is a field that has been so far deemed to be rather science fiction than a real possibility. In particular, the idea of gradual top-down (semi-)automated formalization of mathematics written in books and papers has been so far considered outlandish in a way that is quite reminiscent of how the idea of ATP in large general mathematics was considered outlandish before the first large-scale experiments in 2003. The proposed AI solution should be similar for the two problems: guide the vast search space by the knowledge extracted from the vast amount of problems already solved. It is interesting that already within the QED project discussions, Feferman suggested[17] that large-scale formalization should have a necessary top-down aspect. Heuristic AI methods used for knowledge search and machine translation are becoming more and more mature, and in conjunction with strong ATP methods they could provide a basis for such large-scale (semi-)automated QED project.

References

1. The QED Manifesto. In: Bundy, A. (ed.) CADE 1994. LNCS, vol. 814, pp. 238–251. Springer, Heidelberg (1994)
2. Alama, J., Kühlwein, D., Tsivtsivadze, E., Urban, J., Heskes, T.: Premise selection for mathematics by corpus analysis and kernel methods. CoRR, abs/1108.3446 (2011)
3. Alama, J., Kühlwein, D., Urban, J.: Automated and Human Proofs in General Mathematics: An Initial Comparison. In: Bjørner, N., Voronkov, A. (eds.) LPAR-18. LNCS, vol. 7180, pp. 37–45. Springer, Heidelberg (2012)
4. Althaus, E., Kruglov, E., Weidenbach, C.: Superposition Modulo Linear Arithmetic SUP(LA). In: Ghilardi, S., Sebastiani, R. (eds.) FroCoS 2009. LNCS, vol. 5749, pp. 84–99. Springer, Heidelberg (2009)
5. Bancerek, G., Endou, N., Sakai, Y.: On the characterizations of compactness. Formalized Mathematics 9(4), 733–738 (2001)

---

[17] http://mizar.org/qed/mail-archive/volume-2/0003.html

6. Benzmüller, C.E., Paulson, L.C., Theiss, F., Fietzke, A.: LEO-II - A Cooperative Automatic Theorem Prover for Classical Higher-Order Logic (System Description). In: Armando, A., Baumgartner, P., Dowek, G. (eds.) IJCAR 2008. LNCS (LNAI), vol. 5195, pp. 162–170. Springer, Heidelberg (2008)

7. Bertot, Y., Casteran, P.: Interactive Theorem Proving and Program Development - Coq'Art: The Calculus of Inductive Constructions. Texts in Theoretical Computer Science. Springer (2004)

8. Bonacina, M.P., Lynch, C., de Moura, L.M.: On deciding satisfiability by theorem proving with speculative inferences. J. Autom. Reasoning 47(2), 161–189 (2011)

9. Brown, C.E.: Satallax: An Automatic Higher-Order Prover. In: Gramlich, B., Miller, D., Sattler, U. (eds.) IJCAR 2012. LNCS, vol. 7364, pp. 111–117. Springer, Heidelberg (2012)

10. Carlson, A., Cumby, C., Rosen, J., Roth, D.: SNoW User's Guide. Technical Report UIUC-DCS-R-99-210, University of Illinois at Urbana-Champaign (1999)

11. Claessen, K., Sorensson, N.: New Techniques that Improve MACE-style Finite Model Finding. In: Baumgartner, P., Fermueller, C. (eds.) Proceedings of the CADE-19 Workshop: Model Computation - Principles, Algorithms, Applications (2003)

12. Dahn, I., Wernhard, C.: First order proof problems extracted from an article in the MIZAR Mathematical Library. In: Bonacina, M.P., Furbach, U. (eds.) Int. Workshop on First-Order Theorem Proving (FTP 1997). RISC-Linz Report Series No. 97-50, vol. (97-50), pp. 58–62. Johannes Kepler Universität, Linz, Austria (1997)

13. Grabowski, A., Korniłowicz, A., Naumowicz, A.: Mizar in a nutshell. Journal of Formalized Reasoning 3(2), 153–245 (2010)

14. Hähnle, R., Kerber, M., Weidenbach, C.: Common Syntax of the DFG-Schwerpunktprogramm Deduction. Technical Report TR 10/96, Fakultät für Informatik, Universät Karlsruhe, Karlsruhe, Germany (1996)

15. Hales, T.C.: Introduction to the Flyspeck project. In: Coquand, T., Lombardi, H., Roy, M.-F. (eds.) Mathematics, Algorithms, Proofs. Dagstuhl Seminar Proceedings, vol. 05021, Internationales Begegnungs- und Forschungszentrum für Informatik (IBFI), Schloss Dagstuhl, Germany (2005)

16. Harrison, J.: HOL Light: A Tutorial Introduction. In: Srivas, M., Camilleri, A. (eds.) FMCAD 1996. LNCS, vol. 1166, pp. 265–269. Springer, Heidelberg (1996)

17. Jaskowski, S.: On the rules of suppositions. Studia Logica, 1 (1934)

18. Korovin, K., Voronkov, A.: Integrating Linear Arithmetic into Superposition Calculus. In: Duparc, J., Henzinger, T.A. (eds.) CSL 2007. LNCS, vol. 4646, pp. 223–237. Springer, Heidelberg (2007)

19. Matuszewski, R., Rudnicki, P.: Mizar: the first 30 years. Mechanized Mathematics and Its Applications 4, 3–24 (2005)

20. McCune, W.W.: LADR: Library of Automated Deduction Routines, http://www.mcs.anl.gov/AR/ladr

21. McCune, W.W.: Mace4 Reference Manual and Guide. Technical Report ANL/MCS-TM-264, Argonne National Laboratory, Argonne, USA (2003)

22. McCune, W.W.: Otter 3.3 Reference Manual. Technical Report ANL/MSC-TM-263, Argonne National Laboratory, Argonne, USA (2003)

23. Meng, J., Paulson, L.C.: Lightweight relevance filtering for machine-generated resolution problems. J. Applied Logic 7(1), 41–57 (2009)

24. Niles, I., Pease, A.: Towards a standard upper ontology. In: FOIS 2001: Proceedings of the International Conference on Formal Ontology in Information Systems, pp. 2–9. ACM Press, New York (2001)

25. Nipkow, T., Paulson, L.C., Wenzel, M.T.: Isabelle/HOL - A Proof Assistant for Higher-Order Logic. LNCS, vol. 2283. Springer, Heidelberg (2002)
26. Nonnengart, A., Weidenbach, C.: Computing small clause normal forms. In: Handbook of Automated Reasoning, vol. I, pp. 335–367. Elsevier and MIT Press (2001)
27. Otten, J., Bibel, W.: leanCoP: Lean Connection-Based Theorem Proving. Journal of Symbolic Computation 36(1-2), 139–161 (2003)
28. Prevosto, V., Waldmann, U.: SPASS+T. In: Sutcliffe, G., Schmidt, R., Schulz, S. (eds.) ESCoR 2006. CEUR, vol. 192, pp. 18–33 (2006)
29. Quaife, A.: Automated Deduction in von Neumann-Bernays-Godel Set Theory. Journal of Automated Reasoning 8(1), 91–147 (1992)
30. Quaife, A.: Automated Development of Fundamental Mathematical Theories. Kluwer Academic Publishers (1992)
31. Ramachandran, D., Reagan, P., Goolsbey, K.: First-orderized ResearchCyc: Expressiveness and Efficiency in a Common Sense Knowledge Base. In: Shvaiko, P. (ed.) Proceedings of the Workshop on Contexts and Ontologies: Theory, Practice and Applications (2005)
32. Riazanov, A., Voronkov, A.: The Design and Implementation of Vampire. AI Communications 15(2-3), 91–110 (2002)
33. Schulz, S.: E: A Brainiac Theorem Prover. AI Communications 15(2-3), 111–126 (2002)
34. Sutcliffe, G.: Semantic Derivation Verification. International Journal on Artificial Intelligence Tools 15(6), 1053–1070 (2006)
35. Sutcliffe, G., Puzis, Y.: SRASS - A Semantic Relevance Axiom Selection System. In: Pfenning, F. (ed.) CADE 2007. LNCS (LNAI), vol. 4603, pp. 295–310. Springer, Heidelberg (2007)
36. Sutcliffe, G.: The 4th IJCAR automated theorem proving system competition - CASC-J4. AI Commun. 22(1), 59–72 (2009)
37. Sutcliffe, G.: The TPTP World – Infrastructure for Automated Reasoning. In: Clarke, E.M., Voronkov, A. (eds.) LPAR-16 2010. LNCS, vol. 6355, pp. 1–12. Springer, Heidelberg (2010)
38. Urban, J.: Translating Mizar for First Order Theorem Provers. In: Asperti, A., Buchberger, B., Davenport, J.H. (eds.) MKM 2003. LNCS, vol. 2594, pp. 203–215. Springer, Heidelberg (2003)
39. Urban, J.: MPTP - Motivation, Implementation, First Experiments. Journal of Automated Reasoning 33(3-4), 319–339 (2004)
40. Urban, J.: XML-izing Mizar: Making Semantic Processing and Presentation of MML Easy. In: Kohlhase, M. (ed.) MKM 2005. LNCS (LNAI), vol. 3863, pp. 346–360. Springer, Heidelberg (2006)
41. Urban, J.: MaLARea: a Metasystem for Automated Reasoning in Large Theories. In: Urban, J., Sutcliffe, G., Schulz, S. (eds.) Proceedings of the CADE-21 Workshop on Empirically Successful Automated Reasoning in Large Theories, pp. 45–58 (2007)
42. Urban, J.: MPTP 0.2: Design, Implementation, and Initial Experiments. Journal of Automated Reasoning 37(1-2), 21–43 (2007)
43. Urban, J., Sutcliffe, G.: ATP Cross-Verification of the Mizar MPTP Challenge Problems. In: Dershowitz, N., Voronkov, A. (eds.) LPAR 2007. LNCS (LNAI), vol. 4790, pp. 546–560. Springer, Heidelberg (2007)
44. Urban, J., Trac, S., Sutcliffe, G., Puzis, Y.: Combining Mizar and TPTP Semantic Presentation Tools. In: Proceedings of the Mathematical User-Interfaces Workshop 2007 (2007)

45. Urban, J.: An overview of methods for large-theory automated theorem proving (invited paper). In: Höfner, A., McIver, G. (eds.) ATE Workshop. CEUR Workshop Proceedings, vol. 760, pp. 3–8. CEUR-WS.org (2011)
46. Urban, J., Alama, J., Rudnicki, P., Geuvers, H.: A Wiki for Mizar: Motivation, Considerations, and Initial Prototype. In: Autexier, S., Calmet, J., Delahaye, D., Ion, P.D.F., Rideau, L., Rioboo, R., Sexton, A.P. (eds.) AISC 2010. LNCS, vol. 6167, pp. 455–469. Springer, Heidelberg (2010)
47. Urban, J., Rudnicki, P., Sutcliffe, G.: ATP and Presentation Service for Mizar Formalizations. J. Autom. Reasoning 50(2), 229–241 (2013)
48. Urban, J., Sutcliffe, G., Pudlák, P., Vyskočil, J.: MaLARea SG1 - Machine Learner for Automated Reasoning with Semantic Guidance. In: Armando, A., Baumgartner, P., Dowek, G. (eds.) IJCAR 2008. LNCS (LNAI), vol. 5195, pp. 441–456. Springer, Heidelberg (2008)
49. Urban, J., Vyskočil, J., Štěpánek, P.: MaLeCoP Machine Learning Connection Prover. In: Brünnler, K., Metcalfe, G. (eds.) TABLEAUX 2011. LNCS, vol. 6793, pp. 263–277. Springer, Heidelberg (2011)
50. Weidenbach, C., Dimova, D., Fietzke, A., Kumar, R., Suda, M., Wischnewski, P.: SPASS Version 3.5. In: Schmidt, R.A. (ed.) CADE 2009. LNCS, vol. 5663, pp. 140–145. Springer, Heidelberg (2009)
51. Wiedijk, F.: CHECKER - notes on the basic inference step in Mizar (2000), http://www.cs.kun.nl/~freek/mizar/by.dvi

Author Index

Author Index